# DEATH
## from the heavens

Other Books by Kenneth P. Werrell

*Sabres over MiG Alley: The F-86 and the Battle for Air Superiority in Korea* (2005)

*Chasing the Silver Bullet: U.S. Air Force Weapons Development from Vietnam to Desert Storm* (2003)

*Blankets of Fire: U.S. Bombers over Japan during World War II* (1996)

*"Who Fears?": The 301st in War and Peace, 1942–1979* (1991)

*Archie, Flak, AAA, and SAM: A Short Operational History of Ground-Based Air Defense* (1988)
(revised and updated as *Archie to SAM*, 2005)

*The Evolution of the Cruise Missile* (1985)

*Eighth Air Force Bibliography*
(1981; revised and updated 1996)

## A HISTORY OF STRATEGIC BOMBING

**KENNETH P. WERRELL**

Naval Institute Press
Annapolis, Maryland

Naval Institute Press
291 Wood Road
Annapolis, MD 21402



Library of Congress Cataloging-in-Publication Data
Werrell, Kenneth P.
Death from the heavens : a history of strategic bombing / Kenneth Werrell.
p. cm.
Includes bibliographical references and index.
ISBN 978-1-59114-940-8 (acid-free paper) 1. Bombing, Aerial—History.
2. Strategic forces—History. I. Title.
UG700.W47 2009
358.4'209—dc22

2009009081

Printed in the United States of America on acid-free paper

14 13 12 11 10 09 9 8 7 6 5 4 3 2
First printing

*To Les, Brad and Jen, Rik and Lori, Lin and Eric,*
*Sara, Jeff, Zak, Amanda, Rebecca, Eric, Harry, and Kenny.*
*May they never experience the subject of this book.*

*And as always,*
*Jeanne*

# CONTENTS


| | | |
|---|---|---|
| *Acknowledgments* | | ix |
| *Abbreviations* | | xi |
| *Introduction* | | xiii |
| Chapter 1. | The Early Years through World War I: First Steps | 1 |
| Chapter 2. | The Interwar Years: Bombers, Doctrine, and Combat | 21 |
| Chapter 3. | German Strategic Bombardment: Limited Efforts | 46 |
| Chapter 4. | British Strategic Bombing: Destruction by Night | 68 |
| Chapter 5. | U.S. Strategic Bombing in Europe: Day Bombing | 99 |
| Chapter 6. | Razing Japan: The Zenith of World War II Strategic Bombing | 129 |
| Chapter 7. | The Postwar Era: The End of Propeller-Powered Bombers | 155 |
| Chapter 8. | Between Korea and Vietnam: The Transition to Jets | 180 |
| Chapter 9. | The 1960s and 1970s: Vietnam, New Aircraft, Munitions, Tactics, and Training | 214 |
| Chapter 10. | Missiles: Winged and Ballistic | 237 |
| Chapter 11. | Strategic Bombardment into the Twenty-first Century: A Limited Future? | 277 |
| *Conclusion* | | 297 |
| *Source Notes* | | 301 |
| *Index* | | 325 |

# ACKNOWLEDGMENTS

This study is based on my career-long research, which has focused on aviation history, specifically of the U.S. Army Air Forces/U.S. Air Force, using primary, multiarchival sources as well as a host of studies by authors whom I have credited in the Source Notes.

Over the years I have received considerable assistance from archivists at various locations. At the Air Force Historical Research Agency, Joe Caver, Archie Difante, Jerome Ennels, Dan Haulman, Forrest Marion, and Anne O'Connor were critical. Bill Elliott at the Air Force Materiel Command and Doug Lantry and Brett Stolle at the National Museum of the United States Air Force were also important. I also wish to thank Tim Nenninger at the National Archives, Tom Lubbesmeyer and Mike Lombardi at Boeing, and Alan Renga at the San Diego Air and Space Museum for their help. The reference librarians at the Air University and Radford University, especially Bud Bennett, interlibrary loan librarian at Radford, proved essential through the years. A number of individuals at the Naval Institute Press deserve credit for encouraging and making this project possible, including Mark Gatlin, Rick Russell, Janis Jorgensen, and especially copy editor Karin Kaufman. A special debt of gratitude is due the history faculty at Duke University, who taught, encouraged, and mentored me. Of that noble band, Theodore Ropp was my major advisor, and Irving B. Holley my continuing inspiration.

Pride of place, of course, goes to my wife, Jeanne.

# ABBREVIATIONS

| | |
|---|---|
| AAA | Antiaircraft artillery (flak) |
| AAF | Army Air Forces |
| ABM | Antiballistic missile |
| ACM | Advanced cruise missile |
| ACTS | Air Corps Tactical School |
| AFCE | Automatic flight control equipment |
| ALCM | Air-launched cruise missile |
| AMSA | Advanced manned strategic aircraft |
| ASM | Air-to-surface missile |
| ATA | Advanced tactical aircraft |
| ATB | Advanced technology bomber |
| ATRAN | Automatic terrain recognition and navigation |
| AWPD | Air War Plans Division |
| BMD | Ballistic missile defense |
| BMEWS | Ballistic missile early warning system |
| BSAX | Battlefield surveillance aircraft |
| CBO | Combined Bomber Offensive |
| CCS | Combined Chiefs of Staff |
| CEP | Circle error probable |
| DEW | Distant Early Warning |
| ECM | Electronic countermeasures |
| EVS | Electro-optical viewing system |
| FAS | Federation of Atomic Scientists |
| GAF | German air force |
| GAM | GPS-aided munition |
| GCI | Ground-controlled intercept |
| GD | General Dynamics |
| GLCM | Ground-launched cruise missile |
| GPS | Global positioning system |
| ICBM | Intercontinental ballistic missile |
| IF | Independent Force |
| IFF | Identification Friend or Foe |
| INF | Intermediate-range Nuclear Forces |
| IR | Infrared |
| IRBM | Intermediate-range ballistic missile |

| | |
|---|---|
| JCS | Joint Chiefs of Staff |
| JDAM | Joint direct attack munition |
| JSOW | Joint stand-off weapon |
| LABS | Low-altitude bomb system |
| LGB | Laser-guided bomb |
| LPI | Low probability of intercept |
| MAD | Mutual assured destruction |
| MIRV | Multiple independently targetable reentry vehicle |
| MPS | Multiple protective shelter |
| MRBM | Medium-range ballistic missile |
| MRV | Multiple reentry vehicle |
| MX | Missile experimental (Peacemaker) |
| NACA | National Advisory Committee for Aeronautics |
| NASA | National Air and Space Administration |
| NATO | North Atlantic Treaty Organization |
| NBC | Nuclear, biological, and chemical (weapon) |
| OAS | Offensive avionics system |
| PGM | Precision-guided munition |
| RAF | Royal Air Force |
| RAM | Radar-absorbing material |
| RCS | Radar cross-section |
| SAC | Strategic Air Command |
| SALT | Strategic Arms Limitation Treaty |
| SAM | Surface-to-air missile |
| SCAD | Subsonic cruise armed decoy |
| SDI | Strategic Defense Initiative |
| SIOP | Single Integrated Operational Plan |
| SLBM | Submarine-launched ballistic missile |
| SLCM | Sea-launched cruise missile |
| SRAM | Short-range attack missile; short-range air-launched missile |
| SSBN | Ballistic missile nuclear submarine |
| SSF | Seaplane Striking Force |
| START | Strategic Arms Reduction Treaty |
| TAC | Tactical Air Command |
| TERCOM | Terrain contour matching |
| TFX | Tactical Fighter Experimental |
| ULMS | Undersea long-range missile system |
| USAF | U.S. Air Force |
| USSBS | United States Strategic Bombing Survey |
| USSTAF | United States Strategic Air Forces in Europe |
| VHF | Very high frequency |
| WMD | Weapons of mass destruction |

# INTRODUCTION

Strategic bombardment combines the allure of a new weapon of war with two of the most glamorous and romanticized technological fields of the twentieth century, aviation and space. Its story is a tale of high hopes and great efforts, along with considerable costs. But there is also a dark side to this subject, linked with death and destruction and the well-founded belief that strategic bombardment has the potential to end civilization as we know it. And while the end of the Cold War has dampened many of these fears, the proliferation of nuclear weapons and their delivery systems has mutated and heightened concerns. There is growing apprehension that the new nuclear-armed nations may not be as responsible as the older ones and may not adhere to the established rules of the nuclear powers. This fear for the future, coupled with strategic bombing's record against civilians, has branded aerial bombardment as a weapon and symbol of barbarism.

## The Promise

Homeland targets have been considered much more vulnerable than fielded military forces, and attacking them, rather than the enemy's military, has been likened to a stiletto to the heart instead of an axe to the limbs. Air power advocates and theorists forecasted, if not promised, that this new form of warfare would be cheap, quick, and decisive compared to traditional means of warfare. But to achieve its potential, strategic bombardment had to overcome a number of obstacles. These included the requirements of range, accuracy, bomb load, target intelligence, and the ability to counter poor weather and enemy defenses. The power of strategic bombardment increased as delivery systems improved, first with bombers and later with missiles. Initially guided bombs and missiles bolstered bomber capabilities, but later missiles—ballistic and cruise, air-, sea-, and land-launched—challenged and supplanted bombers as the strategic weapon of choice.

In World War I air power, especially as employed as observation, reconnaissance, and, increasingly, ground support, proved very valuable. Some believed that the airplane could do far more, that the airplane could decide future wars, for limited strategic bombing operations during World War I revealed its promise, encouraged military theorists, fanned the imaginations of proponents, and stoked the fears of the public. The desire to avoid the slaughter and indecisiveness of World War I fostered interest in strategic bombardment. Although these ideas far exceeded the capabilities of the technology of the 1920s and 1930s, during this period they gained considerable currency. World War II would test these concepts.

The midcentury conflict saw two major coalitions fighting a total war, mobilizing their economies and populations, and employing the most modern technologies. This massive war involved the bulk of the industrialized world, devastated countries, and killed perhaps 50 million. Early in the war the Germans and Japanese made effective use of air power, which greatly contributed to their spectacular victories. The Allies responded, slowly but massively, and overwhelmed the Axis powers with superior numbers and competitive technology. Both coalitions made effective and widespread use of air power, but only the Americans and British employed strategic bombardment on a grand scale. This bombing was hampered by problems associated with weather, accuracy, intelligence, enemy defenses, the flexibility and resilience of enemy economies, the stoicism of the population, and the resolve of the enemy's leadership. It involved considerable effort and cost to both attackers and the attacked and left an ongoing controversy as to its efficacy and morality. To this day two questions remain: Was World War II a fair test of the concept and how effective was strategic bombardment in the war? At the time the conflict appeared to validate the advocates of air power and support the views of strategic bombardment enthusiasts now armed with nuclear weapons.

World War II reordered the world balance of power and helped lead to the bipolar world and the Cold War, which pitted the American-led "free world" coalition against the Soviet-led Communist coalition in a rivalry that dominated world events for the next four decades. Both sides fielded large numbers of strategic bombardment systems that could deliver nuclear weapons. The adversaries first deployed bombers, and then, beginning in the late 1950s, technological advances changed the shape of air warfare and strategic bombardment. In short order ballistic missiles were considered superior to bombers for both the nuclear deterrent and nuclear war-fighting roles. Cruise missiles first aided then threatened to replace the penetrating bomber. Meanwhile bombers also were challenged by surface-to-air missiles (SAMs) and smaller aircraft.

Since World War II air power (perhaps more accurately, aircraft and some missiles) has been commonly employed in large and small wars by major and minor powers. At the same time strategic bombardment has seen only limited action and nuclear weapons have not again been employed. The smaller powers could not afford it, while the major powers could not use it fully because of political constraints and limited strategic targets. Although strategic bombardment was greatly enhanced by improving technologies, it was unable to achieve victory by itself. Was this because of a flawed concept or the political restraints?

## This Book

Some definitions are in order to clarify the discussion that follows. First, I use the term "air power" to include all uses of air vehicles as well as missiles and space vehicles instead of the more precise but awkward "air and space power." Second, I use the term "strategic bombardment," a subset of this construction, to encompass both strategic bombing (aircraft) and strategic missiles (both air-breathing cruise and rocket-powered ballistic missiles). Strategic bombardment is perhaps best understood for what it is not. It is neither tactical operations, the attack of enemy troops (close air support), nor interdiction, the attack of enemy supplies that sustain the troops. Instead, strategic bombardment strikes the enemy's homeland, bypasses its armed forces, and directly hits the source of its power, be it physical targets, such as war industry (munitions' plants, for example), economic targets (fuel, transportation, or electricity), or psychological targets (the enemy's civilian morale).

What follows is the story of strategic bombardment from its inception through its development to the present. This study centers on the technology and employment of strategic bombardment. Doctrine, another important element, is mentioned, but it occupies a lesser role, as it merits. Bombers, U.S. efforts, and World War II dominate the narrative, as their significance deserves and demands. Coverage of U.S. aspects is based mainly on primary sources, with the exception of World War I, while coverage of non-U.S. aspects is drawn from secondary sources. The notes for each chapter (see the Source Notes) begin with a bibliographic discussion that pinpoints the most important (but not all) sources used. The notes refer chiefly to quoted material.

So to the basic question: Has strategic bombardment fulfilled its promise?

Chapter 1

# The Early Years through World War I

## FIRST STEPS

In the summer of 1783 the Montgolfier brothers launched an unoccupied hot air balloon, and that September they repeated the feat, this time with a balloon that traveled a mile and a half carrying a sheep, a rooster, and a duck. The first manned free flight took place in November 1783, lasted twenty-five minutes, and crossed over five miles of Paris. A decade later, the French, clearly leaders in this field, established a balloon force, and in 1794 their armies employed captive balloons as a reconnaissance platform. Napoleon, however, disdained new technology and abolished the unit in 1802. The next combat application of ballooning was seen in August 1849, when the Austrians used small hot air balloons to drop 30-pound bombs (ineffectively) on besieged Vienna. During the American Civil War both sides employed balloons but abandoned them before the war's end. During the Franco-Prussian War, the French flew sixty-six balloons out of encircled Paris. Others were used by the British in the Sudan (1882), Bechuanaland (1884), and the Boer War (1899) and by Americans in Cuba (1898).

Heavier-than-air craft did not appear until the twentieth century. These offered numerous advantages over the lighter-than-air craft as well as technical challenges. Power was one of the challenges. Steam and electric power were unsuitable, and only toward the end of the nineteenth century did the internal combustion engine present a viable power source. Control was a more difficult issue. The sustained effort of the Wright brothers broke that barrier and was the last critical step on the road to successful powered flight.

The military quickly employed the heavier-than-air craft in war. When the Italians went to war against the Turks in Libya in 1911, they deployed nine airplanes with the invading force and went on to establish a number of "firsts." In November they dropped the first bombs in combat, despite the two Hague conventions (1899 and 1907) that prohibited such action. (There is no mention in the secondary sources of any protests or international reactions to these violations.) These rather modest efforts are noteworthy coming so soon after the first successful powered flight and as an accurate precursor of what was shortly to follow.

World War I erupted two years later. While there were many more aircraft involved in the war, these were limited in capability and used primarily in the same roles. Strategic bombing, of a sort, was introduced only a few days after the war's beginning in August by the Germans, who dropped a few small bombs on Paris, killing one woman. The British bombed a Cologne railway station in October but achieved more dramatic success that same day when one aircraft destroyed a zeppelin in its shed at Dusseldorf.[1] The Germans dropped their first bomb on England in December 1914. While these were mere pin pricks, they foreshadowed more concerted efforts. Nevertheless, for the most part the airmen supported the ground forces. Fighter pilots attracted the most attention at the time, and since, with the myth and mystique of the "Red Baron," his cohorts, and opponents. The air war was much cleaner and more romantic than the horror and slaughter of trench warfare on the western front, although also quite deadly. Strategic bombing has received much less attention, correctly so, as it had less impact on the war and was much smaller in scale. It did, however, cast a long and dark shadow.

## The Zeppelins

The first sustained strategic air offensive came from a technology that would become a mere footnote in history, the lighter-than-air machines. It is easy from our present position to scoff at these craft, knowing how they fared, going extinct as the dinosaurs that they resembled. It should not be forgotten, however, that in the early years of the war, airships had clear advantages over their heavier-than-air competitors: greater altitude, rate of climb, range, payload, and endurance.

Lighter-than-air machines markedly improved in the nineteenth century. Initially balloons were lifted by hot air, which was quickly seen as an inferior lifting agent when compared with hydrogen, although the latter presented a constant and considerable problem due to its flammability. Balloons evolved from an essentially nonsteerable, inverted-teardrop shape into powered, elongated balloons called dirigibles. Count Ferdinard von Zeppelin was responsible for airship development in Germany and went on to dominate the technology and give the vehicle his name.

The count made his first flight in a balloon during the American Civil War and ten years later conceived of the idea of a large "air cruiser." It was not until 1894 that these ideas took shape in his proposal for a vehicle 384 feet long powered by two 11-horsepower (hp) engines, a project the German government reasonably rejected on technical grounds. Nevertheless Zeppelin persisted. He modified his early design to a more practical cigar-shaped gas bag braced by a rigid structure with an under-slung gondola for

**(Right) In January 1915 the Germans initiated the first strategic bombing campaign employing zeppelins against Britain. Weather, engine unreliability, and highly flammable hydrogen were great hazards, and half of the zeppelins built were lost. The L-12 was hit by antiaircraft artillery on its maiden flight in August 1915 and went down in the English Channel. Note the aircrew on both gondolas and on the top surface. (U.S. Naval Institute Photo Archive)**

the crew. His first airship flew in July 1900. It was scrapped in 1901 after two more flights and a total flying time of just over two hours because it exhibited inadequate control, low speed, and a weak hull.

Zeppelin built and flew other airships that led the German War Ministry to purchase two, but they did not do well in competition against short-range, nonrigid, lighter-than-air craft. The count then formed the world's first commercial airline, which carried almost thirty-three thousand passengers on sixteen hundred flights before the war broke out. Thus the name "zeppelin" and the craft's concept became known throughout the country, the Continent, and the world and became a point of pride for Germans.

The German navy also showed an interest in airships. The navy's chief, Gross Admiral Alfred von Tirpitz, saw its promise as a scouting vehicle equipped with radios as early as 1906 but wanted a craft with longer range and greater speed (to buck the North Sea winds) than what Zeppelin was able to deliver. However, pressure from the kaiser and the public, along with the threat that the airship maker might sell airships to Britain, forced the German navy to order their first machine in April 1912, which initially flew that October.

The Germans began to train crews, carry out long-range flights, and conduct maneuvers with warships. Then the program was stunned in September 1913 when the first navy zeppelin was lost to bad weather. Although previously three airships had been destroyed by high winds or fire, this was the first in which there were fatalities. Just over a month later a second airship caught fire and was destroyed, killing, as in the prior incident, all but one of the crew. The twin disasters almost wiped out the navy's experienced airship personnel. At the outbreak of the war, the navy had one airship and the army six, and in short order the military took over three civilian machines.

Despite great expectations the airships' combat record was disappointing. During the opening months of the war the army lost half of its machines to ground fire. Concurrently German airmen began making long-range attacks, starting with a raid on Antwerp in early September 1914. Although both the German public and military wanted to bomb London, the kaiser was reluctant, possibly concerned about neutral opinion and perhaps fearful of hitting his royal relatives. There is no indication that the Germans gave any attention to the international treaties that banned the bombing of cities. The kaiser vacillated, but after the French bombed Freiburg in January 1915, he authorized attacks on London with certain geographic and target restrictions. After an abortive effort, on January 19, 1915, three naval airships dropped a ton of bombs over eastern Britain, killing four and injuring sixteen. It was not until the last day of May 1915 that an army airship attacked London, dropping three thousand pounds of bombs and killing seven. The first Battle of Britain had begun.

The airships encountered major navigational difficulties during these operations. Map reading and celestial navigation required adequate visibility, which was a problem in the European winter. Dead reckoning was also problematic due to the primitive state of weather forecasting, especially for the Germans, who lacked reporting stations to the west, the source of the prevailing weather.[2] Beginning in April 1915 the Germans employed radio aids to permit navigation in all weather conditions. The airship crew transmitted radio signals to ground stations that used triangulation to determine the craft's position and radioed that information back to the crew. This system encountered problems with the radio transmissions and allowed the British to plot the airship's position. In 1918 the Germans fielded a system that used transmissions from two ground stations to allow an airship's crew to ascertain its position.

A second difficulty was the weather. Poor weather and zeppelin operations were a dangerous mix. The airships were unwieldy under the best of circumstances and difficult to handle in high winds and turbulence. As most only had a top speed in the range of sixty miles per hour, they had marginal speed when flying in high winds.[3] The airship's large structure was also vulnerable to the stresses of turbulent weather.

A third problem was the unreliability of the zeppelin's power plants. Airships had multiple engines that were accessible to the crew in flight, a wise provision because the engines frequently broke down. Engine failure in calm air was not critical, but against high winds it certainly could be. The airships were underpowered when all engines were running at peak performance and certainly had problems when any failed or were not producing full power.

There were also the enemy air defenses. The airships were large and slow and relatively easy to spot in clear weather. They lacked a speed advantage over aircraft, although initially they could climb faster and higher than the defending fighters. The zeppelin's slow speed and large size gave antiaircraft artillery gunners, as well as infantrymen and machine gunners, a much easier target than aircraft, and the use of hydrogen made the airships vulnerable. Of the fifty-three or so naval airships lost during the war, at least ten were downed by fighters and eight by ground fire. Some of the other losses may have resulted from damage suffered from either or both of these agents, but clearly weather and mechanical problems were a major cause of losses.

The new weapon forced the British to field new defenses. The major problem for home defense was that the western front enjoyed top priority on artillery pieces, aircraft, and personnel. One result was to employ, in the words of one officer, the "deaf, the blind and the mentally deficient."[4] In a similar manner, gun tubes and first-line aircraft were desperately needed in France, which left only obsolete equipment at home. Luck and visibility seemed to be the most important factors in the defender's success or failure.

The British deployed ground observers, who visually and aurally detected the aircraft and reported their position. The first technological aids they received employed two horns from a phonograph player attached to a pivoting pole and connected to a doctor's stethoscope. At least some of these devices were manned by blind individuals with acute hearing. Later the defenders built at least four concrete sound reflectors, fifteen feet in diameter, that could detect aircraft at a range of six to eight miles and perhaps as far as twelve to fifteen miles. Sound detection (and tracking) was the best available technology of the day, although it was short ranged, fickle, and unreliable.

Initially British fighters lacked adequate armament to down the airships. The defenders used bombs and darts, but these required the fighter to get above the zeppelin, brave defensive fire, and accurately release the missiles, not an easy task. Grapnels and ramming were considered, while flare pistols and grenade launchers proved useless. The standard British .303-caliber machine-gun ammunition proved ineffective, leading to the invention of incendiary bullets that could ignite the hydrogen. These went into action in the summer of 1916 and became the best fighter weapon.[5] The Germans defended the zeppelins with machine guns mounted on the top of the craft and in the gondolas beneath it. Until the last year of the war the airships carried eight or more machine guns, which then were reduced to two or none to lighten the craft and gain additional altitude.

A Belgian-based British naval aviator downed the first airship in aerial combat in June 1915, and British gunners at Dover downed their first zeppelin that August with a 3-inch shell. The airship fell into the English Channel and, although a German destroyer towed the downed craft to

Ostend, it was written off. That same morning airship gunners shot down a British interceptor, the only such recorded success by a zeppelin. The British-based airmen achieved their first aerial kill in September 1916, when a fighter, using incendiary ammunition for the first time, destroyed an airship. This came during the largest and only joint German zeppelin raid of the war, an attack consisting of dozen naval craft and four army machines. This armada dropped seventeen tons of bombs but inflicted little damage or injury on the British. The one downed airship cost four times as much as the damage inflicted during this raid, while the sixteen crewmembers killed were four times the number of civilians killed. This was the last army airship raid of England, for in June 1917 the army pulled the zeppelins back to Germany and that summer broke up all but two "super Zeppelins" that it turned over to the navy. Meanwhile, the German navy persisted.

The Germans constantly attempted to improve their airships throughout the war. In May 1916 the first of a new type arrived for service against Britain. Powered by six rather than five engines, they were more powerful, more streamlined, better protected (by as many as ten machine guns), and could carry five tons of bombs. Nevertheless these were little faster and flew at about the same altitude as their predecessors.[6] In late November 1916 British aircraft downed two zeppelins. As a consequence the head of the naval airship detachment wrote, "Machine guns are no defence for the airship against aeroplanes . . . [therefore] 'attack only with cloud cover.'"[7] The navy concluded that the best defense against interceptors was altitude. To gain height the Germans ruthlessly cut weight in both new construction and existing airships. They removed the sixth engine, got rid of all the guns, cut the fuel and bomb load, and used lighter hull girders. German priorities are clearly seen in their decision to remove parachutes that had been carried for a brief time in 1917, saving about eighteen pounds per man. The manufacturer tightly doped the outer cover of the craft to reduce drag, reduced the weight of the gas cells by using two layers of silk rather than three layers of cotton fabric. These measures reduced weight and, depending on the particular airship, allowed operations up to eighteen thousand feet. In practice, however, the zeppelins cruised at lower altitudes to aid navigation and increase bombing accuracy. Lighter construction in the succeeding airships (spacing the largest frames farther apart), and new, streamlined gondolas, which were retrofitted into the fleet, further increased performance. New airships could reach almost twenty thousand feet.

Higher altitude operations introduced new problems. They subjected the crew to both the cold and anoxia, and while the airships carried compressed oxygen, they did not carry heating devices. Higher altitude operations made navigation more difficult; caused engine problems; froze instruments, water ballast, and lubricants; shattered celluloid windows; made the craft's fabric brittle; and caused control cables to stick. Nevertheless the increased altitude could render the defending craft impotent as fighters used through 1917 had ceilings below thirteen thousand feet. Not until the end of the war did the British field fighters that could reach these higher operating altitudes.[8]

(Right) The Germans began using Gotha bombers in May 1917 to bombard Britain. Although carrying a lighter bomb load than the airships, they were faster and, flying in formations, better able to defend themselves. However, British defenses inflicted heavy losses on the Germans, forcing them to turn to night operations. (National Museum of the USAF)

The new "Height Climbers" made their first attack in March 1917, during which one airship was lost. The attacks persisted with limited bombing results and continued casualties. On October 19, 1917, the Germans launched eleven airships on the last great airship raid of the war. Although none of the seventy-three British pilots launched that night made contact with the intruders, the raid proved to be a great disaster. Despite a favorable weather forecast, the zeppelins ran into a gale that along with mechanical breakdowns and French antiaircraft fire contributed to the loss of five of the giant craft. Nevertheless the attacks continued.

On August 5, 1918, the Germans launched their first attack on Britain in four months; it consisted of five airships led by the commander of the naval air division in an airship that had first flown only a month earlier. The defenders got thirteen fighters up and in clear weather downed the lead airship. One British aircraft was lost to unknown causes; at least two others were lost in forced landings. Within a week a British fighter downed another airship over the North Sea. These zeppelin losses, but especially the death of their leader, marked the end of the airship as a bomber.

## German Strategic Bomber Operations

Fixed-wing aircraft joined zeppelins in the assault on Britain during the last two years of the war. The bombers raised Germans hopes, although it took six bombers to deliver as much tonnage as one zeppelin. The commander of the German air force proposed thirty Gotha G.IV bombers for these attacks, to be joined later by a second squadron consisting of the even larger R-type Giant bombers.[9] The Germans deployed the new models to Ghent, about 170 miles from London, the first arriving in March 1917. In the words of General Erich Ludendorff, the "purpose [of the bombing] was to shatter the unity of the enemy and dispel their belief in victory."[10] Secondary objectives were to disrupt British industrial production and communications between the ports and London, destroy supply facilities at the ports, and dislocate transportation across the channel. London would be the focus of these attacks. The task of the "England Squadron" was a tall order for a handful of aircraft embarking on a new form of warfare.

Two engines powered the Gotha G.IVs, which had a cruising speed of eighty miles per hour. They usually carried a bomb load of 660 pounds during the day (flying lower at night, the load increased to 1,100 pounds), were armed with two or three machine guns, and were manned by a crew of three. One innovation that increased their defensive capabilities was to cut away a portion of the rear deck, allowing a gunner to fire downward and to the rear of the bomber. Although the Gotha was considered agile in flight, it was tricky to fly and land at light weights, accounting for many landing accidents. For this strategic campaign the Germans fitted the bomber with primitive oxygen equipment and extra fuel tanks, which provided an additional two hours of flying time. These aircraft did not carry heating devices, parachutes, or radios. However, despite the skepticism of the crews, they did carry carrier pigeons in the event the crew went down at sea.

In late May 1917 the Gothas made their first attempt to raid London, but the mission was aborted when the Gothas encountered storms; instead, they bombed the coastal town of Folkestone, killing 95 and injuring another 260, the highest losses yet inflicted by an air attack. In mid-June, six of twenty Gothas reached central London without encountering British fighters. On their withdrawal one Bristol fighter of the ninety-two British aircraft that rose in defense attacked the German formation but broke off its action after its guns jammed and the gunner was fatally wounded. The bombers had attacked the British Empire's capital in broad daylight without a loss, killing 162 and injuring 432. Due to improved bombs and accuracy, this raid inflicted greater damage than any zeppelin attack.

The Gothas proved much more difficult than the zeppelins to counter because they were faster and better able to defend themselves in formations that massed their firepower. While the British could get one hundred or more aircraft aloft in defense, these were a wide mixture (the Royal Flying Corps flew twenty-one different types) of obsolete aircraft that were outmatched by the Gothas. The bulk of the German losses occurred from accidents or lack of fuel.

To respond to the threat, the British shifted a few fighter units from France, equipped other units with more modern aircraft, and increased the number of antiaircraft guns. On a mid-August raid 11 German bombers crossed the coast, inflicted little damage, but lost 1 bomber to 133 British fighters and 4 others in crash landings, a high toll. As with the airships, weather could also cause great problems in the aircraft operations. On August 11, 1917, twenty-eight Gothas launched against Britain encountered stiff winds that reduced their ground speed to fifty miles per hour. Thirteen bombers were destroyed. Four days later the German Gothas made their last daylight attack on Britain. Eleven of the fifteen that took off made it to England, but only to coastal targets. One was lost to antiaircraft fire and two to fighters. In response to these substantial losses, in September the Gothas turned to night operations.

One British reaction to the bomber attacks was to deploy barrage balloons in the fall of 1917, a device already employed by the Germans and Italians. A British innovation suspended an "apron" between the balloons with one-thousand-foot streamers. There were plans to implant twenty of these installations, but fewer than half were in action by May 1918. Although the balloons had little direct impact, damaging only one Giant bomber, they did force the attackers higher.

The Germans modified the Gothas for the air offensive. The Gotha G.V was more streamlined and, unlike its predecessor, had its fuel tanks inside the fuselage, not under the engines. But the use of unseasoned wood and extra equipment added nine hundred to one thousand pounds, and poor-quality fuel decreased engine power and hurt performance, resulting in a bomber with little more speed, an inferior rate of climb, and no more altitude capability than the earlier Gothas. The Germans built eighteen of this model, the first of which reached the combat units in August 1917. The Gotha was replaced by a better aircraft, the larger and aptly named Giant.

The Russians were the first to fly a four-engine aircraft, Igor Sikorsky's "Le Grand," which made its initial flight in May 1913. The Germans quickly followed with a series of large bombers that came to be designated Risen or R-type. These aircraft, much larger than the Gothas, were ungainly looking machines with long, narrow (high aspect ratio) wings that contradicted the old pilot saying, "If it doesn't look good, it can't fly good." Like the zeppelins they had engines that were accessible in flight, and some mounted their engines in the fuselage coupled to external props. The latter arrangement concentrated weight near the center of gravity, allowing good flight characteristics, but this was counterbalanced by the increased complexity, greater weight, and the loss of power absorbed by the transmission system. The decentralized engine placement proved lighter and more reliable. The Germans built between fifty-five and sixty-five R-planes, of which about thirty saw action.

The Staaken R.IV was the first of these that engaged in combat in the west. It was powered by six engines, two geared to a tractor prop in the nose and two in each nacelle geared to a pusher prop, and carried seven machine guns. It first flew in August 1916 and saw service on the eastern front before deploying to Belgium in August–September 1917 for operations against Britain. The R.V had five engines, four in two nacelles, and one in the nose, which turned three tractor props. Unlike the R.IV, the engine transmissions caused numerous problems. The R.V was accepted in September 1917, flew with a crew of eleven, and had a higher top speed and ceiling than the R.IV.

The R.VI was the largest aircraft to be produced in numbers during the war. The first was completed late in 1916, and after testing it was delivered to the air arm in June 1917. This aircraft dispensed with the coupled engine arrangement and mounted four engines tractor-pusher fashion turning four props in two nacelles. Two pilots sat

The German R.VI could carry more than two tons of bombs and truly lived up to its nickname (Giant) with a wing span only slightly less than that of the largest World War II bomber, the B-29. It flew almost a tenth of the German bomber sorties against Britain, on which it suffered no combat losses. (San Diego Air and Space Museum)

side by side in a completely enclosed cabin with an open position in the nose for one crewmember who could function as aircraft commander, bombardier, or forward gunner. The Giant had a span of 138 feet, mounted as many as seven machine guns (although usually three), and generally was manned by a crew of seven. The Giant's crews, unlike those in the Gothas, were equipped with parachutes and a radio. The craft could exceed eighty miles per hour and carried just over two tons of bombs. At least eleven of the eighteen bombers built were destroyed in the war, although only one clearly fell to enemy action.

Three R.XIV aircraft followed the R.VI model. They had the earlier five-engine arrangement, with one tractor prop in the nose and a tractor and pusher arrangement in each nacelle, and they reverted to an open cockpit configuration. The aircraft was heavier than its predecessors and thus had slightly less performance. The R.XIV saw service on the western front, with one shot down by a night-flying Sopwith Camel near the front lines.

The increasing problems encountered by the zeppelins pushed the Germans to make greater use of airplanes. On September 28, 1917, six Giants participated in their first bombing operation against Britain. In one week in September 1917, the Germans launched ninety-two Gothas, fifty-five of which crossed the coast; however, fewer than twenty reached London, while only one of five Giant sorties hit the city. The Germans lost thirteen Gothas. Nevertheless, these raids forced as many as half a million Londoners to move at night into the underground, the countryside, tunnels, and government buildings. The raids had an adverse impact on the war effort; for example, the production of the great arsenal at Woolwich for one period fell to 60 percent of capacity. To inflict this damage the Germans paid a considerable price. On December 5, 1917, British antiaircraft guns got two Gothas of sixteen that made it to England, and in all, six were destroyed on this raid. The bombers caused little damage; in fact, of eighteen casualties in London, eight resulted from antiaircraft shell fragments.

By the end of the year the British had increased their defenses to include 224 fighters, although one-third were obsolete types. They flew aerial patrols at night above the balloon aprons that by this time had been raised to ten thousand feet. The British also increased antiaircraft gun defenses, deploying 150 fixed guns around London, another 100 in other areas, and a number of mobile units.

The largest attack on London came on May 19, 1918. The Germans thrust thirty-eight Gothas, two C-types, and three R-planes toward Britain, of which only eighteen Gothas and one Giant reached the British capital. The defenders had a good night: The eighty-eight fighters that got aloft claimed three Gothas, while antiaircraft guns fired over thirty thousand rounds and claimed three more. A seventh bomber crashed near its airfield. The attackers did unload eleven tons of bombs on Britain that killed 49 and injured another 177. This was the last assault on London . . . during this war. The Germans then directed their bombers to support its army's offensive in France.

The Germans also raided French cities. During the war Paris was attacked by fifty-eight airplane and airship sorties that dropped thirty-three tons of bombs that killed 270 Parisians. The French deployed about nine hundred guns and forty thousand men to defend their capital.[11]

German aircraft operations against Britain in World War I were modest by later standards. The Gothas completed 292 sorties and dropped ninety-three tons of bombs, while the R-planes, for all the attention they received, completed a mere 28 sorties and dropped only thirty tons of bombs. The Giants carried a much greater bomb load, proved more reliable, and had a lower loss rate than the Gothas. The Germans lost sixty Gothas, the bulk destroyed in crashes in Belgium (thirty-six), with British fighters claiming eight and guns twelve. British home defenses did not get any of the R-planes, although two were destroyed in crash landings in Belgium. The German bombers killed 860 Britons and injured 2,100. In comparison the airships delivered 220 tons of bombs on two hundred sorties that killed 560 and wounded 1,400. (London antiaircraft guns accounted for 24 of the dead and nearly 200 of the injured.) The aerial bombardment caused $15 million in property damage, about equally credited to bombers and airships. The Germans losses, excluding those at their bases, totaled 160 airship and sixty Gotha crewmembers; the British lost twenty-eight airmen in the defensive effort.

At its peak British home defenses consisted of over seventeen thousand personnel, 470 antiaircraft guns, 620 searchlights, and up to 380 aircraft. The zeppelins and airplanes might not have inflicted much damage on the British homeland or military, but their operations eroded British civilian morale and diverted considerable forces from the western front. Even allowing for its novelty, the bombing of London had a disproportionate impact on the population and, as shall be seen, on the bombardment theorists.

## Allied Strategic Air Attacks

The Allies also engaged in strategic bombing, albeit to a lesser degree than the Germans. This was a result of a number of factors, but certainly not of restraint. One reason was geography. London and Paris were closer to German bases than Berlin was to Anglo-French bases. The Germans also had zeppelins and more sophisticated, longer-range bombers sooner than did either the British or the French. Other factors were the desperate need for Allied air power on the western front and the objections of both the British army and navy to independent strategic bombing. In fact, Hugh Trenchard, best remembered as the father of the Royal Air Force (RAF) and as a strategic air power advocate, early on characterized strategic bombing operations as a "great waste." Nevertheless the British pushed this effort in response to public and political pressure. The French, much closer to German bases and with a less well defended capital, were more reluctant to take this course of action.

The Royal Navy took the lead in strategic bombing. While at first glance this seems odd, it made sense at the time because the British army was tied down in the trenches on the western front, there was no independent air force, and the Royal Navy had been assigned the task of home defense. In April 1916 the Admiralty's director of

(Right) The Allies flew their first strategic attack in October 1916. The two-seat De Havilland 4 (DH-4) carried only a 400-pound bomb load but proved to be an outstanding aircraft and formed the core of the British and American day-bombing force. (Air University Press)

Air Services wrote a memo for the War Cabinet noting that a strictly defensive posture was inadequate and urging instead a "vigorous offensive." Such an effort would restore the initiative to the British, restrict zeppelin operations, divert German forces from the western front, and batter German morale. The British were not the only ones thinking in these terms, for the next month the French proposed a combined bombing effort under French control and from a French base against German industrial targets. Representatives of the two countries in July 1916 agreed on a list of four target areas; however, due to limited forces only one was practical: blast furnaces in the Saar-Lorraine-Luxembourg area that produced almost half of all German steel. The airmen were also authorized to conduct "specific reprisal" attacks on selected towns and cities much deeper in Germany.

Compared to later operations the force allocated for this offensive was amazingly small. The planners hoped to equip a force with thirty-five bombers and twenty escort fighters by July. In fact the airmen had an average of forty-three pilots and thirty-five aircraft available for action during the time of greatest bombing activity (October 1916 to March 1917). The bombing effort was further undermined by an exceptionally low serviceability rate of approximately 20 percent. Most of these aircraft were Sopwith 1½ Strutters serving as both bombers and fighters. Configured as a bomber the Strutter could carry four 65-pound bombs at a cruise speed of nearly one hundred miles per hour with a combat radius of almost 150 miles. The plan called for an expansion to approximately one hundred aircraft, a number later doubled. The army opposed this concept. In June 1916 Trenchard asserted that observation was aviation's primary objective. The British ground commander in France was more forceful when he wrote the War Office in November 1916 that his forces should have top priority and that the strategic bombing effort was a "very serious interference with British Land Forces, and may compromise the success of my operations."[12]

On October 12, 1916, British and French bombers accompanied by escort fighters, including some piloted by Americans in the soon-to-be-named *Escadrille Lafayette*, attacked the Mauser armament plant at Oberndorf. While eighteen of the thirty-one bombers hit the city with just under two tons of bombs with unknown results, German fighters rose and downed nine Allied bombers with no German losses. One of the deepest raids was on May 18, 1918, when thirty-three British aircraft bombed Cologne, killing 110. There were also reprisal raids. On April 14, 1917, the bombers attacked Freiburg in response to the sinking of two hospital ships.

Between July 1916 and April 1917 the Royal Naval Air Service launched eighteen attacks against targets behind German lines, some flown along with French aircraft. These operations were hampered by weather, aircraft, and supply problems. Nevertheless the unit flew almost 170 sorties in 1916 and 1917, 59 percent against blast furnaces. While the two allies bombed together, their objectives were fundamentally different. The French only saw morale as a target in retaliatory raids, while the Admiralty considered morale a much higher priority.

Some other aspects of this small bombing campaign deserve notice. The airmen relied on crew reports and bombing effort (e.g., sorties and bombs dropped) to measure bombing results. Separate postwar American and British bombing surveys found much less damage than the wartime reports and demonstrated the shortcomings of this method. Bomb damage assessment is a problem that has persisted throughout the history of bombing. Another problem was that perhaps as many as one-quarter of the bombs failed to explode. One aspect that wartime intelligence may have gotten wrong, as have some postwar historians, is the assertion that the Germans were forced to deploy fighter units to defend against this bombing. Overall the claims of bombing effectiveness were overblown, another perennial aspect of aerial warfare.

German attacks on Britain provoked a public and political outcry that the British military could not resist. Trenchard appeared before the British Cabinet in June 1917, where he rejected politicians' call for an airborne defensive patrol belt and instead argued for a ground offensive in Belgium to seize the Gotha bases. Trenchard's next suggestion was to attack the German airfields. The British airman saw retaliatory attacks as both counterproductive (the Germans would respond) and impractical (the British lacked suitable aircraft). Public pressure forced the government to react and therefore recall some units from France and divert new aircraft to home defense. The prime minister also formed a committee, consisting of himself and South African general Jan Smuts, to investigate British air policy. This led to a separate and coequal Air Ministry. Smuts's second report also observed that aircraft were capable of independent operations against the enemy's heartland. On April 1, 1918, the Royal Air Force was formed.

The push for an offensive air force gained momentum fueled by political pressure, the Smuts report, and the anticipated surplus of aircraft. The British adopted a policy of retaliation to satisfy the British public and undermine enemy morale, although the British field commanders thought morale bombing was irrelevant and reprisal attacks had to be carefully considered. The French were restricted by their aircraft and adopted a practical concept well within their means, targeting the German rail network as well as iron and steel installations.

In October 1917 the Royal Flying Corps formed the 41st Wing with the mission of strategic bombing of the German homeland. The unit consisted of three squadrons, one

The Handley Page O/100 first flew in December 1915 and became Britain's night strategic bombing mainstay. Its 1-ton bomb load was fitted in an enclosed bomb bay. One of its unique characteristics was its folding wings, which permitted it to use existing hangars. (San Diego Air and Space Museum)

day-bomber unit flying de Havilland DH-4s and two night-bomber squadrons flying Handley Page O/100s and F.E.2bs. Unlike the earlier British strategic bombing unit, the 41st made little effort to coordinate with the similar French bombing offensive. Ninety-five of 118 bombers dispatched on the day raids dropped eleven tons of bombs on blast furnaces, while 40 of 59 aircraft launched on the night attacks dropped nineteen tons of bombs. Postwar allied bombing surveys and German officials led one historian to conclude that "material results were incommensurate with the effort" and at best disappointing.[13] The attacks on railroads, consisting of 33 percent of the tonnage, proved no better. The overall cost during this period was twelve aircraft destroyed, eight pilots killed, and ten other crewmembers missing in action.

The British were of two minds as to the primary mission of the bombing. In June 1918 the new chief of the Air Staff, Sir Frederick Sykes, presented a plan to the War Cabinet to attack Germany's root industries, of which chemicals were the primary target.[14] Meanwhile the public, members of Parliament, and the cabinet minister for air urged the bombing of cities. The French, who supported British operations flying from French bases, had increasing reservations about bombing German cities, because they feared retaliation, and thus wanted control over such operations. (The British formed a unit based in Britain equipped with the long-range Handley Page V/1500 to free them from French restrictions, but the war ended before these went into action.) The key player in this, Trenchard, straddled the fence on targeting,

(Left) Handley Page modified the O/100, designated it O/400, and built 430. In September 1921 this Handley Page O/400 was flying over Aberdeen, Maryland, with a 2-ton bomb. (National Museum of the USAF)

noting industrial targets but adding that "the moral effect at present is far greater than the material effect."[15] Despite what London wanted, Trenchard went his own way. While he ostensibly planned to hit industrial targets, his bombers primarily bombed tactical targets, with airfield attacks rising from 13 percent in June 1918 to 50 percent in August. At the same time railroads were the targets of 55 percent of the effort in June and 31 percent in August.

In February 1918 the 41st Wing was redesignated Eighth Brigade, which in June 1918 became the core of the Independent Force (IF), sometimes referred to as the Independent Air Force. Trenchard took command despite his objections and reservations concerning strategic bombing, for he was Britain's leading airman and was highly thought of by both the Americans and French. The purpose was "direct action against the heart of the German industrial system."[16] This strategic force grew in numbers and capability during the year, and by September the British were fielding nine bombing squadrons, four flying DH-4/9/9As in daylight operations and five flying Handley Page O/100s in night operations. One step backward was the introduction of the DH-9, which was "generally unsatisfactory—and unwanted."[17] Although it looked like the famous DH-4, its ceiling and cruising speeds were inferior and its engine inadequate and unreliable. The main change was to move the fuel tank from between the pilot and observer forward of the pilot. The next version, the DH-9A, featured a larger wing and a more powerful engine. The increased power gave the newer aircraft about equivalent performance with the DH-4.

Handley Page provided the backbone of the IF's bombers. The first of these, the O/100, had been designed for sea patrol duty in response to a December 1914 Royal Navy specification. The aircraft was manned by a crew of three and featured an enclosed bomb bay that could carry a maximum of two thousand pounds, which was three times that of the Short bomber and six times that of the DH-4, Britain's other bombers of the day. It also had three gun positions that mounted up to five machine guns. Initially the manufacturer armored both the nacelles and crew cabin, however, it later deleted the nacelle armor and enclosed cabin. The operational O/100 had a span one hundred feet (from which the designation was derived) and wings that folded back, outboard of the engine nacelles that housed two 250-hp (later 285-hp) engines. Handley Page built forty-six O/100s, the first making its maiden flight in December 1915. In one of those awkward twists of history, the pilot of one of these bombers became lost in the fog on a delivery flight in January 1917, prior to its first bombing mission in March, landed behind German lines, and was captured in his brand new bomber. Pilots considered the O/100 a pleasant aircraft to fly, although it was heavy on the controls and slow to respond.

The manufacturer modified the aircraft by moving the fuel tanks from the nacelles into the fuselage, substituting props rotating in the same direction for opposite rotation, and using higher power engines, which merited its redesignation as the O/400. The

new model began to replace the O/100 in early 1918 and went on to form the bulk of the IF. The fastest version could reach a maximum speed of ninety-eight miles per hour at sea level, had an endurance of eight hours, a combat radius of over 220 miles, and a service ceiling of eighty-five hundred feet. In all, the manufacturer delivered 430 before the war's end. Although a large and unwieldy machine compared to the single-engine bombers, the Handley Page was little criticized by its crews, and during the summer of 1918 it had a lower accident rate than the single-engine DH-4s and F.E.s.[18]

The aircraft that would have carried the bombs to the German capital was the Handley Page V/1500, which first flew in May 1918. This bomber had greater capability than its predecessors, with a bomb load of seventy-five hundred pounds, over twelve hours of endurance, and up to five machine guns in three defensive positions. It had a

(Left) The first four-engine British bomber was the Handley Page V/1500, which initially flew in May 1918. Each nacelle mounted a puller and pusher engine. Built to bomb Berlin from Britain, it did not see combat, although three were standing by when the Armistice was concluded. (Air University Press)

span of 126 feet and was powered by four 375-hp engines configured tractor-pusher fashion in two nacelles. The manufacturer delivered the first three in November and apparently they were fueled and bombed up for an attack on Berlin when the Armistice was concluded.

In the last five months of the war the IF dropped 550 tons of bombs, 390 tons at night. The British bombed targets during the war as deep as 150 miles within Germany, hitting Bonn, Cologne, Frankfurt, and Stuttgart. The cost was high: 109 aircraft were lost, according to one source, 104 day and 34 night bombers according to another. For the period October 1917 through to the Armistice, for every British strategic bomber lost to enemy action, three were lost to accidents, close to the ratio of losses endured by the British airmen in the last eighteen months of the war.

While the British had a grand idea of an Allied strategic force consisting of some sixty American, British, and French squadrons, the field commanders feared a drain on their forces as well as a lack of control. Tepid French support was a problem because the logical place for the bases was in France. It was not until late October 1918 that Marshal Ferdinand Foch, the Allied commander on the western front, agreed to the organization of an Allied air force, only shortly before the war ended. It should be noted that there was tension among the Allies over control and targeting, Trenchard in fact admitted that he considered autonomy more important than cooperation.

During the course of the war the German airmen inflicted about twice the casualties on British civilians (1,400 killed and 3,400 injured) as did the British (750 killed and 1,300 injured). Allied bombing caused about $6 million in damage on Germany, while the British suffered about $15 million in damage. Two factors help account for this difference: German air defenses were better at warning civilians than the British system and the Germans had better strategic bombers and more experienced aircrews than did the IF.

## Conclusions

Clearly the World War I air war was overshadowed by the ground war. Aircraft had more impact on the public than on the battlefield, and more on the imagination than the reality. Air power showed greater results in tactical operations (certainly observation and reconnaissance and at times in direct support) than in strategic operations. Strategic bombing was a small portion of World War I air operations, certainly compared to what was to follow. The combatants conducted raids with only a handful of aircraft that carried meager bomb loads. With essentially no navigational aids, weather and night were serious obstacles, and it is little wonder the results were limited. At least one-fifth of the IF's bombs hit an unintended target or no useful target, and 5 percent of British and 6 percent of French bombs were duds. Study of the British bombing reveals that day bombing caused more property damage and casualties per ton of bomb dropped than did night bombing.

What was the result of the World War I bombing experience? For one thing, the British, the prime target of the strategic bombing, came to emphasize the psychological impact of bombing rather than its physical effects. This is probably an understatement, as this belief has been described as "evolv[ing] into a widespread obsession" and this attitude "became the predominant justification for the RAF."[19] While the British experience in the trenches on the western front was similar to that of the other combatants, its experience with strategic bombing was not. The bombing of Britain, along with the acknowledged vulnerability of London, made the British public and politicians extremely sensitive. Aviation had drastically changed Britain's unique security position. As one Briton noted shortly after the first powered flight, "England is no longer an island."[20] Yet the results of World War I bombing were overestimated: The physical impact of the strategic bombing was minimal, and while the bombing on occasion shook civilian morale, its impact tended to diminish over time. A third aspect is that strategic bombing in World War I was hampered by small forces further reduced by low in commission rates, restricted by the limited capabilities of bombers and zeppelins, and limited by bad weather. The small numbers of airships and airplanes that made it to the target areas delivered small bomb loads, and not very accurately. In addition losses were high, mostly due to accidents. In view of the crude equipment and very limited training, it is remarkable that losses were not higher. The most significant influence of World War I strategic bombing was to divert resources from other operations and, in the long run, the impression it left on the minds of the airmen and the civilian decision makers. The air war in general and the strategic bombing campaign in particular demonstrated more promise than performance. Certainly the strategic effort was spectacular and dramatic, but in truth it had minimal effectiveness and influence on World War I.

Chapter 2

# The Interwar Years

## BOMBERS, DOCTRINE, AND COMBAT

In the 1920s and 1930s aviation made rapid strides. These were the golden years for aviation, as new aircraft, flashy barnstormers, record-breaking flights, and seemingly endless promises for the future dazzled the press and the public. In these two decades new technology appeared that not only changed the appearance of aircraft but also greatly improved aircraft performance.[1] In military aviation bombers were the first to take advantage of new technological developments so that between the late 1920s until the mid-1930s they achieved at least performance parity with fighter aircraft.

### The U.S. Superiority in Bomber Aviation

America emerged from World War I with a large air force, albeit one that flew foreign-designed and -manufactured aircraft. In short order the postwar period brought a drastic draw-down in the U.S. military. Nevertheless, bolstered by a vibrant interest in flying and simulated by growing commercial aviation, in roughly a decade U.S. industry brought the country up to the state-of-the-art in long-range aviation.

In January 1918 the Army contracted with the Martin Company to design a bomber superior to the World War I Handley Page. The resulting MB-1 (Martin's designation; GMB was its Army designation) first flew in August 1918. Smaller than the Handley Page, it was a twin-engine biplane fitted with five guns in three positions, manned by a three-man crew, with a top speed of 105 mph and a service ceiling of 10,300 feet. It was followed by the MB-2 (NBS-1, night bomber short range), ordered in June 1920. Although it looked the same, the NBS-1 had a larger wing and greater weight and could carry a heavier bomb load. The Army ordered about 130. While

it could carry three thousand pounds of bombs, its flying performance was less than its predecessor. Nevertheless the NBS-1 was significant as the aircraft that participated in Billy Mitchell's bombing of surface ships in July 1921, made up the Army's entire bomber force with ninety aircraft in 1925, and equipped all eight Army bomb squadrons until the late 1920s, when it was replaced by the Keystone series.

## The Barling Bomber

Of all the aircraft of the 1920s, none was as large or as bizarre as the Barling bomber (NBL-1, night bomber long distance). Its three wings, four rudders mounted in a biplane box tail, multiple wheels on its fixed landing gear, and birdcage of wires and struts made it an aeronautical monstrosity. The aircraft's poor performance further undercut the Barling's reputation. It is not surprising, then, that one author includes the bomber in his book *The World's Worst Aircraft*.[2] This aircraft is the poster child for the airman's mantra, "If it don't look good, it can't fly good."

The Air Service wanted a multiseat night bomber for long-range operations. More precisely this aircraft was pushed by Billy Mitchell, who in 1919 asked Walter Barling, who had worked on a similar aircraft (the British Tarrant Tabor), to design an aircraft to carry bombs heavy enough to sink a battleship. (And thus the aircraft was later dubbed "Mitchell's Folly.") In May 1920 the Army Engineering Division issued ambitious specifications that included six machine guns, a five-thousand-pound bomb load, a twelve-hour endurance at ten thousand feet, and a maximum speed of one hundred miles per hour at sea level. The Air Service hired Barling to design the bomber, and Wittemann-Lewis won a contract to build two of the aircraft, later reduced to one because of cost. The NBL-1 made its first flight in August 1923 and proved underpowered, restricted by the requirement to use the Liberty engines of which the United States had a surplus. Nevertheless, the Barling set a number of records for duration, load, and altitude with load.

The NBL-1 was the largest aircraft in the world. It was powered by six engines mounted in four nacelles, the inboard nacelles enclosing two engines that turned tractor and pusher props, the six producing a total of 2,400 hp. The Barling had a crew of a least four: two pilots seated side by side with dual controls and two flight engineers, along with a bombardier and gunners who manned seven guns mounted in five

(Top) The MB-2 (NBS-1) equipped all of America's bomber forces in 1925 and served into the late years of that decade. It could carry three thousand pounds of bombs and, like the Handley Page O/400, featured folding wings. The MB-2 is best remembered for its role in the sinking of surface ships led by Billy Mitchell in July 1921. (Historical Research Agency, USAF)

(Right) America's Barling bomber (NBL-1) was a giant disappointment. The largest and heaviest aircraft of its day, it was an aerodynamic monstrosity with its six engines, three wings, and boxlike empennage with four rudders. This massive airframe coupled with inadequate power produced poor performance. (National Museum of the USAF)

positions. At an empty weight of almost twenty-eight thousand pounds and takeoff weight (without bombs) of just over thirty-two thousand pounds, it was the heaviest aircraft flying for a number of years.

The bomber proved to be a failure. An official test report indicated that the Barling could not meet performance specifications even when flying without armament or bombs. These tests demonstrated a maximum speed of ninety-six miles per hour at sea level and a service ceiling of seventy-three hundred feet. (This meant that it could not cross the eastern U.S. mountains with a heavy load or with a safety margin in the event of engine failure.) Its range was also disappointing: only 170 miles with 2.5 tons of bombs. The test pilot noted that considering its size it was not difficult to land, but it was "somewhat different" than normal size aircraft, referring to the craft's inertia and slow response to the controls. The NBL-1's approach speed was fast; the pilot opined it should not be slower than

ninety miles per hour, which was only slightly less than its top speed.

The Barling appeared at several air shows in the mid-1920s, where its giant size impressed the crowds. However, its limited performance dissatisfied the airmen, and the bomber made its last flight in May 1925. For its day it was an expensive machine, costing $.5 million along with an addition $.7 million for a hangar. In 1928 it was burned, on the orders of Henry "Hap" Arnold, who would command the Army Air Forces (AAF) during World War II.

## From Barling to B-10

For decade and a half after the end of World War I, U.S. Army airmen used twin-engine, short-range aircraft as their standard bombers. And these were little improved over what had been in service in the last year of World War I: The DH-4 had a top speed of 124 mph, whereas ten years later the Curtiss Condor B-2 could reach 130 mph. These bombers looked very much the same, with any increase in performance due mainly to greater engine power, which increased from the 420 hp of the famous Liberty engine to the 575 hp of the Wright R 1820, which powered the B-6A.

During the 1920s the airmen's conception of bombers was in flux, as seen in their shifting aircraft designation systems. In September 1919 the Air Service adopted an aircraft designation system of fifteen aircraft types, including three bomber categories: day bombardment (DB), night bombardment short distance (NBS), and night bombardment long distance (NBL). In 1923 a board of officers recommended that for the purpose of requirements and specifications aircraft be categorized into seven groups, which included light bombers and heavy bombers. Light bombers would carry twelve hundred pounds or less, and heavy bombers more than that. Heavy bombers were to have three or more engines, be armed with four machine guns in two turrets, and be capable of achieving a service ceiling of ten thousand feet. The system was changed the next year to include categories for attack (A), bombardment (B), bomber long range (BLR), heavy bombardment (HB), and light bombardment (LB). In 1928 a number of airmen pushed for two specialized types of bombers: a long-range night bomber capable of carrying heavy loads and a short-range, high-speed day bomber with a lighter bomb load but heavier armament. This concept was supported by two air organizations (the Bombardment Board and the Air Corps Tactical School) that also endorsed the specification for the high-speed bomber of 160 mph, a service ceiling of eighteen thousand feet, and a radius of 250 miles, with a bomb load of twelve hundred pounds and an armament of six machine guns. In March 1930 the Air Corps Tactical School recommended the substitution of the day and night system with one that again categorized bombers by their bomb load: the light bomber with a maximum of twelve hundred pounds and the heavy bomber with a minimum of twenty-five hundred pounds. Other airmen saw range as the critical distinction between bomber types, a view aided by the rapidly improving aeronautical technology.

The Army, that is, the non-airmen, however, wanted a single general-purpose bomber to save money. This led in 1928 to the conversion of two observation aircraft, the Fokker (XO-27) and Douglas (XO-35), into fast bombers. Both had retractable landing gear and were powered by 600-hp engines. Fokker delivered the XB-8 to the Air Corps in February 1931; the Douglas XB-7 appeared slightly later. The Fokker could reach 160 mph, the Douglas 169 mph, despite open cockpits and wing struts. The XB-7 proved superior, leading the Air Corps to order seven Y1B-7s for service tests. Douglas made some minor changes, including installing 675-hp engines that pushed top speed to 182 mph. This was sixty miles per hour faster than the Keystone bombers and competitive with the Air Corps fighters of the day, as they demonstrated during the 1933 Fort Knox maneuvers, where they outran pursuing P-12s. Due to the rapid advances in bomber aircraft, the B-7 was quickly superseded and the airmen ordered no further examples. At about the same time Boeing came up with a comparable aircraft.

The Boeing bomber was a derivative of its single-engine, all-metal Monomail transport. The XB-901 bomber first flew in April 1931, powered by two 575-hp engines and featuring retractable landing gear that helped it reach a maximum speed of 163 mph and a service ceiling of over nineteen thousand feet. The aircraft had four open cockpits (with the pilot and copilot in tandem positions) that were criticized for the difficulty they posed to the front gunner and copilot if they had to abandon the bomber in flight. The test pilot also noted the heavy rudder pressure required at high speed, even though it was the first American aircraft to have rudder servo tabs that employed aerodynamic forces to assist the pilot. The Air Corps ordered a prototype and six service test models designated YB-9, with supercharged 600-hp engines turning three-bladed props. These improvements increased top speed to 188 mph. The airmen also bought a second version, designated Y1B-9, which was powered by liquid-cooled engines that also produced 600 hp but was 14 mph slower than the YB-9. (There is no explanation for this disparity other than a small difference in weight and props.) The Air Corps fitted five other aircraft (Y1B-9A) with the radial engines, allowing them to match the YB-9's top speed. It was defended by two machine guns and could carry twenty-two hundred pounds of bombs. The bombers now had speed comparable with the best fighters of the day. The B-9 seemingly had a bright future with the Air Corps as the airmen preferred it to the B-7 because of its greater range (540 versus 410 miles) and heavier bomb load (2,260 versus 1,200 pounds). It should be noted, however, that its engines vibrated excessively, the metal skin wrinkled, and the fuselage tended to twist. Meanwhile a more impressive aircraft put an end to Boeing's hopes for a large production order.

## The Martin B-10

While neither a strategic nor long-range bomber, the Martin B-10 demonstrated U.S. technological prowess and leadership in bomber aircraft, which the United States has maintained since this aircraft's first flight. The Martin Company produced a brilliant design that incorporated most of the technological advances of the day—all-metal construction,[3] twin engines, monoplane configuration, an enclosed bomb bay, and

retractable landing gear—with only one regressive element: three open cockpits. The XB-907 initially flew in February 1932 and went off to Dayton for testing that summer, where it registered a top speed of 197 mph and a service ceiling of twenty thousand feet. This performance was impressive, offset by problems with severe engine vibrations that shook loose rivets, wires, and brackets and a higher landing speed and longer takeoff roll than of existing aircraft. Martin went back to the drawing board and made some significant additions that included a larger wing, more powerful engines, an enclosed manual nose gun turret, and full engine cowlings. These changes increased the top speed to 207 mph, the world's first 200-mph bomber, and reduced landing speed from 81 mph to 71 mph. The Army ordered forty-eight production aircraft which incorporated two sliding canopies, one over the pilot and the other over the rear gunner. These YB-10s were followed by the B-10B, the major production model that featured manually deployed wing flaps and controllable pitch propellers and, with greater engine power, increased top speed to 213 mph. In short, the Martin B-10 was the first bomber to incorporate all of the technological improvements of the era and was rewarded with outstanding performance.

The B-10 handled well and was considered a very "forgiving aircraft," a quality highly regarded by pilots. On the negative side, the crew could not exchange positions in flight, the copilot would have great difficulty landing the aircraft due to his restricted forward visibility, and the aircraft had a tendency to float when landing at light weights. Because of its chunky profile and despite its outstanding performance, the Martin bomber earned the unflattering nickname "Flying Whale." The Army bought 154 B-10s and its derivatives, the most it acquired of any bomber type between the DH-4 and the Douglas B-18.

The B-10 was the fastest bomber of its day and garnered much publicity, many

(Left) The Martin B-10 incorporated the latest aviation technology of the early 1930s and was the breakaway bomber that clearly demonstrated U.S. dominance in that field. Despite its chunky appearance and unflattering nickname ("Flying Whale"), it flew well and fast (it was the first two-hundred-mile-per-hour bomber) and more than 270 were built. (National Museum of the USAF)

records, and frequent awards. In 1932 Martin received the prestigious Collier Trophy, awarded for the outstanding annual achievement in aviation. Henry "Hap" Arnold led ten B-10s on an eighty-three-hundred-mile flight to Alaska and back in the summer of 1934, which attracted considerable press attention. The Martin bomber also made distant flights to Argentina and the Dutch East Indies.

The government prohibited Martin from selling the B-10 to foreign governments until 1936. Then a number of air forces bought the bomber, including those of Argentina, China, Holland, Russia, Siam, and Turkey. The Chinese Martins saw action in 1937, with two dropping leaflets over Nagasaki, Japan, in May 1939. (It later would be one of the first and one of the last American targets in the Pacific war.)

The B-10 dominated the bomber class in the early 1930s but in turn fell victim to rapidly advancing aviation technology. Martin entered it in the 1936 Air Corps bomber competition in which the Boeing B-17 was the overwhelming winner. The B-10 would see action early in World War II, but was quickly shot out of the skies.

## The Really Big Bombers

The issue of desired aircraft type, fueled by rapid technical progress, became embroiled in a roles and missions fight in the United States. Traditionally, and understandably, the Navy had the responsibility for the sea-based protection of the U.S. coast, while the Army had the responsibility for land-based defenses. The increasing range of aircraft upset this tidy division as the Navy put aircraft aboard ships and ashore while the increasing range of land-based aircraft allowed the Army to extend its operations well beyond the shore line and the previous line of demarcation, the range of its coast defense artillery. This situation led to duplication and conflict between the services.

The Air Corps pushed the coast defense mission, which fit in with its quest for longer range bombers. In November 1931 the Air Corps advised the Army chief of staff that the airmen should develop a long-range reconnaissance aircraft, and in 1933 the Air Corps chief defined its coast defense responsibility in terms of the maximum range of aircraft. That same year Brig. Gen. Oscar Westover, then assistant chief of the Air Corps and later the chief, stated that "high speed and otherwise high-performing bombardment aircraft, together with observation aviation of superior speed and range and communications characteristics, will suffice for the adequate air defense of this country."[4] In that same report Westover further noted that the high speed of bomber aircraft combined with its ability to fly in close formation meant that "no known agency can frustrate the accomplishment of a bombardment mission."[5] The connection between what kind of bomber and the missions of coast defense and strategic bombardment were entangled by the rapid technological changes taking place. This is not the place to sort out the relative importance of these factors but only to observe that they all were present and all played a role in what transpired.

(Left) The Boeing B-15 was an ambitious giant that exceeded the technology of the day. It had a greater wing span than its offspring (the B-29) but fell short of its range requirement, had limited performance, and was difficult to handle. (National Museum of the USAF)

Concurrently an Air Corps engineering study concluded that one aircraft could combine the reconnaissance and bombing function and reach out five thousand miles, with a bomb load of two thousand pounds and a speed of two hundred miles per hour. In December 1933 the Air Corps proposed to build such an aircraft, "Project A," which was quickly approved. In May 1934 the chief of the Air Corps authorized negotiations with Boeing and Martin for designs for such a machine with the stated mission of destroying distant land or naval targets and reinforcing Hawaii, Panama, and Alaska. In June 1934 Boeing won a contract that led to an order for one aircraft.

Originally designated XBLR-1 (experimental long-range bomber), the aircraft was redesignated XB-15 before its initial flight in October 1937. It had taken three and a half years to get the giant into the sky, at this point the largest and heaviest American aircraft to fly. Following tests at Dayton it made a number of notable flights ferrying medical supplies to Chile and setting a number of records. Similar to the Barling bomber, the XB-15 was large, underpowered, and lacking in flying performance . . . and only one of each was built. Probably its chief claim to fame was its size; its 149-foot wing span was longer than the World War II B-29s (141 feet) and the huge wing permitted inflight inspection and maintenance of the engines.[6] It was initially defended by three .30-caliber and three .50-caliber guns mounted in manually operated gun positions, could carry a maximum of six tons of bombs, and with a ton of bombs could fly thirty-four hundred miles. However, the aircraft fell short of the range requirement, was heavy on the controls, and was hard to handle. In Curtis LeMay's words, "It was a brute of an airplane to fly."[7] The big plane was overshadowed by more practical (and smaller) types that fought in World War II.

One author wrote, "The story of the XB-15 is one of high hopes, disappointment, quiet accomplishment, and finally unmerited consignment to oblivion."[8] Another concluded that the B-15 was "born out of despair, raised on a succession of bad breaks, and cheated by time."[9] These statements are a mixture of romanticism and rhetoric, hyperbole and fact. Less poetic but more to the point, the B-15 was just too ambitious for the technology of the day (specifically, the power plants). The rapidly advancing technology outmoded the aircraft even before it took to the air, as demonstrated by another Boeing product, the famous B-17 Flying Fortress. Meanwhile the airmen put another giant into the sky.

The B-19 was even larger than the B-15 and thus a greater challenge. In 1935 the Air Corps invited both the Douglas and Sikorsky companies to compete in "Project D" to build the aircraft, funding both to design it, then picked Douglas to build the flying article. But by the summer of 1938 Douglas had lost interest in the project, and in an unprecedented move the company recommended that it be cancelled because the design was outmoded, its weight was dramatically increasing, and Douglas needed personnel for other designs more certain of production. Nevertheless the Air Corps pressed on, albeit at a slower pace.

The XB-19 finally made its first flight in June 1941. Testing revealed brake problems, and although the aircraft could reach a top

Even larger than the B-15 was the Douglas B-19, which was obsolete before its maiden flight in June 1941. It employed two innovations, reversible pitch propellers and powered gun turrets, but it proved to be another lackluster performer. As with the other abortive giants, it is remembered for its size and lack of success. (National Museum of the USAF)

speed of 224 mph for a limited time, overheating of its R 3350 engines forced the use of the engine coolers, which reduced the maximum speed to 204 mph. The last major change came in January 1944, when the more powerful Allison liquid-cooled engines replaced the Wright radial engines, which led the airmen to redesignate the bomber XB-19A. With the Allisons, top speed increased to 265 mph, and service ceiling rose from thirteen thousand to thirty-nine thousand feet. However, the giant aircraft needed much greater power to propel its enormous bulk at competitive speeds.

Unlike the B-15 the Douglas bomber did not set any records nor make any memorable flights. However, its contribution was similar; it was an exercise in designing, building, and operating a giant aircraft, as was its claim to fame, its size. It was in a line of development of giant aircraft that featured long-range and heavy defensive armament, rendered obsolete by later bomber and fighter technology. As we shall soon see, other bombers would wage the strategic air war for the United States, specifically the B-17 and B-24 in the European theater and mainly the B-29 in the Pacific theater.

## French and Soviet Programs

Although only three countries engaged in strategic air warfare during World War II, two additional nations built strategic bombers in the interwar years. The Soviets deployed a large and successful air force and more heavy bombers than any other nation during this period. The French also built heavy bombers, but as with their main defense preparations, these were too little, too late.

The Russians were the first to field a four-engine bomber, although they did not engage in strategic bombing during World War I. In the civil war that followed, aviation did not prove significant. After the war, however, the new Soviet regime put considerable emphasis on aviation as a symbol of modernism and communism. The Russians built a number of large bombers. The most famous large aircraft, although a propaganda machine rather than a bomber, was the eight-engine *Maxim Gorky*, which had a 207-foot wing span and first flew in June 1934. Another was the five-engine ANT-14.[10]

Our interest is more narrow: the Soviet heavy bombers. The four-engine ANT-6 began in December 1926, first flew in December 1930, and was redesignated as TB-3. The aircraft was like others of the day, a low-wing, all-metal monoplane with open crew positions and fixed landing gear. It was defended by ten machine guns in five positions, carried a bomb load of twenty-two hundred to forty-nine hundred pounds, and had a top speed of 186 mph. Later versions reduced the defensive armament and crew. The aircraft went into service in 1932 and flew in the various Soviet conflicts of the late 1930s and in World War II as a transport. The significance of the bomber is that about 820 were built between 1932 and 1937, the largest fleet of long-range, heavy (strategic) bombers fielded during the interwar years. But events in the second half of that decade, specifically the Spanish Civil War and the purges, dampened if not killed strategic bombing ideas in Russia. So ended what one historian has described as "the Soviet's brief but serious flirtation with the strategic bombing ideas of Giulio Douhet."[11]

In 1934 the Soviets put forth a specification for a better long-range bomber that became the ANT-42, later known as the TB-7 and then as the Pe-8. It first flew in December 1936. An unusual modification mounted a fifth engine in the fuselage to power the superchargers that doubled its service ceiling from fifteen thousand to over thirty-five thousand feet and gave it a top speed of 250 miles per hour, which enabled it to outrun German Bf 109Bs and He 112s

The Soviet TB-3 first flew in December 1930. It is notable not for its flying or operational performance but because more were built than any other strategic bomber during the interwar period. Note the parasite fighters hung underwing. (San Diego Air and Space Museum)

at altitude. But by adding weight (two tons) and reducing volume, it cut both range and bomb load. The bomber was protected by twin 7.62-mm guns in the nose, a single 12.7-mm gun in the rear of each of the inboard engine nacelles, and one 20-mm cannon each in a tail and a dorsal turret. Versions built during the war included one with diesel engines, an arrangement that increased range but decreased speed and ceiling, and another that went into production in 1943 without the nacelle guns. The latter could achieve a maximum speed of 280 mph and a range of thirty-seven hundred miles. An indication of the Soviet focus is that they built fewer than 150 Pe-8s, removed the bomber from production in 1944, and that long-range bombing consumed only 4 percent of their air force effort during World War II.

The French did less well. While they turned out a profusion of aircraft models, some with competitive performance, these arrived too late to be of any help in World War II. Development of heavy bombers was even slower due to lack of funds and a belief in multipurpose aircraft. The Bloch 162 developed from a line of four-engine transports and was smaller, shorter legged, and faster than the B-17. But it did not make its first flight until June 1940. The C.A.O. 700 was about to make its maiden flight when the Armistice grounded all French aircraft.

## British, German, and American Doctrine

World War I promised that airplanes would be useful in future wars, a lesson acknowledged by most and doubted by few. The question was how to use this new weapon. The postwar consensus emphasized air power's role in the support of armies and navies. Nevertheless, some airmen believed the airplane could do much more and insisted the war had ended before their ideas could be demonstrated. A few even claimed that attacks on the enemy's rear areas could, by themselves, be decisive—a concept that became the basis of strategic bombing theory.

A number of individuals in different countries laid out this theory, the most famous being a Briton, Hugh Trenchard; an Italian, Giulio Douhet; and an American, Billy Mitchell. Although these men and their ideas received considerable attention, the basic concept of bombing cities and civilians was most memorably put forth by British prime minister Stanley Baldwin, who summed up the concept of strategic air warfare in 1932 when he told members of Parliament, "There is no power on earth that can protect [the civilian] from bombing, whatever people tell him. The bomber will always get through. . . . The only defense is the offense, which means you have got to kill more women and children than the enemy if you want to save yourselves."[12] Thus the ideas that formed the core of the strategic bombing theory emerged during and soon after World War I: There was no way to stop the bombers, cities and civilians would be their targets, and these attacks on the enemy's homeland would be decisive. Airmen in numerous countries accepted and espoused these ideas.

If land warfare has its Carl von Clausewitz and sea warfare its Alfred Thayer Mahan, air warfare theory is dominated by

Giulio Douhet (1869–1930). Like his fellow theorists he is more quoted and cited than read, but in contrast, his influence is much more in dispute. It appears that the American and British airmen constructed their theories independent of Douhet. Nevertheless, it is correct to comment that his writing, and especially his classic book *The Command of the Air*, "remains the most eloquent, elaborate, and comprehensive theory of air power in the interwar period."[13] One prominent military writer, however, has called Douhet the "prophet of the ridiculous."[14]

Douhet trained in the artillery arm and went on to serve on the Italian General Staff early in the twentieth century. After Italy's entry into World War I, Douhet pushed hard for a massive aviation build up, but his criticism of Italy's floundering war effort led to his court-martial and a one-year prison sentence. However, Italy's poor wartime performance allowed him to gain a high position in Italy's air arm during the conflict, although his court-martial was not reversed until the war's end. He was promoted to general and later became a supporter of Benito Mussolini, who gave him a high position after coming to power, although in short order he resigned. He died in 1930, only a few years before his ideas would be tried and tested in the largest war in man's history.

*The Command of the Air* appeared in print in 1921, although it was not published in English until 1942. Earlier Douhet materials were published outside of Italy, in both Britain and France in 1933 and Germany in 1935. However, Douhet had the greatest and earliest influence in the United States. An article on Douhet by the Italian air attaché was published in an American aviation periodical in 1922 and excerpts of his work reached the American airmen in 1922, followed shortly by a translation of *The Command of the Air*. A translation of the second edition of the book was in circulation in the Air Corps in 1933, and Douhet was mentioned in Air Corps Tactical School lectures in the 1930s.

Douhet wrote that the next war would be a total war with no boundaries in regards to geography or targets. He saw no defense against aircraft. "No fortifications can possibly offset these new weapons, which can strike mortal blows into the heart of the enemy with lightning speed," he noted, adding that "with the air arm it is easy to strike but not to parry."[15] He maintained that the "air arm is eminently offensive," and went on to hold that it is "completely unsuitable for defensive action."[16] Because there was no way to stop bombing operations, all effort must go into offensive operations and combatants must absorb the bombing attacks and endure losses to inflict more damage. Therefore the airmen should concentrate on independent (read: strategic) operations, not auxiliary tasks such as support of armies and navies. (In 1926 he went further, writing that "auxiliary aviation is worthless, superfluous, harmful.") The first task in an air war was to defeat the enemy's air force, which should be attacked not only in the air but also on its bases and in its factories. He opined that the new weapon should be organized into a coequal and independent military branch. Douhet also asserted that bombers were not intended for air-to-air combat, that defensive armament was only for crew morale, and that "pursuit planes must clear the sky of enemy interference before the bombers can accomplish their mission."[17] He astutely saw that air warfare's biggest difficulty, one that remains to this day, was targeting.

His most controversial beliefs, and what usually passes for the shorthand summary of Douhet's views, are that air war could be decisive and that civilian morale was the key target. He believed that all "citizens will become combatants [as] . . . there will be no distinction any longer between soldiers and civilians."[18] In his words, "By bombing the most vital civilian centers [air attack] could spread terror through the nation and quickly break down [the enemy's] material and moral resistance."[19] He further held that "mercifully, the decision will be quick in this kind of war . . . [and] these future wars may yet prove to be more humane than wars in the past in spite of all, because they may in the long run shed less blood."[20]

Douhet made a number of statements that would prove to be in error. For example, he believed the power of antiaircraft artillery was limited, that firepower not speed would determine the aerial battle, that civilian aircraft could be converted into effective bombers, and that poison gas would be widely used. In addition he grossly overestimated bombing destruction.[21] In an article published only a few days after his death in 1930, Douhet wrote that a future war between France and Germany would be over in a day.

In brief, Douhet had broad, bold, imaginative ideas on strategic bombing and put these into print early. Working on the limited experience of World War I, he saw the great impact bombing would have in the future. Some of his assumptions and conclusions proved in error because the new technology had not yet matured, and perhaps because he was not a flyer. Nevertheless he set the agenda. His direct influence remains in dispute and may well have been limited, but he was frequently mentioned by aviators throughout the world and used to bolster their equally far-reaching ideas. Certainly his core ideas on bombing's decisiveness, morale as the critical target, and air power's offensive nature were believed by many during the interwar years, and for that matter since then. And while World War II did not confirm Douhet's ideas, and may in fact have refuted some if not many of them, the emergence of the long-range bomber delivering nuclear weapons during that conflict certainly gave new life to them. These early and far reaching predictions of the future of air power merit the position that Douhet has achieved as the most important advocate of strategic air power.

British airmen held essentially the same positions as Douhet. Hugh Trenchard (1873–1956), who led the RAF (1918–30) during its infant years and is credited with being its father and savior, wrote in 1916 that the airplane was an offensive weapon that could not be stopped by defensive aircraft. Thus he favored building as many bombers and as few fighters as possible. Also like Douhet he saw air superiority as the first goal of air forces. During World War I he pushed interdiction targets as the focus of British bombing, although he did see the importance of a number of strategic targets. He is best known for his emphasis on the effect of bombing on morale, which he

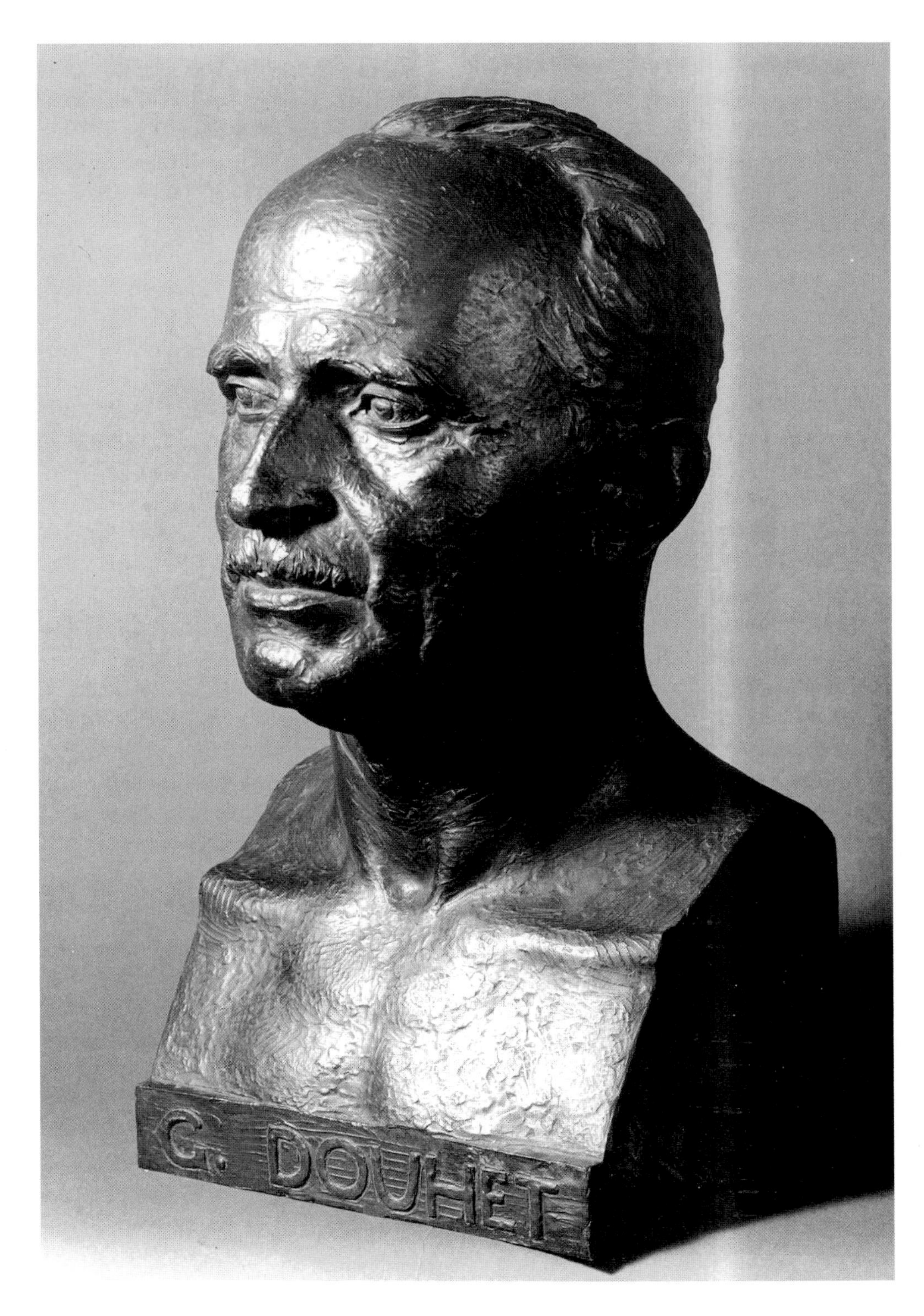


The Italian Giulio Douhet (1869–1930) remains the eminent strategic bombardment theorist. His classic book *The Command of the Air* appeared in 1921 and held that bombing could be decisive and civilian morale was the key target. This bust is appropriately located in the Air University Library in Montgomery, Alabama, the second site of the Air Corps Tactical School and the present location of the USAF's professional schools. (Air University Press)

Hugh Trenchard (1873–1956) *(left)* commanded British strategic bombers during World War I, fathered the RAF, and preserved it following the conflict. He saw bombers as an offensive weapon that were unstoppable and believed morale was the primary target. (National Museum of the USAF)

considered greater than on material targets. (This may have originated from bombing's limited power against physical targets as contrasted with its terrifying impact on morale.) "Using a subjective and unprovable statistic that earned him much (largely undeserved) ridicule," one author notes, "Trenchard stated that the psychological effects of bombing outweighed the material effects at a ratio of 20 to one."[22] However, in 1928 Trenchard wrote of the need to target military objectives and avoid "indiscriminate bombing of a city for the sole purpose of terrorizing the civilian population."[23]

The RAF survived during the lean interwar years, due in large measure to Trenchard's skill in fighting and manipulating the bureaucracy. During this period the RAF took two roads, one in the field and one in theory that enabled it to maintain its independence. In practical terms the RAF carved out a niche for itself by helping pacify colonial uprisings with the tactic of "air control," which used bombing to selectively harass, intimidate, and coerce those who got out of line. Air control proved effective as well as far cheaper than conventional punitive expeditions, which fit the British quest for reducing their military budget. However, operations against tribesmen armed with rifles did not prepare an air force for an air war against a major power.

At the same time the airmen developed a strategic bombing concept. Britain feared an aerial assault, the so-called "knock out blow," more than any other country. In 1924 the RAF estimated that based on World War I experience, one ton of bombs would inflict fifty casualties and that the French, the posited enemy of the day, had the ability to inflict more than twenty thousand casualties in the first week of hostilities and twice that with a maximum effort. These beliefs led the British to downgrade the importance of the air superiority battle and to emphasize strategic bombing with an air force composed of twice as many bombers as fighters. In the mid-1920s the RAF also held that long-range fighters were unnecessary for bomber protection. On one key point the British concept differed from Douhet's. While the British held that victory could be had by breaking enemy morale, this was not to be done by bombing people but by bombing the factories that employed them. The RAF explicitly rejected the idea of indiscriminate terror bombing, and unlike Douhet, who thought the chemical weapons ban would be quickly broken, did not foresee the use of poison gas in a future war. However, the airmen did not make clear how the factories could be destroyed without killing the workers and their families who lived nearby. Ironically, in view of what would transpire in World War II (and his role in promoting that tactic), Winston Churchill noted during World War I that he doubted "that any terrorization of the civil population which could be achieved by air attack would compel the Government of a great nation to surrender."[24]

The British feared a "knock out blow," a "bolt out of the blue" bombing offensive against London, which was not just their capital but also their economic and social center. London was seen to be more vulnerable than any other European city because of its population concentration and its short distance from the coast and to potential enemy air bases on the Continent. In Churchill's words, London was "a tremendous fat cow . . . tied up to attract the beasts of prey."[25] In fact, British leaders had sought an international prohibition of bombing even before the Wrights's first flight.[26] In the early 1930s the world community discussed general disarmament, and the British made several attempts to abolish aerial bombardment. These efforts got hung up on questions of converting civil transports into bombers and the desires of France and Italy to retain bombers for use in their African colonies and then were dashed when the Germans bolted from the conference.

During the same time frame, the development of strategic bombing in the United States was hindered by the Army's tactical support requirements, restricted funds, and public belief in a "defensive war only." The interwar years were lean for the U.S. military, especially for the airmen who were developing a rapidly evolving and expensive technology. Nevertheless great strides were made with both hardware and doctrine. But before focusing in on the mainstream of American doctrinal development, a discussion of the most famous American military aviator of the interwar period is in order.

William "Billy" Mitchell (1879–1936) enlisted in the Army during the Spanish American War and later learned to fly at the relatively advanced age of thirty-six. He was in France when the United States entered World War I and soon became the top-ranking American airman in Europe. During the war he proved to be an excellent combat leader, commanding large air forces, and came into contact with Hugh Trenchard. In 1921 he planned the bombing operations that sank captured German warships, which dramatically demonstrated the vulnerability of navies to aircraft and the power of the air arm, an event that was the high point of his career. Because his ideas for an independent air force and personal ambitions were thwarted, a frustrated Mitchell made some harsh, intemperate public remarks, was court-martialed, and consequently resigned from the Army in 1926. A prolific writer, he pushed for a greater role for the air arm and its independence until his death in 1936. He became a symbol of the maverick standing up against the system, a mistreated visionary and towering air force leader. He inspired many airmen, and some refer to him as a prophet and martyr of air power.

What were his ideas? It is clear that Mitchell did not contribute new concepts. He was familiar with Douhet and went along with the major thrust of his ideas, although Mitchell was less extreme than the Italian. His thinking evolved over the years from first emphasizing control of the air and fighter aircraft to later views of attacks on "vital centers" with an air force built around the bomber. In January 1920 Mitchell wrote of hitting nerve centers at the start of a war, the next year he included bombing industry for both physical and psychological purposes, and later he held that population centers would be attacked. He predicted that in future wars entire cities would be destroyed by aerial gas attacks. As early as 1922 he held that a proper air force could win a war by itself and later (1930) that the air arm could bring "quick decisions." Mitchell believed that the bombers would require fighter escort and that antiaircraft artillery was futile. He is notable as a symbol and polemicist rather than as a thinker and theorist. So where did the Americans form their strategic bombing doctrine?

World War I sowed the seeds. An aeronautical commission sent to Europe to survey the scene (headed by Maj. Raynal Bolling), and strongly influenced by Mitchell, called for a force consisting of over 60 percent bombers and the bombardment of Germany and wrote that such bombing might be decisive.[27] This was rejected by the War Department, which in August 1918 approved a plan for a force in 1919 of about 20 percent bombers. Despite the limited use and impact of strategic bombing in the war, the airmen developed a full-blown and independent American theory of strategic bombing at the Air Corps Tactical School. The Tac School occupied this central position because it was the airmen's highest educational establishment and was where most of the United States' top air leaders of World War II served as either instructors or students during the interwar years. William

(Right) William "Billy" Mitchell (1879–1936) is best known for bombing tests that sank battleships, his 1935 court-martial (seen here), and as a strident promoter of an independent air arm. Mitchell was a prolific writer who advocated an air force built around bombers and the attack of "vital centers." (U.S. Naval Institute Photo Archive)

Sherman, one of the instructors at the school, set out the thinking of the American airmen in a book based on his lectures at the school. He noted approvingly that Marshal Ferdinard Foch, the top French general of World War I, saw "the possibility of bringing such pressure to bear on civilian populations as to end war through the action of the air force alone."[28] Sherman wrote that aircraft were purely an offensive weapon and that a vigorous offensive was the best defense. Reflecting on the lessons of the air war, he observed that while night bombing was done in comparative safety, day bombing, despite tight formations and improved gunnery, required escort. He observed that antiaircraft guns were ineffective and nothing more than a drain on resources. As for the bombing of civilians, he held that "among peoples in whom the spirit of sport has been strongly inculcated [Western Europeans and former British colonials?], it is peculiarly abhorrent to contemplate the waging of war on unarmed civilians of all ages and sexes."[29]

The American strategic bombing theory was formulated at the Air Corps Tactical School, where Kenneth Walker (1898–1943) was an instructor and a prime contributor to this theory. He was killed in action early in World War II leading a bombing mission and was awarded the Medal of Honor. (Air University History Office)

He had no confidence that international law would prevent such attacks but wrote that "a potent restraint will always be exercised by the fear of reprisals."[30] Sherman also noted that industry was made of a complex system of interlocking factories, so that only key plants had to be destroyed to cripple the entire economy, what later would be called the "bottleneck" or "strategic web" theory. The last point was the crux of what became the distinctive U.S. strategic bombing theory.

The American strategic bombing theory appeared in the early 1930s concurrent with rapidly advancing aviation technology that produced the equipment needed for strategic operations and that gave the bomber temporary ascendancy over the fighter. Technical developments obviously influenced doctrine, and the officers at the Tac School thought mostly in terms of the future. They added two distinctive elements to create the American strategic bombing theory.

First, the American bombing advocates maintained the bombers could reach their targets without escort and without suffering undue attrition. This was a break from the past because it had been neither their World War I experience nor what appeared in their writings and lectures during the early 1920s. While escort was considered useful, most if not all airmen around the world believed it was technically impossible to build an aircraft with both fighter performance and bomber range. It was late in the 1920s when the "unescorted" bombing theory evolved with American airmen insisting that bombers maintaining tight formations, mounting heavy defensive firepower, and flying at high speed could nullify enemy fighters. Escort was considered desirable but unobtainable and unnecessary. Certainly the American airmen attempted to provide fighter escort, with most attention focused on fielding a heavily armed, multiseat, multiengine aircraft—a concept advanced as early as 1920. While such an aircraft offered the advantages of greater range and heavier firepower, efforts to build one with performance competitive with defensive fighters failed. The airmen also discounted antiaircraft artillery, positing that high-altitude operations would lessen or neutralize its impact. In brief, and in words that serve as a military analog to Stanley Baldwin's famous quote, American bomber advocates believed in the phrase attributed to Tactical School instructor Kenneth Walker (1898–1943) that the "well-organized, well-planned, and well-flown air force attack will constitute an offensive that cannot be stopped."[31]

The absence of a practical escort did not upset the bombing enthusiasts. As we shall see, the advent of the B-17 confirmed the bomber advocates' faith in unescorted bomber operations as here was a aircraft that could outfly, outgun, and thus outfight fighters—at least when it first took to the air in the mid-1930s. The American bombing theory emerged during 1930–35 at a time when the bomber had achieved at least technological parity with the fighter. Little wonder that the "dominating echelon [in the Air Corps], both numerically and in terms of rank, firmly believed that a bomber, through applying proper formation and mutual defense, could whip opposing fighters, penetrate to the target and destroy it, and that therefore there was no need for fighters."[32] Bomber proponents failed to foresee that the defense would also benefit from advancing technology (especially radar) and that the fighter would regain its ascendancy.

A second idea that made the American bombing theory unique was what came to be called the "industrial web" concept: Destroying key elements of a nation's economy would be decisive. Early on American airmen shared the views of their European contemporaries, who saw cities, people, and factories as correct targets. In 1926, however, an airman instructing at the Tactical School wrote that only key plants had to be destroyed, an idea that came to dominate. The lure of greater efficiency in terms of greater results that required fewer aircraft, crews, and bombs along with lesser overall casualties were critical factors. If these bottleneck targets could be identified and then destroyed, the bombing proponents believed a nation's ability to fight would be broken and the civilian life so disrupted that an enemy would be forced to surrender. This concept was reinforced when deliveries of a new aircraft were delayed. Investigation revealed that a flood at the sole manufacturer of the springs used in the controllable-pitch propeller had stopped production, an example instructors at the Tactical School used in their lectures during the early 1930s. The airmen attempted to identify these bottleneck targets during the interwar years and, for example, noted such targets as transportation, steel, and electric power in the lectures given during the 1933–34 school year at the Tac School. As the airmen studied the U.S. economy for other clues to economic bottlenecks, they added finance, utilities, raw materials, oil, and the food supply to the target list. In 1939 an Air Corps lecture listed electric generating plants in the United States and synthetic oil refineries in Germany as examples of bottleneck targets.

The key to this bombing doctrine was accuracy. The American emphasis on high-altitude and high-speed formations of heavily armed aircraft necessitated compromises in other performance. For example, as altitudes increased to lessen the impact of enemy antiaircraft artillery, bombing accuracy decreased. Other areas of compromise concerned bomb load and range. The airmen favored range and heavy defensive firepower at the expense of bomb load, which further increased the need for accuracy. Hence the importance of the highly sophisticated and accurate Norden bombsight, which emerged in the 1930s. Because of the location of the United States and its potential enemies, range was one of the most difficult and obvious problems for the airmen. To overcome range limitations, they discussed technological developments such as air-to-air refueling and the use of foreign air bases. By the mid-1930s they had developed their own strategic bombing theory: Unescorted, heavily armed aircraft flying in formation could economically and accurately bomb and destroy industrial targets from high altitude in daylight and thus win wars.

## Combat Operations between the World Wars

Conflicts continued in the years following World War I, the "war to end all wars." The limited use of aviation in these fights was in stark contrast to the extraordinary pace of aircraft development of the period. One of the most important conflicts was the Russian civil war, which raged between 1917 and 1922, in which aviation played no significant role. Aircraft were sparsely employed, but certainly not strategic aviation. The same was true in the Allied interventions in Russia (1917–22), the Russo-Polish War (1920), and the Greco-Turkish War (1920–22), and the one major war in the Western Hemisphere, the 1933–35 Chaco War.[33] Air power was more important in the one-sided Italian war against Ethiopia (1935–36). The war's image of spear-wielding Ethiopians fighting against poison gas obscures the effective Italian use of air power, especially air transport, in difficult conditions.

The conflict in Spain in 1936–39 was not as important in shaping the twentieth-century world as the revolutions in China and Russia, yet from the point of view of air power it was the most important of the interwar conflicts. Because of the large number of aircraft engaged, their quality, and the war's location, this conflict was closely watched by the militaries of the major powers. In July 1936 a military coup failed to topple the government in Spain leading to a prolonged and costly civil war. Germany and Italy quickly sent men and machines to assist the rebels (known as Nationalists). The airlift of Spanish regular troops across the Straits of Gibraltar and over the Spanish navy, which remained mostly loyal to the government (known as Republicans), by German Ju 52s probably saved the revolt and was the most decisive use of air power in the war. The initial aircraft employed were upgraded versions of World War I aircraft, albeit with better performance. While the bulk of the Spanish air force remained loyal to the government, German and Italian aid gave the rebels the edge in the air. Then beginning in October 1936 the Soviets weighed in on the side of the government, sending not just numbers of aircraft and personnel but also two of the most modern aircraft of the day, the I-16 monoplane fighter, which was far superior to the German and Italian biplane fighters, and the twin-engine SB monoplane bomber, which could easily outrun the Nationalist fighters. This shifted the air balance back to the government. The Germans responded with the best they could produce (which would form the core of the World War II Luftwaffe), including Bf 109 fighters and the Do 17, He 111, and Ju 87 bombers. By mid-1937 the Nationalists had regained air superiority, an advantage they maintained for the remainder of the war.

Almost all of the aerial operations supported the ground troops. The Nationalists did bomb Republican ports, as well as the major cities of Madrid in 1936, killing an estimated one thousand civilians, and Barcelona in 1938, killing about thirteen hundred. In all perhaps twenty-five thousand civilians of the more than four hundred thousand killed in the war fell to bombing. However, the most memorable attack of the war, and for that matter, perhaps of all time, was the April 1937 bombing of Guernica. Two dozen German bombers, mostly Ju 52s, dropped forty-five tons of high explosives and incendiaries followed by He 51s, which strafed the survivors. Most accounts put the death count at sixteen hundred, but because this event was highly publicized as an example of Nationalist (and especially German) ruthlessness, highlighted by Pablo Picasso's vivid painting, this figure may well be inflated; regardless, the attack and the casualty toll are controversial.

Some of the lessons drawn from the Spanish Civil War relate to strategic bombing, although little such bombing was conducted there. Observers noted that bombing accuracy was poor but that Germans did better than their allies and foes. This may have resulted from the superior Nationalist (specifically German) antiaircraft fire that forced Republican aircraft to bomb from higher altitudes than Nationalist bombers. Antiaircraft artillery was effective, with the

Germans' rated as excellent. One significant conclusion was that bombing cities did not break civilian morale; in fact, the observers concluded that civilian morale hardened. Simply put, some observers believed that terror bombing had failed and Douhet had been repudiated.[34] The observers concluded that converted transports were not adequate bombers. Perhaps more noteworthy for American airmen, the consensus was that bomber operations required fighter escort. One American observer reported that the theory of bomber invulnerability "no longer holds." The Air Corps' own organ emphasized that the "comparison of an airplane to a flying fortress is possible only in the minds of the theorists."[35] The observers pushed for greater speed for both fighters and bombers, some asserting that additional guns were no substitute for more speed. One report held that many believed that future bombing would be done at night.

There were those who questioned the validity of lessons drawn from Spain. George Kenney, one of the United States' top air generals during and after World War II, wrote that there was no legitimate test of air power in Spain as both sides lacked a true air force and the numbers to fight a real air war. Other American officers concurred. One noted that only small air forces were engaged, bombs were light, no gas was used, and there was restraint. Another wrote that while aircraft had not lived up to expectations, Douhet's theories had not been tested. A fourth warned that a future war might not be like the Spanish Civil War. In a 1937 speech at the Army War College, Henry Arnold, wartime chief of the Army Air Forces, stated, "The powers, capabilities and limitations of bombardment aircraft were not properly tested in Spain. So let us be careful not to draw lessons about heavy bombardment from air work in that theater."[36] Civil wars are not conductive to strategic bombing because both sides wish to preserve the economic base. Franco did just this, on occasion criticizing his allies for bombing Spanish cities. Thus air power was restrained: There were no concerted attacks on strategic targets.

One last note about the Spanish Civil War is that the Germans gained much knowledge from the conflict. They were able to not only combat test some of the hardware that would be important in the coming conflict but also to train their men and work out tactics. This combination, more than Luftwaffe numbers, was to be the most important factor in the early German victories in World War II. The Soviets also should have benefited from the experience in Spain. However, Joseph Stalin's purges of the late 1930s decimated the Russian military, especially those who had served in Spain. One lesson, that superior German equipment and tactics made clear, was the need for modernization. Certainly the Soviets learned that tactical support was more effective than strategic bombing. So the country that had been the leader in building long-range bombers for most of the decade turned its back on that concept.

The Soviets were involved in other air operations preceding World War II. The Japanese invaded Manchuria in 1931 and then other parts of China in 1937. Although the Japanese fielded a superior air force, the Chinese did raid Formosa in December 1938 and get B-10s over Hiroshima, but they dropped only leaflets. The Russians began supplying aerial aid to the Chinese in September 1937 and prior to 1941 provided the Chinese with the most aid of any outside power. They may have sent more than eight hundred, and perhaps as many as one thousand, aircraft to the Chinese, but they had no significant impact on the war.

More significant to air power were the battles waged between the Soviets and Japanese. In the summer of 1938 there was a border clash between the two at Lake Khasan and the next year at Khaklhin-Gol (Nomonhan). In the latter there was extensive aerial combat with over one hundred aircraft battling each other on numerous occasions. Both sides employed heavy bombers, the Japanese the Italian built B.R. 20s and the Soviets their tried and true TB-3s. There were, however, no strategic targets in the war zone so these were employed in other roles. The Russians made significant gains on the ground, but there is considerable controversy about the air battle with each side claiming a major victory. The Russians clearly did better against the Japanese in China than they had against the Germans in Spain. However, their next action a few months later was a severe jolt to Soviet pride and competence.

The Russians made territorial demands on the Finns in 1939 to bolster the western defenses of Leningrad but were rebuffed. Consequently the Soviet initiated the conflict in November 1939 that lasted into March 1940. During the "Winter War" the Soviets employed overwhelming numbers and superior equipment, which the Finns met with courage, skill, and determination. The Finns initially fielded about 145 aircraft against about 900 Soviet aircraft, and during the course of the conflict the former received another 200 aircraft and the latter increased their strength to 1,500 to 2,500. Despite their numerical inferiority and hodgepodge of mainly obsolete aircraft, the Finns inflicted heavy and embarrassing losses on the Soviets, downing 600 to 900 aircraft for the loss of 60 to 70. Although the majority of the air action was in support of the ground battle, Soviet aircraft did bomb rear areas and Finnish cities with little success and considerable problems. The Finns also launched a number of raids on Soviet bases in Estonia. Thus the Soviet air force, which began the 1930s as one of the premier air forces in the world, and which had led the way with modern fighters and bombers, ended the decade in obsolescence, highlighted by being outfought by a much smaller power. This conflict may have led the Germans to underestimate the strength of the Soviets at the same time alerting the Communists to their own weakened state.

chapter 3

# German Strategic Bombardment

## LIMITED EFFORTS

The Treaty of Versailles prohibited Germany from maintaining the key weapons that emerged from World War I: poison gas, submarines, tanks, and military aircraft. The treaty also forced the Germans to disband their air force and turn over aircraft to the Allies. Although slowed, German ambitions and skills along with the negligence and apathy of the Western democracies allowed the Germans to end the 1930s with the strongest air force in the world. The revival of the Luftwaffe took three paths: manufacturing, commercial air, and the secret development of a military air force.

### The Rebirth of the German Air Force

The first to break the bonds on German aviation was the manufacturing sector. By early 1920 it was exporting aircraft, although the Germans were initially hobbled by restrictions on the performance of commercial aircraft, the numbers manufactured, and the size of the work force. Nevertheless German firms built aircraft at home and abroad. In short order the German aviation industry was competitive with its European rivals in terms of production and technology, with the notable exception of aircraft engines.

Germany had even greater success with civil aviation. The Germans pushed commercial aviation, along with flying and gliding clubs, to encourage aviation and by the mid-1920s Germany was the most air-minded nation in Europe. In 1926, when the last fetters on commercial aviation were lifted, the Germans formed the national airline Lufthansa, which enjoyed both a monopoly within Germany and government subsidies. It emerged as the best-equipped and -operated airline in Europe and was soon known as well for its extensive overseas activities. As the decade ended

Germany had the largest airline industry in Europe, flying almost twice as many passenger miles as France, their closest competitor, a position it held until the war. Civil aviation allowed the Germans to train aircrews, develop modern aircraft, introduce radio navigation aids, pioneer aircraft instruments, and push forward in the areas of night and bad weather flying, all areas critical to a strategic bombing force.

The reemergence of the German air force (GAF) took a longer understandably clandestine path. The Germans used Lufthansa flying schools as well as a secret training center at Lipetsk, Russia, to circumvent treaty restrictions that limited them to training a mere half a dozen pilots a year. The Germans operated in Russia between the summer of 1925 and fall of 1933, training personnel as well as testing tactics and aircraft. Even before the rise of Adolph Hitler, Germany had made progress toward building a first-class air force.

Hitler took the wraps off of the German military and rapidly expanded it after his accession to power in January 1933. The next year he announced conscription, and in March 1935 he proclaimed the existence of the Luftwaffe. It consisted of approximately 1,800 aircraft and included 370 bombers (Do 11, Do 23, and Ju 86) along with 250 Ju 52 transports, which could also serve as bombers. The year 1935 saw other evidence of the growing GAF strength, as aircraft output increased from an average of 190 per month in the first half of the year to 300 in the last months of the year.

In March 1935 the Germans were conducting final trials of a number of new fighters, dive bombers, reconnaissance aircraft, and medium bombers. This air force, especially the bombers, was intended to impress the world with its numbers and act as a deterrent to shield German rearmament. The 1935 program called for the Do 17, He 111, and Ju 86 to equip bomber units. The German buildup was aided by the coincidental rapid advances in aviation technology after a decade of relative stagnation. Thus the Luftwaffe was not burdened with a large inventory of obsolete aircraft in the wake of the rapidly changing aviation technology.

The one major area in which the Germans trailed was piston aircraft engines. The terms of the Treaty of Versailles, the lack of raw materials, and the neglect of this field by German companies forced airframe manufacturers to rely on imported and licensed engines. The Germans also trailed in the adoption of higher octane fuels, which increased engine power.

## The Dornier and Junkers Bombers

During the 1930s the Dornier company dominated the development of bomber aircraft in Germany. In March 1930 the four-engine Do P first took to the air, manufactured in the Dornier factory in Switzerland in response to a 1927 army requirement for a four-engine long-range bomber. It was tested, along with the Do 11 and Rohrbach Roland, at Lipetsk. The breakthrough aircraft, however, was the Do F, which made its maiden flight in May 1932. This twin-engine medium bomber incorporated such modern features as Townend engine cowlings, metal skin fuselage, and retractable landing gear, enabling it to reach a maximum speed of 155 mph. The Germans changed its designation to Do 11 in 1933, and along with the Ju 52/3m, the aircraft formed the core of the initial GAF bombing units. It suffered a number of design flaws, however, specifically, wing vibration, poor stability, and landing gear difficulties that forced it to operate with the landing gear locked in the extended position. One prominent aviation historian writes that the Do 11 "acquired so unenviable a reputation as a result of accidents stemming from both structural weaknesses and execrable handling characteristics that it had been dubbed the *Fliegender Sarg* (Flying Coffin) by crews unfortunate enough to fly it."[1]

Dornier modified the aircraft, which it redesignated Do 13. It was a simplified design that used flaps and fixed landing gear, along with more powerful engines, but it was no better, and perhaps worse than the original. Therefore Dornier stopped production and made further modifications. The improved version was redesignated Do 23 and began to supplant the Do 11 in the summer of 1935. The GAF received over 200 of the Do 23s, along with 150 Do 11s and 12 Do 13s. These Dornier bombers were disappointing at best, certainly slow, sluggish, and dangerous. Therefore they were quickly phased out of front-line service into training and support duties.

In contrast the Junkers aircraft had a long and distinguished career. The Junkers Ju 52 is best remembered as a transport *Tante Ju* (Auntie Ju), the German counterpart to the classic Douglas DC-3/C-47 "Gooney Bird."[2] The Ju 52/3m made its maiden flight in April 1931 and was distinguished by its three engines, fixed landing gear, and corrugated metal fuselage. It was designed as a transport and only intended to be the interim bomber until the Do 11 became available. But the latter's problems, and the former's virtues (sterling flying characteristics, simplicity, ruggedness, and ease of manufacture), gave the Junkers craft a much greater role, so that at the beginning of 1936 it equipped two-thirds of the GAF's bomber force. The bomber version was armed with one 7.9-mm gun in an open top position and a second gun in a semiretractable "dustbin" mounted just aft of the landing gear and between the aircraft's two bomb bays, which could accommodate a maximum of thirty-three hundred pounds of bombs. The bomber had a tactical radius of 310 miles at a cruising speed of 153 mph.

Other aircraft in the GAF's early inventory also served as both bombers and transports. Early in 1934 the Germans issued a requirement to the Junkers and Heinkel companies for an aircraft that could be fitted as either a fast transport or a bomber to replace the Do 23 and Ju 52, which resulted in the He 111, Do 17, and Ju 86. The first two were successful, the last was not. One last medium bomber deserves mention before returning to our focus. The Ju 88 first flew in December 1936, and one of the prototypes reached an astonishing speed of 323 mph. The aircraft entered service only shortly before the outbreak of the European war and because of its performance and adaptable design became the mainstay of the GAF bomber force, with almost fifteen thousand built.

The German experience with heavy bombers was much less satisfactory. The GAF was interested in the concept of strategic bombing and did make efforts to introduce a strategic bomber, but it failed. This constitutes one of the intriguing questions of the war: What if the Germans had fielded a satisfactory long-range bomber?

In the 1920s the Germans built a few four-engine bombers, the Rohrbach Roland and Dornier P, and in the 1930s the Ju 89 and Do 19. The GAF began plans for a long-range bomber in 1934, and gave the program top priority in their development program. In the summer of 1935 the Luftwaffe issued Dornier and Junkers a requirement for a four-engine heavy bomber that became known as the "Ural Bomber" because it was to have sufficient range to reach east of Russia's Ural Mountains. This plan called for full-scale production by 1938. While the Luftwaffe chief was a recent convert to the idea of strategic bombing, there were many in the GAF who opposed the concept, fearing it would overextend the already stretched German training establishment and war industry.

The Junkers response was to use its Ju 86 airframe and substitute gasoline engines for the diesels. The resulting Ju 89 was a low-wing monoplane powered by four liquid-cooled engines, featuring twin vertical stabilizers and defended by two 7.9-mm guns, one in the cockpit and one in the tail, along with single 20-mm cannons mounted in dorsal and ventral turrets. In April 1936 the Germans issued new specifications for the aircraft that almost doubled the range and speed requirements, clearly beyond the state of the art, leading some to speculate that this was a move to kill it and the competing Dornier (Do 19) project. The death of the program's patron, GAF chief Walther Wever, in June 1936 and the emergence of a more advanced heavy bomber project ("Bomber A," which spawned the He 177) put the "Ural Bomber" program in jeopardy. In addition, German air staff thinking shifted to the belief that medium bombers could satisfy Germany's tactical and strategic needs. Factors behind this change included the success of the dive bomber in Spain, the inaccuracy of horizontal bombing,[3] the fact that two and a half twin-engine bombers could be built at the same cost as one four-engine bomber, and Germany's shortage of materials. The GAF put its hopes and money on the Ju 88. Meanwhile the Ju 89 made its initial flight in December 1936, only months before its termination in April 1937. The Germans continued to test the aircraft before converting it to a transport. The bomber could reach a maximum speed of 242 mph without armament and was designed to carry a bomb load of thirty-five hundred pounds.

The Junkers' rival, the Do 19, was no improvement, and in fact it was slower. It too had twin vertical stabilizers but was powered by four radial engines mounted on a wing set midway on the fuselage. The aircraft's defensive armament consisted of a single 7.9-mm gun in the nose and tail and a single 20-mm cannon in dorsal and ventral turrets. The aircraft first flew in October 1936. The bomber is credited with a top speed of 196 mph and was designed to carry thirty-five hundred pounds of bombs.

The Luftwaffe was a powerful tool in Hitler's march to power. It helped cow the Western nations into allowing the remilitarization of the Rhineland in March 1936, the annexation of Austria in March 1938, and the shameless sellout of Czechoslovakia that September. The threat of war, particularly the fear of bombing of cities, bolstered by Hitler's rhetoric and backed by the Luftwaffe, paved the way for Germany's bloodless aggression.

## German Strategic Bombing Doctrine

The development of German air theories has been distorted by writers who have taken the events of World War II and folded them back into the prewar years. From this has evolved the myth of the Luftwaffe as only a close support air force with a bomber arm designed for terror bombing. This is convenient and simple, but just plain wrong.

Immediately following World War I, the German military ignored strategic air operations. In 1921 German doctrine held that air superiority was the primary goal of the air force, and because of bomber vulnerability in daylight, bombing operations would be conducted at night against railroads and supplies. Nevertheless, in 1924 war games German players "attacked" French industry. Two years later the Germans issued a directive that called for two air forces, one tactical and one strategic. The latter would attack the enemy's homeland and focus on civilian morale. "By attacks against the enemy's major cities and industrial centers," the directive held that "it will be attempted to crush the enemy's moral resistance and will to fight by targeting his armaments industry and food distribution."[4] This document was less radical than Douhet and other air power theorists in that it called for escort fighters, espoused a strong air defense setup, and, most significant, did not assert that air power alone could win a war.

In early 1933 Robert Knauss, the leading German air theorist of the day, advanced a more detailed and daring scheme. He concurred with two tenants of American strategic bombardment doctrine that fighter escort was unnecessary because bombers could protect themselves and that modern nations could be "crippled" by a few selective attacks on electric power, transport, and fuel. But he also echoed Douhet by advocating attacks on enemy air forces, mobilization centers, and terror bombing of major cities. While there were some German officers who objected to Knauss's ideas, these views were well received by German airmen.

The most important German airman in the interwar years was Walther Wever. An outstanding staff officer, he was bright, hard working, charming, forceful, and regarded by his superiors as a possible candidate for command of the army. In 1933 he was transferred to the air arm, learned to fly at age forty-six, and became the first chief of the Luftwaffe. Wever shared the view that strategic bombing was primary, but he was more moderate than some air advocates as he did not hold that air power alone could win wars. Wever believed that command of the air and air defense were two of the air force's primary tasks. He laid out five objectives for the air arm: first to defeat enemy air forces, the next three to support army and navy operations, and fifth to influence the battlefield by stopping enemy armament production. Wever implored his subordinates to "never forget that that the bomber is the decisive factor in aerial warfare."[5] And in late 1934 war games Wever used deep strikes.

That same year Wever ordered the formation of an air doctrine that was drawn up in 1935 as Luftwaffe Regulation 16: The Conduct of the Aerial War. It stressed air power's offensive nature and that the primary mission was to break the will of the enemy's armed forces. The document called for supporting both the other services and independent operations, emphasizing that the first and most important mission was to destroy the enemy's air force. However, the core of air operations was to hit the sources of military power, production, food, power, transportation, and administration. The German airmen rejected Douhet's aim of terrorizing enemy civilians on moral and military grounds, although retaliation was another matter. This document also denied the need for fighter escort. Clearly the Germans gave strategic bombing considerable thought and were thinking in terms of attacking the enemy's industrial base. They were influenced by Douhet, although with reservations. In the words of one historian, "The theories of Douhet enjoyed far more thorough scrutiny and acclaim in Germany than in many European nations."[6]

## The Luftwaffe in Action

The war in Europe broke out on September 1, 1939, when Germany invaded Poland. To the German's surprise, the fear of the GAF did not keep the British and French from fulfilling their agreements and joining the war against Germany. The Germans planned to bomb the Polish capital and principal city on the first day of the war because Hitler wanted to crush Polish resistance as quickly as possible to deny his partner of convenience, Joseph Stalin, a larger stake in a defeated Poland. Weather thwarted this effort, delaying the attack for two weeks when the Germans hit the city with 183 aircraft, a mission notable primarily for the confusion among the attackers. The Germans also flew numerous leaflet-dropping missions over Warsaw, but the city would not surrender. The Soviets joined the invasion of Poland on September 17.

On September 25 the Luftwaffe flew 1,150 sorties and dropped five hundred tons of high explosive and seventy-two tons of incendiaries on Warsaw. (Thirty Ju 52s were involved, with aircrew literally shoveling incendiary bombs out of the aircraft.) The attack killed thousands of civilians, destroyed 10 percent of the city's buildings, and cost the GAF three aircraft. Although Warsaw was a legitimate military target because it was defended and already under German artillery fire, the attack was elevated to far more. The Germans made a propaganda film, *Baptism by Fire*, that played up the ferocity and efficiency of the air attacks, which stoked the fears of Western politicians and the public. On their part, the Allies and sympathetic neutrals wrote of German barbarity in attacking the city and civilians.

After a brief lull, in April 1940 the Germans quickly overran Denmark and Norway. The Luftwaffe was again effective in supporting the ground troops, although more important in battering the Royal Navy and in deploying German forces. The next month the Germans attacked the Low Countries and France and won a stunning victory despite the Allied numerical advantage in troops and tanks (the Germans had superior numbers of aircraft).

One action in the western campaign is significant to our story. With their ground forces stalled outside of Rotterdam and fearing imminent British landings, the Germans wanted to take the city as quickly as possible. They threatened to bomb the city, and a German general gave the order "to break the resistance at Rotterdam by every means" with the caveat to "use all means to prevent unnecessary bloodshed amongst the Dutch population."[7] On May 14 the GAF launched one hundred He 111s against the city, which the ground commander attempted to halt as negotiations with the defenders were underway. Unfortunately fifty-seven bombers failed to receive the recall signal because of a breakdown of radio communications and poor visibility that negated the backup signals (flares). As a consequence ninety-seven tons of bombs hit the city and created a fierce fire in the old wooden buildings and factories, killing nine hundred to one thousand civilians. Again the propagandists on both sides played up the attack to the debit of history. Another consequence of the attack was that the British saw no further need to restrain their bombing raids to attacks west of the Rhine.

## The Battle of Britain

The quick and overwhelming German victory on the Continent left Britain without allies, with a battered army facing the experienced and triumphant German military, and with the prospect of defeat. Hitler assumed the British would show good sense and surrender to, or at least compromise with, the inevitable. However, the Germans faced a situation unlike any they had mastered so easily in the past, for this was more than a "mere" river crossing; to get at the British army it would have to defeat (or neutralize) both the Royal Air Force and Royal Navy. The German failure against the former was clearly one of the decisive battles of World War II and probably the most significant air campaign of all time.

The Battle of Britain began with a German advantage in overall numbers of combat aircraft but, contrary to conventional wisdom, almost equal numbers of the key weapon, modern single-engine fighters. (In early August the GAF had 805 serviceable Me 109s facing 715 RAF Spitfires and Hurricanes.)[8] These fighters were essentially equivalent in quality, each having advantages and deficits. Initially the GAF had an edge in experience and tactics versus the RAF's advantage of a well-organized air defense system built around radar, which provided early warning of the attackers and control of the defenders. In addition the British were fighting over their own soil, allowing higher pilot and aircraft survival.[9]

The Luftwaffe was better prepared for conducting a bomber offensive than either the U.S. or British air forces were when they began their later efforts. They had a large number of bombers, bases relatively close to their targets, night and bad-weather bombing aids, and escort fighters. Nevertheless these proved inadequate. The GAF shifted from attacks on the RAF to attacks on London for a variety of reasons, including faulty intelligence that indicated the RAF fighter force was spent. The former resulted from a German navigation error that led to the mistaken bombing of London on August 24. The British quickly responded, although only twenty of the eighty-one bombers launched attacked Berlin. In retaliation the Germans began the infamous Blitz of London on September 7. This was a critical mistake.

In short order heavy daytime losses forced the GAF to switch to night attacks. The Luftwaffe was prepared for night attacks fielding a Pathfinder Force trained to mark targets for the main force using a bad weather/night navigation system (*Knick-ebein*) that was accurate within five hundred yards. But it was not enough. Carl Spaatz, an American air observer in London, and later commander of U.S. strategic bombers assaulting Germany, noted at the time, "The British are winning. . . . The Germans can't bomb at night—hell, I don't think they're very good in daylight—they haven't been trained for night bombing. Nope, the British have got them now."[10] The Luftwaffe pounded London for sixty nights and during the course of the year killed forty thousand Britons. The air battle along with end of summer led Hitler on September 17 to "postpone" the invasion, although the GAF night attacks continued against weak RAF air defenses.[11]

In the end the fierce German attack failed to break British morale (it actually stiffened) or defeat RAF fighter command. Although both combatants submitted exaggerated victory claims of enemy aircraft destroyed (the GAF 3,058 and the RAF 2,698), it appears that the RAF downed almost twice as many German aircraft as they lost (1,733 and 915, respectively). These were heavy losses for both air forces, especially in terms of aircrew. It should be emphasized that experienced pilots rather than numbers of aircraft proved to be the crucial element in air combat in World War II.[12]

Both sides made serious miscalculations and proved ill prepared for the battle. In fairness, wars and battles never go according to plan. The British based their defense on the assumption that the Germans would be operating from German bases and therefore would only launch light bombing attacks. RAF equipment, pilot experience, and tactics were inadequate at the beginning of the battle, but both improved. On its part the Luftwaffe was better prepared, equipped, and experienced, albeit mainly as a tactical air force. The GAF had no plans for an air offensive against London and its successes against inferior air arms did not help it in a conflict against an equivalent one. German aircraft proved unsuited for the task at hand. Their best fighter, the Me 109, was range limited.[13] Their stellar dive bomber, the Ju 87 Stuka, which had performed so well over Poland and Western Europe under an umbrella of air superiority, proved too vulnerable over Britain and was pulled out of action. Likewise the Me 110, designed as a bomber escort, failed against the Hurricanes and Spitfires and had to be escorted. The Germans did not have a heavy bomber, and their medium bombers lacked sufficient defensive firepower, armor, and bomb load to operate successfully against a first rate defense. And the GAF, hampered by poor intelligence, shifted targets. As the battle unfolded the British demonstrated great adaptability as they made cool, calculated, and correct moves to preserve their fighter force, modifying both their tactics and aircraft while inflicting heavy casualties on the Germans. In brief, the RAF played its hand well and improved as the fight unfolded; the GAF did neither.

Thus RAF Fighter Command broke the German string of overwhelming victories and their aura of invincibility. The British remained unconquered. The prewar air power theorists had been disproved—defense against bombers was possible—and while the bombers might "always get through," they did so without achieving decisive effect while suffering unsustainable losses. The Battle of Britain was the primary factor in why the Germans did not attempt a cross-Channel invasion and therefore clearly deserves credit in changing the course of World War II. Finally, it is no discredit to the British people, to those who served in the air defense, particularly those "few" Fighter Command pilots, to note that the GAF did not have the edge it enjoyed in its previous victories. I concur with the view that "barring some gross and irreversible error by [Air Chief Marshal Hugh] Dowding [commander of Fighter Command] and his senior commanders, the Luftwaffe could not have won the battle."[14]

The night bombing of Britain continued, albeit on a lesser scale. Another city martyred by bombing was Coventry. It clearly was a legitimate target because of its numerous factories, but its long history (complete with Lady Godiva and Peeping Tom) and historic buildings made it something else in the public's mind. On November 14–15, 1940, the Germans launched more than 550 aircraft at Britain, and 450 of 510 destined for Coventry unloaded about 550 tons of bombs on the city. British defenses were weak, twenty-four heavy and twelve light antiaircraft guns and 135 sorties by a wide variety of ill suited aircraft that downed only two GAF aircraft. The bombing was remarkably accurate due to excellent navigational aids,[15] limited British resistance, and low bombing altitudes. The attack destroyed or severely damaged one-third of the city's factories, killed 570, and seriously injured another 860 of the city's population of about 230,000. The six-hundred-year-old cathedral was among the buildings destroyed. Allegations of prior notice of the attack and Coventry's deliberate sacrifice to preserve the secret of broken German codes still abound, but are untrue.

The principal significance of the Coventry bombing was its impact on public opinion. It joined the list of cities that symbolized the power, frightfulness, and immorality of strategic bombing in World War II and enhanced the impression of Hun inhumanity. The list would grow, joined by cities devastated by the Allies. Two other points are worthy of note. Coventry was studied by British Bomber Command as a test case for its operations against Germany. And although most of the city's factories were back in production within six to seven weeks and civilian morale was not seriously shaken, the British believed that similar attacks on German cities would create decisive results. Second, despite the conventional wisdom, there were no immediate calls from the survivors of Coventry for retaliation.

## The Heinkel 177

In contrast to the great German aviation success of getting a jet fighter into operations before the Allies, the Germans failed to field anything comparable in numbers or performance to the heavy bombers

**(Right) One of the worst aviation failures of World War II was the Heinkel 177. To achieve range and diving requirements, the Germans mounted four engines in two nacelles that turned two propellers. This arrangement was a fiery disaster, and consequently the Germans never fielded an effective long-range bomber force as did their foes. (USAF)**

employed by the British and Americans. The one GAF long-range bomber that did get into service is instead better known for engine fires, structural failures, and lackluster operational results. One aviation historian calls this program "the most dismal chapter in the wartime record of the German aircraft industry,"[16] while another notes that "no bomber of W.W. II suffered more unpopularity with its crews or produced so little tactical results for its total cost in effort and exchanges, than the He 177."[17]

The Germans began to seriously investigate a long-range bomber beginning in 1934, an effort encouraged by the charismatic chief of the GAF, Walther Wever. The next summer the Luftwaffe sent specifications to Dornier and Junkers that produced the unsuccessful Do 19 and Ju 89. Even before the Do 19 first flew the GAF decided it needed a more capable bomber. In June 1936 the Luftwaffe instructed five manufacturers to design a heavy bomber that could employ dive bombing tactics, a key, controversial, and critical requirement.

(The reader will note that this preceded the German experience in Spain.)

The German Air Ministry issued a design requirement to Heinkel, only months before it cancelled the Ju 89 and Do 19 in 1937, that called for a top speed of 335 mph and the ability to carry at least two thousand pounds for 4,160 miles. The company's major challenge was the lack of suitable power plants, which it met by mounting two engines side by side and coupling them through a transmission to a single propeller shaft, thus combining the power of two engines with the drag of one propeller. The He 177 was a four-engine bomber with two props, an arrangement that Heinkel disliked but found necessary to achieve the required performance and dive bombing capability. A second innovation to reduce drag involved a surface evaporative cooling system instead of conventional radiators. This worked on the He 100 fighter, but not on the larger He 177, and was deleted by the spring of 1939, a change reducing speed from 311 mph to 286 mph. Another innovation intended to reduce drag was to use remotely controlled gun turrets instead of manned turrets. The bomber was to be defended by three of these along with a manned tail turret.

If these innovations did not challenge the design sufficiently, the dive bombing requirement of 60 degrees surely did. Operationally this would improve bomb accuracy, but decreasing altitude would expose the bomber to increasingly lethal antiaircraft artillery. The design implications were also significant because dive bombing produced greater aerodynamic loads that required a stronger airframe and wing that in turn increased aircraft weight and decreased performance. In short the He 177 was an aircraft that demonstrated German ambition and ingenuity well beyond the capabilities of German industry.

The prototype flew for the first time in November 1939 and quickly exhibited significant problems as the flight was cut short due to alarming engine temperatures. In addition the aircraft performed poorly with its top speed 15 percent and range 25 percent below requirements. More dramatically, crashes took a terrible toll, destroying four of the first five prototypes between April 1940 and June 1941 and nine preproduction aircraft during trials. Engine fires dogged the aircraft throughout much of its service, prompting the label *Luftwaffenfeurzeug* (Luftwaffe's lighter), along with a parcel of other unflattering nicknames. Engine lubrication proved inadequate, engine placement dangerous, and transmission problems troubling. The plan for remote-controlled guns failed. Little wonder the program suffered continued delays.

In the summer of 1941 the Germans deployed thirty-four He 177s to a French based antishipping unit, but within weeks teething problems forced their withdrawal. In a similar manner a dozen He 177s pressed into transport service during the abortive German airlift of Stalingrad proved a failure. The GAF transferred these aircraft to bomber duties with little better results, as seven were lost to noncombat causes on only thirteen missions. In 1943 the He 177 returned to the sea war, inflicting limited damage while suffering heavy losses. One notable aspect of these operations was the employment of guided bombs. Daylight operations with the Hs 293 missiles began on November 21, 1943, but proved expensive, forcing the GAF to switch to less costly and less effective night operations. The He 177s continued antishipping operations over the Mediterranean into 1944, employing the radio-controlled FX 1400 glide bomb, a 3,500-pound armor-piercing bomb fitted with wings and a cruciform tail.[18]

The He 177 also engaged in strategic attacks. In December 1943 Hitler ordered a response to the Allied bombing of Germany with raids on British cities. The GAF massed its bombers, including thirty-five to forty-six He 177s among the four hundred to five hundred aircraft involved. The "Little Blitz" began on January 21–22, 1944, with 270 sorties, including a dozen flown by He 177s. RAF night fighters claimed two of the latter, and German flak a third. The He 177s logged 5 of the 125 sorties on the last large bomber attack on London (April 18–19). These raids faded out by the end of May.

The Germans launched missiles from He 177s against the Normandy invasion. The cost was high as on its first operation, one He 177 unit lost half of twenty-six bombers committed and the next day another six. In short order, the bombers were directed to sow mines that were somewhat more effective—sinking nine warships and seventeen others by the end of the month—and less costly. At this point the Luftwaffe redeployed the surviving antishipping units from the battle to bases in Germany and Norway.

The major action of the bombers was now on the eastern front. In contrast to the German operations against Britain and the Anglo-American bombing of Germany, there was essentially no strategic bombing in the Russo-German war. The December 1940 plan for the war excluded Soviet industry from attack, and once the war degenerated into a bloody slugfest, the Germans needed all the air power they could muster to support their beleaguered troops. The only example of strategic bombing was a brief campaign in June 1943, when the Luftwaffe flew almost one thousand sorties dropping about 1,000 tons of bombs on a tank factory, 324 tons on a rubber plant, and 181 tons on an oil refinery. In May 1944 the GAF transferred He 177s to join newly formed units in East Prussia to mass ninety bombers for attacks on Soviet supply and assembly targets. The He 177s operated at high altitudes against light resistance and thus suffered minimal losses. Even the longstanding problems with engine fires seemed to be solved. However, when a Russian offensive erupted in June 1944, the Germans thrust the He 177s into action as ground support aircraft and took heavy losses, losing more than ten of twenty-four bombers during the first low-level attack. None of the various imaginative German schemes to hit Soviet power plants and dams (bombers, commandoes, guided bombs, mines, suicide attacks, and Mistel weapons) were employed.[19]

By the summer of 1944 the Allied bombing offensive had quickly reduced the amount of aviation fuel available to the GAF. To conserve what they had, the Luftwaffe grounded the He 177s and literally parked the bombers out in the open. Of 3,100 built, probably only 200 saw service. The GAF was not short of aircraft, it was short of trained crews and fuel.

Reportedly the He 177 was an easy aircraft to fly and as maneuverable as the smaller Ju 88, although, unlike its foreign rivals, it was difficult to takeoff and land. The main version produced, the 565 A-5, had a maximum speed of 303 mph and with two Hs 293 missiles was able to fly thirty-four hundred miles. It was manned by a crew of six and defended by numerous machine guns and cannon mounted in six positions. One writer contrasts the German heavy bomber with the Avro Lancaster and the Boeing B-17 commenting that the He 177 "was over-complex, unforgiving of mishandling and vulnerable to battle damage. Soon after its maiden flight, it gained notoriety as a killer of its crews, and it never shed that image."[20] The Germans did better with strategic missiles; at least this effort cost fewer German lives.

## The V-1 (Buzz Bomb) Campaign

The German strategic bomber effort was much smaller and much less effective than the one waged by the British and American air forces. Yet the Germans surpassed the Western Allies by introducing two new strategic weapons into combat, winged and ballistic surface-to-surface missiles.

The V-1 in a terminal dive over London. The cheap, pulsejet-powered missile proved ineffective militarily, but it had a great psychological impact and cost the Allied defense four times more than the German's offensive. (USAF)

The V-weapons allowed bombardment from a distance, day or night, fair weather or foul. Both missiles proved to be spectacular and innovative, but neither affected the outcome of the war. Decades later both concepts would develop into bomber rivals as well as bomber munitions.

The first German V-weapon was a small, pilotless aircraft with various designations, although best known as the V-1. Such weapons had been proposed as early as World War I, but problems with control, accuracy, and the cost of a conventional engine were barriers. The invention of the cheap pulse jet engine in the late 1920s reenergized the concept, although various technical issues (the engine's high fuel consumption, short life, and severe vibrations along with the missile's problems of control and accuracy) required solution. From the German army's viewpoint, the issues of range, accuracy, and cost kept the project from receiving priority in the mid-1930s.

The conquest of France opened an opportunity for this weapon because it reduced the severity of the range problem, and because accuracy was directly related to range, the accuracy problem diminished as well. The war also stretched the Luftwaffe thin, which along with the GAF's lack of a suitable strategic bomber made a cheap bombardment device appealing. Allied bombing of Germany created a demand for retaliatory weapons while interservice rivalry nudged the GAF to match the army's emerging V-2 ballistic missile program.

The V-1's development was rapid. In April 1941 the Luftwaffe tested the engine in the air, in June 1942 it gave it the highest priority, and in December 1942 it glide tested the V-1 airframe, followed that same month with a ground-launched powered flight. In May 1943 the Germans ordered both the V-1 and V-2 into full production. The V-1 showed its potential that July by impacting half a mile from its target over a 152-mile range, a fluke in view of later performance. Problems would be one of the costs of the missile's rapid and concurrent development; by the end of July only one-third of eighty-four air- and ground-launched missiles tests were successful. Another issue was accuracy. The manufacturer had expected that 90 percent of the missiles would impact within 6 miles of the target and achieve a circle error propable (CEP) of 1.85 miles, but the results were disappointing.[21] Of thirty missiles fired in April 1944, nine crashed shortly after launch and the rest hit within eighteen miles of the aiming point, but only ten within six miles.

The V-1 had a small winged airframe half the size of a FW 190 and one-quarter the size of an He 111. What distinguished it from an aircraft was its lack of a pilot and its long engine housing mounted above the aft third of the fuselage and vertical stabilizer.[22] The pulse jet engine used cheap gasoline (75 octane) and a louvered Venetian blind–type device in the intake which rapidly opened to admit air and closed for combustion, giving it its distinctive ("putt, putt") sound. The missile required a boosted catapult launch because it could not operate at speeds less than 150 mph. It could carry a 1-ton warhead on average about 130 miles and crossed the British coast at 340 mph, reaching about 390 mph at the target. Most of the V-1s flew between three thousand and four thousand feet. A small propeller in the nose armed the warhead after it traveled a set distance and then after a preset number of revolutions, activated two detonators that locked the elevators and rudder in a neutral position and actuated spoilers on the tail to push the V-1 into a shallow terminal dive. This maneuver usually cut off the fuel flow, stopping the engine and thus warning those nearby.

The British collected bits and pieces of information and by August 1943 knew of both German missile programs. That month they received photographs of the wreckage of a V-1 that crashed in Denmark, the French underground provided information on the construction of launch sites in France, broken codes revealed additional intelligence, and in November 1943 aerial reconnaissance photos captured an image of a V-1 at the Peenemünde test faculty. In May 1944 the British examined a V-1 that crashed in Sweden. Although the Allies received considerable information about the V weapons, for every item that proved valid, they received numerous other pieces that were in error or exaggerated. As in all matters of intelligence (and history), the story is much clearer in retrospect than it was at the time.

The Allies took active measures to thwart the German missile programs. They bombed launch sites and some production facilities in what one student of the subject has called "an ill-conceived campaign against an ill-defined threat."[23] This effort was costly, achieved little payback, and appears to have been motivated more by political than military factors. Both American and British airmen complained that it diverted the bombers from both the strategic campaign and support of the cross-Channel invasion. An RAF strike on the Peenemünde facility in August 1943 may have delayed the V-2 program six to eight weeks, but it did not impact the V-1 program. Later that month U.S. bombers began attacking seven massive structures built to house German secret weapons (four for the V-1s) near the French coast. Hitler favored these huge structures, although some of his subordinates pushed for smaller, simpler, cheaper, and more readily concealed launching positions. The airmen detected these smaller launch facilities they called ski sites and began a major bombing campaign against them in December 1943. By June 1944 the airmen estimated they had destroyed eighty-two of ninety-six identified sites. Although none of the ski sites launched a missile that reached Britain, the effort cost the Allies 154 aircraft and 771 crewmembers. Some see this battle of the ski sites as a successful diversionary campaign as the Germans responded by constructing even smaller, easier built, and less visible (modified) launch sites. Hitler was reported to have said that every bomb dropped on a V-1 site was one less bomb on Germany. Between August 1943 and August 1944, 12 percent of the Allied heavy bomber sorties and 15 percent of the medium bomber sorties were involved in this campaign. For a number of reasons the British and Americans attacked only one of the modified sites so that by June 1944 the Germans had about half of the 150 sites ready for combat.

The V-1 campaign was delayed less by the bombing than by other factors. The original German plans called for the V-1 offensive to begin in mid-December 1943 with the monthly production rate reaching eight thousand V-1s by September. But by April

1944 the Germans had only three thousand missiles because of production and technical difficulties. The Germans fired the first V-1s against Britain on the night of June 12–13, a week after the D day invasion. The pace of the attack quickly picked up, with the five hundredth launched by June 18, the one thousandth by June 21, the two thousandth by June 29, and the five thousandth by July 22. The V-1 assault battered the war weary Britons of (mainly) London. Production fell by 25 percent, and perhaps as many as 1.5 million Londoners evacuated the city, more than during the Blitz, leading some to believe that the V-weapons had a greater impact on civilian morale than the bombings of 1940–41. These attacks continued until the first days of September, when the Allied ground advance forced the last V-1 unit to withdraw from France. One sure way to prevent missile attacks is to push their bases outside the range of their targets.

Intelligence was only of limited importance to the Germans in the V-1 campaign because of the missile's poor accuracy. One British effort to confront the missile attack involved attempts to shift the mean point of impact six miles, which analysts calculated would decrease casualties by twelve thousand per month. This deception was possible because the British had captured and turned enemy agents and fed the Germans false information that the missiles were hitting beyond their aim point (the Tower Bridge), inviting the Germans to shorten the missile's range, when in fact they were already falling short. This tactic raised issues of who would live and die, a grave moral responsibility, but it appears to have had little effect because there is no evidence that the Germans readjusted their aim. The British shut off a potential source of intelligence to the Germans in July by banning addresses of the deceased in obituary notices because clusters of deaths would indicate missile impacts. The Luftwaffe used other methods to locate the missiles' impact points. Efforts with seismographic and underwater microphones failed. The GAF also fitted 7 percent of the V-1s fired between June 15 and August 1 with radio devices that transmitted during the last thirty-five miles of flight and could be tracked by direction finding units. (By March 1945 half of the V-1s were so equipped.) While this information may have been too crude to adjust the firings, it did correctly indicate that the impact points were mainly east and south of Tower Bridge.

The Allies continued to bomb German facilities. The Allied supreme commander, Dwight Eisenhower, set the V-weapons priority as second to only the urgent requirements of the invasion. Thus the Allies bombed V-1 targets despite the protests of the American and British strategic bomber commanders and their insistence that the destruction of other targets would have greater military impact. Between June 12 and September 3 they flew twenty-six thousand sorties that dropped seventy-three thousand tons of bombs on launch sites, supply depots, and seven German factories at the cost of 197 aircraft and 1,412 aircrew. During July and August 1944 the Allies devoted one-fifth of their total heavy bomber attacks against V-weapons targets with little effect, as the V-1 launch rate did not decline. In all the airmen dropped 122,000 tons on V-1 targets on sixty-nine thousand sorties, compared with the total of seventy-one thousand tons of bombs that the Germans dropped on Britain during the entire war. Nevertheless missile accidents killed more Germans than the number attributed to these air attacks. An extensive American postwar examination of the bombing campaign concluded that the bombing probably delayed the V-1s three to four months but did not delay the V-2s.

Technical problems proved to be a more important hindrance to the missile campaign. One source states that a fifth of the fifty-one hundred V-1s fired at Antwerp crashed, while another asserts that the failure rate was at least 35 percent. Another major problem that undercut missile effectiveness was a lack of accuracy. The overall CEP for the V-1s launched from ground sites in France was about eight miles.

To defend Britain, the Allies employed fighters, antiaircraft guns, and barrage balloons. While the V-1's small size, relatively high speed, and low altitude operation made it difficult to spot, and its simple construction made it tough to down, the missile could only fly straight and level. Although fighters were important, over time the defense evolved to depend mainly on flak deployed at the British coast that by early August numbered six hundred heavy and fourteen hundred light guns. Technical devices such as better radar, gunfire directors, and proximity fuses aided the defense. By the end of the campaign the guns were downing the majority of the incoming V-1s.

The Luftwaffe modified its V-1 offensive in two ways. One effort involved air launching approximately sixteen hundred V-1s from He 111 bombers, beginning in early July and continuing into January 1945. This system had the potential to outflank and spread out the defenses as well as extend the range of the missiles. However, the air-launched missiles had less than a third the accuracy of the ground-launched versions because the problems of He 111 night navigation, low-level flight, and evasion of British fighters magnified the accuracy errors of the missile. (The Germans did deploy a radio buoy to aid the navigation of the missile carriers, but this proved ineffective.) The carrier aircraft also proved very vulnerable to British night defenses. Only 388 of the air-launched missiles got to Britain, and in fact the Germans lost more carrier aircraft (seventy-seven) than the number of V-1s (sixty-six) that reached London. Certainly the British night defenses had improved greatly since1940.

The Germans also employed a longer range, ground-launched version of the missile, firing the first against Britain in March 1945. It had a lighter wing and warhead, enabling it to carry more fuel and fly 220 miles, 90 more miles than the standard version. The Germans fired 275 of these, but only 13 reached London with a CEP of twelve miles.[24] German plans to install more powerful engines to increase speed and range, radio control to improve navigation and accuracy, along with submarine launch and a piloted version were never put into service.

The Germans fired 10,500 V-1s at Britain, 4,300 of which were downed by the defenders, with some 2,400 missiles reaching London. The missiles killed sixty-two hundred civilians, 90 percent of them in London. On a ton-for-ton basis the V-1 killed approximately three times the number of Britons as did the bombing.

The Germans also fired 12,000 V-1s against targets on the Continent, mainly at Antwerp. The city was only defended by flak that destroyed 64 percent of the 2,800 missiles that threatened the highly important port area. Only 211 missiles hit within that zone, of which 150 hit the dock area. The continental attacks killed 4,700.

The Germans built between thirty thousand and thirty-five thousand V-1s, of which perhaps two-thirds were fired in anger. Clearly the Germans might have used the resources devoted to the V-1 in a better fashion. For example, the missile absorbed half of the army's explosive consumption in the third quarter of 1944 and forced the Germans to supplement explosives in artillery shells with rock salt. It is nevertheless true that the V-1 offensive cost the defenders much more than the attackers. A wartime British study put the cost to the Allies (accounting for damage and loss of production as well as British defenses) at 3.8 times the cost to the Germans; however, the Allies could afford the cost, the Germans could not.

The V-1 had little military impact other than diverting resources from other action. It was chiefly a psychological weapon that shook the war-weary Britons but failed to break their morale. Deficiencies of accuracy, numbers, intelligence, pay load, and vulnerability prevented the weapon from achieving decisive results in World War II and seemed to leave the weapon little future.

With the overrunning of the German V-1 launch sites in France, the British thought the bombardment of their homeland was over, and on September 7, 1944, the minister of home security ended the evacuation of London. That same day Duncan Sandys, chief of investigation of long-range German weapons, announced to reporters that "except for a last few shots, the Battle of London is over."[25] Churchill's son-in-law spoke too soon.

## The V-2 Ballistic Missile Campaign

The other German missile program was similar to the V-1 in terms of accuracy, range, goals, and effect, although it was based on an entirely different technology. For in contrast to the cheap, simple, winged V-1, the V-2 (designated A-4 by the Germans) was a ballistic missile that was much larger, more complicated, and more expensive, with heretofore unprecedented performance.

The Treaty of Versailles restricted the German military in many ways but not in the field of rocketry. In 1929 the German army decided to begin rocket research that by mid-1936 encouraged the army to build an experimental facility at Peenemünde, where both V-weapons would be developed and tested. Although there were some efforts in the field of rocketry by Americans and Soviets, the Germans had the lead by the end of the decade. The Germans drew up specifications for a missile that came to be known as the V-2 to carry a 1-ton warhead over 160 miles (twice the range of the longest range artillery piece fired thus far, the infamous Paris gun of World War I) with a CEP of .3 to .45 miles. The rocket had to be transportable and able to move through European railroad tunnels and small villages. The first static test took place in March 1940, and the first attempted launch in June 1942 failed, but on the third try in October, the missile flew 120 miles. This project was

Compared to the V-1 the rocket-powered V-2 ballistic missile was more costly, inflicted lighter damage, and produced less psychological impact. However, in contrast to the V-1, once launched the V-2 could not be stopped. The V-1 delivered 7 percent of the total German tonnage rained on Britain and the V-2 just over 1 percent. (San Diego Air and Space Museum)

on the technological cutting edge, required a number of breakthroughs, and suffered numerous failures and setbacks along the way. Overall, it is estimated that the Germans expended three thousand V-2s in their experiments, testing, and training, one-quarter of which failed.[26] By the summer of 1943 the Germans had begun training operational units at Peenemünde.

In addition to severe technical obstacles, the program encountered bureaucratic and resource problems. Hitler shifted priorities and there was infighting over control of the project. In March 1944 Wernher von Braun, a key scientist, was imprisoned by the SS on the grounds that he was sabotaging the program and was more interested in space travel than developing a military weapon. (This of course would help the German mitigate his wartime connections when he came to the United States after the war to work in the U.S. Army rocket program.) There were also shortages of steel and liquid oxygen, trained personnel, and labor. And although Hitler approved the program for production in November 1942, it was not until summer 1943 that the V-2 received a top priority. At this point Hitler was enamored by the project and wanted a rocket with a 10-ton warhead and a production rate of two thousand per month, both entirely out of the question. One argument used to advance the V-2 was that the Allies were working on a similar program. This was untrue, so just as the Allies pursued an essentially nonexistent nuclear program, the Germans chased a nonexistent missile program.

The V-2 was a large (46 feet long and 5.4 feet in diameter) missile fitted with four fins on its aft end and weighing 13.6 tons at liftoff with its 1-ton warhead. It required thirty-two vehicles and trailers and four to six hours to ready the device for firing (an hour and a half according to another source) and about half an hour to withdraw. Paradoxically, the larger and more complicated V-2 was mobile, whereas the V-1 was not. The V-2 could be launched from any small piece of firm, flat ground, while the ground-launched V-1 required a ramp and catapult. In Peenemünde tests the rocket was controlled during its powered flight by a radio-controlled device (deleted in the operational missiles because of a fear of Allied jamming but restored in the last quarter of production). It had a maximum range of 230 miles (if all the fuel was burnt and no engine cut-off occurred) and a nominal range of 195 miles.[27]

As noted above, the British received fragmentary information about the German's missile programs that initially confused the two programs. In June 1943 Allied photo reconnaissance of Peenemünde revealed the V-2. A year later the British were able to examine the wreckage of a missile that landed in Sweden and retrieve two tons of wreckage. While this gave the Allies some valuable information, it also led them astray. For the crashed rocket was a V-2 testing radio-controlled equipment for the Wasserfall, a surface-to-air missile, leading the Allies to believe that the V-2 would be radio controlled. In July 1944 the British airlifted out drawings and 110 pounds of wreckage from a test shot that had crashed in Poland in May. Thus, with the exception of radio control, the British formed a fairly accurate estimate of the missile's basic characteristics and performance before the first V-2 impacted in Britain.

The Allies took direct action against the V-weapons program beginning with the August 1943 attack of Peenemünde. This bombing targeted the German engineers, missed the V-1 facilities, but killed a key German scientist in the V-2 program along with 178 German technicians. However, the bulk of the more than 700 killed were foreign workers and prisoners. The bombing delayed the program four to six weeks. Allied bombing also knocked out the large launch sites of German V-weapons, pounded a few manufacturing plants, but mainly hit V-1 launch sites. These raids encouraged the Germans to move testing to Poland and manufacturing from three above-ground factories to a massive tunnel complex in the Harz Mountains. Here foreign workers labored under extraordinarily horrible (and lethal) conditions and built approximately six thousand rockets between January 1944 and March 1945.[28] The Allies considered, and rejected, a number of drastic means to counter the V-weapons, including poison gas and reprisal raids on Berlin.

The Allied advance from Normandy interfered with the V-2 deployment; nevertheless, on August 29 Hitler ordered immediate operations. On September 6, 1944, the Germans launched two V-2s against Paris, but both malfunctioned. Two days later they fired one against Paris and two against London. Initially the British government explained the V-2 hits as gas explosions; it was not until two months later that Churchill acknowledged the V-2 bombardment in a speech to Parliament. Unlike the V-1, there was neither a close-in defense against the ballistic missile nor warning of its attack.[29] Allied air attacks did not catch a single V-2 launch crew, although the bombing did slow and hinder supply. Missile malfunctions were a significant problem, with 8 to 10 percent failing at launch, causing more casualties than did Allied aircraft. London was pounded until Allied troops forced the German missile unit to withdraw out of range.

The German missile campaign was hindered by quality control, air burst, and production difficulties, as well a shortage of liquid oxygen. The missile's accuracy was poor, at least 20 times worse than the 1936 goal of a less than one kilometer. In all the Germans fired twelve hundred rockets at London, 45 percent of which hit the city or nearby area, killing 2,800 and injuring 6,500.

The Germans also used the V-2 against continental targets. While the attacks on Britain were strategic, attempting to break British morale, the attacks on the Continent were tactical. The Germans fired 1,600 V-2 against Antwerp, 1,300 of which impacted in the city area, 600 in the city, and 150 in the harbor area (the designated target). These killed 5,100 and injured 22,000. The V-2 attacks against Britain and Antwerp ended in late March 1945.

The Germans did not introduce any improved models of the V-2 into combat as they had with the V-1. Two changes were incorporated, however, the first of which pushed the range out to 220 miles. A second change saw service beginning in January 1945, when an SS battery employed a radio guide beam that enhanced accuracy. A shortage of equipment and technical factors limited this modification's use to the one launch unit. The Germans conducted experiments with larger tanks on the missile that would extend range to three hundred miles and considered alternative launch platforms (canisters towed by submarines and railroad cars), two-stage missiles, and winged versions.

The cutting edge V-2 proved less effective militarily than the cruder V-1; it was a technological wonder yet a military flop. It caused less physical damage than the winged missile, yet cost much more—resources that the Germans could ill afford. The V-2 caused less destruction than the V-1 because it carried a less powerful explosive (to lessen sensitivity to reentry heat) and its impact at Mach 3 tended to bury it deep in the ground, whereas the V-1's blast occurred higher in buildings and thus its effect was broader and more destructive. Because the V-2 gave no warning, there was less psychological impact compared with the Buzz Bomb. Finally, because the ballistic missile could not be countered, the Allies did not expend any resources in direct defense.

The V-weapons were technological innovations that had essentially no effect on the course of the war. (As with so many aspects of the war, there are many "what ifs." For example, Eisenhower wrote that had the Germans employed the V-weapons six months earlier, the cross-Channel invasion "would have proved exceedingly difficult, perhaps impossible.")[30] It was not that the V-weapons came too late in the war, but that they came too soon, before they were mature enough to be effective military weapons.

## In Summary

During World War II German bombers rained just over seventy-one thousand tons of high explosive on Britain; V-1s, fifty-eight hundred tons; and V-2s, eleven hundred tons. These weapons inflicted 147,000 civilian casualties, of which 61,000 were killed, 84 percent killed by GAF bombers.

Strategic bombardment was not as central to German strategy in World War II as it was to the Western Allies. Consequently, compared to the Anglo-American strategic bombing campaign, the German bombing effort was puny. It is ironic that early in the war the GAF was better prepared to conduct strategic bombing than were the Western Allies, employing navigational equipment and tactics not matched by their foes for years. However, at that point they lacked the numbers and aircraft to conduct an effective offensive. A few years later the Luftwaffe effort was greatly surpassed by what the AAF and RAF were able to mount. Unlike the British and Americans, the main air power effort of the Germans (and Russians) was tactical. The only area of strategic bombardment in which the Germans eclipsed the Americans and British was with missiles, and these proved premature and futile.

The impact of the V-weapons was clearly more psychological than physical, more political than military. These could only be a terror weapon because they were deficient in reliability, accuracy, destructive capacity, range, and numbers. It is true they scared the Allies, tied down resources, and altered Allied plans, but at this stage in the war the British and Americans had a surplus of power, in sharp contrast to the Germans. Once the United States entered the war, and as long as the Soviets were engaged full time, Germany was doomed; no weapon aside from a nuclear bomb could possibly save them from defeat. A wiser allocation of resources that emphasized, for example, in the air war, jet aircraft, surface-to-air missiles, and proximity fuses, could have made the Allied air offensive more expensive and less effective, but that would not have changed the outcome of the war. For the Germans could not withstand the attrition of the multifront ground war. In brief, then, the V-weapons were innovative, foreshadowing the air weapons of the future, but mere footnotes in the course of World War II. But as we shall see, the V-1 and V-2 were the forerunners of much more effective strategic weapons.

Chapter 4

# British Strategic Bombing

## DESTRUCTION BY NIGHT

War came swiftly and disastrously for the Western Allies. By mid-1940 Britain stood alone against a triumphant Germany, while the Soviets looked on and fed the German economy and Americans observed from afar, sympathetic but uninvolved. The British survived, although for the next number of years they were besieged by U boats, bombed by the Luftwaffe, and beaten up by German ground forces around the Mediterranean. Between the evacuation from Dunkirk and the D day invasion, air power provided the only major British offensive action in northern Europe.

The air war that ensued was not the quick, cheap, decisive war prophesized by the interwar writers and theorists; instead, the conflict devolved into a battle of attrition, a long, costly, high-tech version of the trench warfare of the previous world war. Tens of thousands of military and hundreds of thousands of civilians died in this air battle before the Germans went down in defeat.

### The Interwar Years

The new British air service, the Royal Air Force, had contributed to the final victory and ended World War I as the world's largest air force. However, the airmen's strategic bombing effort had been limited, although it would have been much greater had the war continued. The RAF declined drastically following the Armistice, and for a time its very fate was in doubt.

The RAF emphasized bombardment aviation, the ability to deter any enemy with a bomber force or attack an enemy's homeland if necessary. In practical terms the new service proved its worth by colonial policing, which was cheaper than the long used punitive expeditions: Economy was the watch word during the interwar years. The ten-year rule, adopted by Britain in 1919, forecast that there would be no major European war for ten years, allowed the military to be reduced, hindered the development of the RAF, and continued for more than a decade. Nevertheless the status of the bomber is clearly seen in the Salisbury committee report approved by the Air Ministry in 1925, which recommended that two-thirds of the RAF's fifty-two squadrons be equipped with bombers. These policies, along with the need to economize, led the British to field a number of single- and twin-engine biplanes that looked and performed like their World War I predecessors.

Thus in the early 1930s British bombers were considerably outmoded by the more modern bombers taking to the air in the United States, especially the B-10 in 1932. This is seen in the RAF adoption of the Handley Page Heyford as a night bomber, its last biplane heavy bomber. It first flew in June 1930, entered service in 1933, and left front-line duties in 1939. The RAF procured three twin-engine bombers to bridge the gap between these older types and modern heavy bombers. All three, the Armstrong Whitworth Whitley, Handley Page Hampden, and Vickers Wellington, first flew in 1936, entered service late in the decade, fought the early battles, and served until 1942–43, when the RAF's four-engine bombers arrived.

Hitler's rise to power and the remilitarization of Germany changed the balance of power in Europe and led to a general rearmament. The British effort was aimed at obtaining parity in air power as a deterrent to Germany. By late 1937 these plans called for the conversion of Bomber Command to four-engine bombers by 1943, plans that raised issues of both money and manufacturing capability. In December 1937 the British revised their offensive plan to one that emphasized air defense. In any event the RAF was ill prepared when Germany invaded Poland in September 1939 and World War II in Europe began.

## Opening Rounds

The outbreak of war found Bomber Command with seventeen of its thirty-three operational squadrons classified as heavy bomber units. Strategic bombing plans called for attacks on the GAF, the aircraft industry, and military transportation, as well as strikes specifically on oil and industry in the Ruhr, Rhineland, and Saar. Bomber Command realized it lacked the power to destroy all these targets but believed that power, coking plants, and oil were special German vulnerabilities. In brief, the airmen were thinking in terms of a bottleneck approach using daylight, unescorted precision bombing tactics. Events would show that while the RAF had solved most of the problems of defensive air warfare early in the war, it faced a long, difficult, and costly struggle to mount an effective offensive campaign.

Disastrous daylight, unescorted bombing missions quickly demonstrated how unprepared Bomber Command was for war. On three missions in September 1939, the British lost twenty-five of sixty-four bombers. These operations revealed the Germans had adequate defenses against unescorted day attacks even around its periphery. In contrast, Bomber Command was deficient in numbers, aircraft, equipment, tactics, training, and plans—just about everything except courage and determination. Bombers burned easily, were poorly defended by their .303-caliber machine guns, used inadequate bombsights, lacked self-sealing fuel tanks, had no navigational aids, dropped inefficient bombs, and flew loose formations. At the same time night leaflet missions operated over the Continent with few losses. (Between May 9 and June 4, 1940, the RAF flew 1,700 night sorties and lost thirty-nine aircraft, a 2.3 percent loss rate, compared with 856 day sorties and a loss of fifty-six bombers, 6.5 percent.) As a consequence Bomber Command turned to night bombing.

The British carefully chose their targets, fearing German retaliation. (In late April the vice chief of the Air Staff, Richard Peirse, noted that it would be foolhardy to provoke a German response because the GAF could deliver a blow four times greater than Bomber Command.) Events quickly changed the RAF's position. On May 15 Bomber Command was authorized to attack east of the Rhine following the German invasion of the Low Countries, and that night it launched one hundred bombers against oil and rail targets in the Ruhr. But competing priorities intervened after the fall of France; most of all French ports that sheltered invasion barges. The Battle of Britain escalated matters. The GAF's accidental bombing of London on August 24 led to the bombing of Berlin the next night and the countering assault on London that began on September 7. The British maintained their policy of precision bombing, now at night, and continued to shun indiscriminate attacks.

Bomber Command believed it was having an impact with its bombing based on the claims of its crews and information that leaked out of Germany. It was not until November 1940 that the RAF formed Spitfire reconnaissance units that could obtain poststrike photography and positive evidence of its bombing. The airmen had neglected aerial cameras and reconnaissance, much to the detriment of the bombing campaign. It bears repeating that intelligence was, and is, a critical element of strategic bombardment. To be clear, this includes both the identification of critical objectives and assessment of damage (bomb damage assessment) of these targets.

Aerial photographs forced the RAF to question the reporting of its aircrews. For example, on the December 13, 1940, raid against Mannheim, 102 aircrews reported that they delivered their bombs in the target area and left the center of the city in flames on this first nonprecision area raid. Photographs, however, indicated a wide dispersal of bombs and showed that the mission had failed. Late in the month Spitfire photos of two oil facilities at Gelsenkirchen revealed that although the majority of 196 aircrews claimed they attacked these targets, there was no indication of important repair work and little sign of the 260 tons of bombs. The realization of these shortcomings led Bomber Command in the first quarter of 1941 to switch from precision attacks to attacks on cities, specifically designating the "industrial centers" of large German cities as the aiming point.

Nevertheless the command's oil plan of January 1941 assumed that under the best conditions of moonlight and visibility, the bombers could achieve an average bombing error of three hundred yards. In April 1941 Bomber Command more than doubled this figure and acknowledged that the error was probably closer to one thousand yards. But even this concession was overly optimistic. In the March 12, 1941, attack of the Focke-Wulf plant in Bremen, fewer than one-third of the bombs hit within six hundred yards of the plant. If that was the situation with moonlight and good weather, bombing accuracy could be expected to be far worse under less favorable conditions. The severity of the problem was driven home by the Butt Report based on 650 photos taken in June and July 1941 during one hundred separate attacks. The photos revealed that only one-third of the crews that reported attacking their targets got within five miles, varying greatly due to location, moonlight, weather, and enemy opposition. Clearly Bomber Command's navigational skills were inadequate.[1]

Although the RAF espoused precision bombing, it certainly had considered attacking German morale. In the fall of 1940 both the prime minister and commander of Bomber Command discussed retaliatory bombing of German towns in response to indiscriminate German bombing, a conversation that raised issues of feasibility, desirability, and morality. The Air Staff believed that morale along with transport were German weak points. British experience with the bombing of London should have given the decision makers pause concerning bombing to affect morale, but that does not seem to be the case. By the end of 1940 the bombers' primary aim was oil in favorable conditions, otherwise morale attacks against Berlin and cities in central and western Germany. In December 1940 the British Ministry of Information opined that the Germans could not take one-quarter of the bombing that the British had withstood. Apparently more than the propagandists believed this, for in July 1941 both the British Foreign Office and prime minister were forceful proponents of bombing civilians.[2]

The Air Staff concluded that different methods were required for city attacks. A comparison of photographs of British and German towns bombed by approximately equal numbers of bombers indicated much greater damage inflicted by the GAF. A September 1941 memo attributed this to the greater proportion of incendiaries carried by the Luftwaffe bombers. In addition German bombs were more powerful, pound for pound, than British bombs.[3] Bomber Command adapted.

In September 1941 the RAF sent a plan to Churchill that called for the total destruction of forty-three German cities with a population of fifteen million people, requiring four thousand bombers and six months. Churchill, despite his support of morale bombing and Bomber Command, was very skeptical of such claims. During World War I, as already noted, Churchill rejected the idea "that any terrorization of the civil population which could be achieved by air attack would compel the Government of a great nation to surrender."[4] Nevertheless, Britain, fighting alone and on the defensive, decided to make the commitment to a major bombing offensive.

The year 1942 was a watershed for Bomber Command. First, it turned to city busting as a strategy. Second, the unit's two major aircraft of the war, the Lancaster and Mosquito, went into service, along with a number of electronic devices that would greatly enhance the offensive bombing effort. These developments, long in the making, are overshadowed in some narratives of the bombing offensive by a third event, the arrival of Britain's most famous air leader of the war.

## Arthur Harris, the Personification of RAF Bombing

The editor of Arthur Harris's wartime dispatches begins with the comment that "there are few more controversial figures in modern British military history than Sir Arthur Harris" and goes on to write that "perhaps only Sir Douglas Haig [of World War I] has attracted so much vilification and public hostility."[5] Harris has come to personify not only the British bombing campaign of World War II but also strategic bombing. Feelings still run high regarding bombing in general and Harris in particular, as demonstrated by the 1992 protests in both Germany and Britain over the unveiling of a statue to Harris and a critical Canadian Broadcasting Corporation TV show that same year. As the most renown German historian of the air war writes, in German eyes Harris "stands for all the horrors of the bombardments and firestorms . . . [and] became the incarnation of violence."[6]

To be clear Harris neither conceived the policy of city bombing nor did he play a role in the innovations that made Bomber Command such a powerful force. In fact he opposed one of the hallmarks of British strategic bombing (the Pathfinder Force) as well as targeting precision targets that he derisively labeled "panacea targets." What he provided for more than three years was firm leadership that overcame obstacles presented by the Germans, weather, geography, and, on occasion in his view, his superiors. Although some would describe him as cold blooded, his men saw him as a forceful, ruthless, single-minded leader. What set Harris apart from his contemporaries was his willingness to make tough decisions with unshakable determination. In addition, Harris illustrates the rewards of being at the right place at the right time (as we shall see with Curtis LeMay's experience with the B-29); he came to Bomber Command after his predecessors had been unable to accomplish very much and just as numerous factors were coming together to increase the effectiveness of British bombing.

Harris assumed command of Bomber Command in late February 1942. He had orders issued just nine days earlier to rain incendiaries on cities to break German morale. Harris found a small force to carry out this task: 374 aircraft available with crews, only 44 of which were heavy bombers. The force grew, and a year later the overall strength was 593 bombers of all types with 363 heavy bombers, most of which were the new and best British bomber of the war, the Lancaster.[7] Bomber Command also incorporated new equipment and tactics during this time period.

Arthur "Bomber" Harris (1892–1984) assumed leadership of RAF's Bomber Command in February 1942. Although not the creator of night area bombing, he became its superb practitioner. His wartime position and the controversy surrounding the bombing of Germany led to him being snubbed after the war as a symbol of the immorality of strategic bombardment. (Air Historical Branch [RAF] Ministry of Defense)

## Navigational Devices

The RAF encountered considerable navigational problems, as the Butt Report made strikingly clear. The British were slow to develop navigational aids and development took some time. What first appeared was Gee, a radio device service tested in May 1941 that went into operations the following March. Three ground stations transmitted signals to the bomber that were displayed on a cathode-ray tube and then translated onto a special chart to indicate the aircraft's position. Gee had three limitations: range was no more than three hundred to four hundred miles, accuracy was relatively poor and declined with range, and the system could be jammed. After August 1942 jamming rendered it essentially useless east of the Dutch coast, however, it continued to assist the bombers on their outbound and inbound legs. By this time 80 percent of Bomber Command carried the device, and by January 1943 the unit was fully equipped.

Oboe was Bomber Command's first true blind bombing aid.[8] Equipment positioned in Britain emitted radio signals, reinforced by onboard equipment that guided aircraft along a parabolic track. Oboe went into operational service in December 1942. Like Gee, the device was limited in range by aircraft altitude and, in fact, had less range than Gee. However its accuracy was far better, measured in hundreds of yards so it was mainly used by the Pathfinder Force. The device doubled Bomber Command's bombing accuracy between 1942 and the spring of 1943. It had two major drawbacks: one pair of stations could handle only one bomber at time, and it could be jammed.

A third aid to navigation was G-H, which combined much of Gee and Oboe and has been described as Oboe in reverse. The bomber carried both a transmitter and receiver that employed ground stations. While its range was limited, its accuracy did not decline with distance and was about that of Oboe. It did have three disadvantages, however: it required a highly skilled navigator, it was limited to handling fewer than one hundred aircraft at the same time, and it could be jammed. The British initiated the program in July 1942, began service trials in June 1943, and started operations in October. Bomber Command did not make extensive use of G-H until the next year in daylight attacks.

A fourth device was navigational radar, airborne maritime search radar adapted to overland duty. Known as H2S, it transmitted radio signals, some of which were reflected back to a cathode-ray tube that gave a general representation of the terrain.[9] Radar took skill to interpret as its presentation varies from visual appearances, and while land-water contrasts were easiest to determine, presentations of open country (few returns) or large urban areas (massive returns) were much more difficult. Because all the radar equipment was carried aboard the bomber, it was not limited in range. Because of a lack of industrial capacity H2S experienced serious production problems until after 1941, nevertheless it went into service in January 1943 and by the end of the year more than 90 percent of the command's bombers were so equipped. H2S proved to be the most useful navigation and bomb aiming device used by Bomber Command. One major liability was that GAF night fighters could home in on radar emissions.

## New Tactics

Better navigational aids along with improved German defenses encouraged Bomber Command to modify its tactics. The unit pressed for concentration of bombers on their routing as well as over the targets, tactics to counter German linear defenses by flooding German radar and antiaircraft artillery (AAA) units. These advantages outweighed the risks of collision, friendly fire (be it machine guns or falling bombs), and greater vulnerability to interceptors. In March 1942 Bomber Command made a major tactical change with a tactic (Shaker) consisting of aircraft dropping illumination to guide in bombers that placed target markers for the main force that followed. Improved target marker flares become operational in January 1943; later the RAF used air markers. Despite the use of these tactics and Gee, success was illusive.

One major innovation instituted by Bomber Command was to organize the Pathfinder Force, which led RAF bombing raids and played a major role in increasing bombing accuracy. The idea of forming a unit of select crews emerged as early as February 1941, and the Shaker tactics and the shortage of Gee equipment helped push its formation. The unit did not enter combat until August 1942 because Harris opposed the idea, contending that it would create a *corps d'élite* at the expense of the morale of the rest of his command and was unnecessary. The British commander was forced to institute a system that he bitterly resented and described in his postwar dispatches as "yet another occasion when a Commander in the field was over-ruled at the dictation of junior staff officers in the Air Ministry."[10] Instead of one Pathfinder unit, Harris wanted each group to have such an organization as tried toward the end of the war.[11] Harris had little interest in such a scheme and no use for the concept of precision bombing or, in his words, attacking panacea targets, because he aimed to devastate urban areas.

The men of the Pathfinder Force supposedly were selected for their abilities, although, according to the official history, "it remains doubtful whether they were always the best crews which Bomber Command had to offer."[12] In addition the Pathfinders never had priority in equipment. This was dangerous work because these men were to fly an extended tour of sixty missions and as the vanguard of the bombing raids would be first in.

The Pathfinders and the new tactics bore results, albeit modest at first. From March to August 1942 night combat photography indicated that 35 percent of the bombers bombed within three miles of the center of the concentration of bombing, and from August 1942 to March 1943 this increased to more than 50 percent. Now the task was getting that concentration over the intended target for during the first period only 32 percent of the bombers were within three miles of the aiming point, while during the second the figure was 37 percent.

## Offensive Bomber Equipment and Methods

The combatants engaged in a lethal race to best use electronic equipment while not giving the opponent an advantage that could be exploited. The interests of the offensive (Bomber Command) had to be balanced against the interests of the defense (Fighter Command and the Royal Navy). The British tended to be conservative, emphasizing how the device might be used against them over the offensive advantages, despite the growing gap between the offensive capabilities of the RAF and the GAF. This combined with a late start meant that Bomber Command did not get significant electronic countermeasures (ECM) into combat until the end of 1942.

In December 1942 the bombers began to use their Identification Friend or Foe (IFF) equipment to jam the German plotting radar (Wurzburg), a process known as Shiver. (In turn the Germans homed in on British IFF, until the RAF learned of the German technique through broken German codes, Ultra, and terminated Shiver in January 1944.) At that same time the British began active measures (Tinsel) to interfere with German radio traffic. The Luftwaffe responded with very high frequency (VHF) radios in March 1943, to which the British answered with Cigar in July, a ground-based system effective out to a range of 140 miles. In October 1943 Bomber Command introduced Airborne Cigar, which overcame the range limitation. In April the British also began using a ground-based system (Grocer) to jam German interceptor radios, later supplemented by an aerial version known as Airborne Grocer. These devices forced the GAF to change its control procedures from the box system and ground-controlled intercept (GCI) to the "running commentary" system, which brought new challenges and new countermeasures. In October 1943 British ground stations began transmitting (Corona) false instructions in idiomatic German and were prepared when the German began using female operators. The GAF then used commercial radio stations (Anna Marie) to transmit announcements and broadcast music that designated cities under attack (for example, waltz music indicated Vienna).

The best known, longest lasting, and simplest ECM devices employed was called Window by the British, Chaff by the Americans, and Duppel by the Germans. Aircraft dropped stripes of aluminum foil cut to half the wave length of enemy radar (similar to Christmas tree tinsel) to produce radar images that complicated the task of radar operators. Its development and delayed employment clearly illustrates the dilemma of balancing offensive and defensive aerial needs. Both the British and Germans knew of the device's impact on radar, but both resisted using it, fearing that their opponent would benefit more than themselves. Although the British Chiefs of Staff agreed in April 1942 to use Window, concerns over its impact on British night fighter radar delayed its employment until the Hamburg raids of July 1943.

Another electronic aid was Boozer, which the British thrust into action in late 1942. It gave a visual indication when the bomber was under airborne radar surveillance (yellow light) or ground radar (red light). Monica was a tail warning device introduced in the spring of 1943 that could detect enemy night fighter radar emissions up to one thousand yards. However, the interceptors were able to home in on Monica, more than nullifying its usefulness. (The British learned of this in July 1944, when a Ju 88 night fighter mistakenly landed in Britain.) In June 1943 Fighter Command employed Beaufighters with a device called Serrate, which detected German night fighter radar emissions and indicated their bearings out to eighty to one hundred miles. In three months British fighters made twelve hundred contacts but only were able to engage twenty in combat and claim the destruction of thirteen. Harris wanted more than ten squadrons of night fighters assigned to the defense of Britain but received only three, which claimed the destruction of 257 Luftwaffe fighters. Both Bomber and Fighter Command flew intruder missions equipped with Serrate, which became increasingly ineffective as the war progressed.

The increasing importance of ECM and growing size and weight of airborne equipment led the RAF to dedicate aircraft strictly to jamming and form a specialized unit with this responsibility. It flew its first ECM mission in November 1943, although the unit did not hit its stride until after the June 1944 Normandy invasion.

In addition to electronics, Bomber Command also used various tactics to complicate the defender's task. To deceive the night fighters the British used diversionary forces, marked cities other than the target, and employed evasive routing. Mosquito bombers were especially important in these efforts but were limited in numbers with only an average of fifty-eight available to Bomber Command as late as March 1944. Because the target area proved most deadly for the bombers, there was a premium on getting the force in and out as quickly as possible. In October 1943 compressing the time over the target was "drastically stepped up."

Bomber Command made other improvements that increased its impact. The British learned that German bombs packed more punch than theirs due to better explosives and a greater proportion of explosive resulting from lighter bomb cases. This was remedied and the RAF went on to drop heavier conventional bombs than any other air force, putting into combat a 4,000-pound bomb ("block buster") in March 1941, an 8,000-pound bomb in April 1942, the 12,000-pound "Tallboy" in summer 1944, and the 22,000-pound "Grand Slam" in March 1945. The British also put a better bombsight into its bombers beginning in September 1942. Another basic change was Bomber Command's reorganization of the bomber crew in early 1942, reducing it from two pilots to one, which certainly eased their training and staffing problem. Two new positions were added: navigator and bombardier (bomb aimer).[13]

## Britain's Other Heavy Bombers

The heavy bomber identified with Bomber Command's air offensive against Germany is quite rightly the Lancaster, but until it appeared in 1942, the RAF used twin-engine medium bombers and two other aircraft, the four-engine Stirling and Halifax. The Short Stirling is remembered at best as

The Short Stirling heavy bomber was designed with a one-hundred-foot wing span to accommodate RAF hangar doors, a bomb bay limiting the size of bombs, and a fuselage sized to carry standard cargo crates. The compromised result led to a disappointing aircraft with limited performance. (U.S. Naval Institute Photo Archive)

a stop-gap bomber and, in the understatement of the official history, "a disappointment." It came to life from a July 1936 specification that generated proposals from eight companies of which Supermarine and Short were awarded developmental contracts. From the start the Short design was compromised in a number of ways. First, its wing span was limited to less than one hundred feet to allow it to fit through RAF hangar doors. This resulted in a low-aspect-ratio wing, which gave it unusual maneuverability for a bomber but seriously restricted its ceiling. Second, the bomb bay design precluded the carriage of bombs larger than 2,000 pounds. Third, the bomber featured a midwing configuration that necessitated a long landing gear. This arrangement tended to cause swings on takeoffs that created safety problems. Finally, the fuselage was sized to accommodate standard packing crates. Understandably the resulting aircraft left much to be desired.

The Stirling made its maiden flight in May 1939 and was destroyed on landing, not a bright omen. Production difficulties and German air raids delayed the aircraft's combat debut until February 1941. The initial design was armed with four .303-caliber machine guns in the tail and twin .303-caliber machine guns in a power nose and retractable ventral turret. (The manufacturer removed the latter on operational bombers, but added twin guns in a dorsal turret and two hand-held beam guns.) The Stirling proved able to absorb punishment, but it was outperformed by both the Halifax and Lancaster. The Stirling was at a great disadvantage because it could only operate at twelve thousand feet loaded, while the other British heavy bombers were flying at twenty thousand feet. By mid-1943 the Stirling was only attacking less defended targets, and it flew its last mission with Bomber Command in September 1944. The British built just under twenty-four hundred Stirlings.

Bomber Command's second heavy bomber, the Handley Page Halifax, was more successful. It began as a two-engine design; however, questions as to the availability of the Rolls Royce Vulture engine led to the use of four Rolls Royce Merlin engines. The bomber made its initial flight in October 1939 and entered combat in March 1941. The production model was defended by nose and tail turrets, mounting respectively two and four .303-caliber machine guns. (The paradox is that the British led the world with power gun turrets but armed them with the inadequate rifle caliber .303 machine guns, which were markedly inferior to the .50-caliber machine guns and cannons they faced.)[14] Its light armament led to the addition of a two-gun dorsal turret in the Halifax II series. However, the added drag of the top turret combined with limited engine power adversely affected flying performance. More serious was the aircraft's tendency to enter uncontrollable spins at heavy weights, which increased accidents. Harris described the Halifax as a "virtual failure," citing its vulnerability, poor performance, vicious handling, and unsatisfactory exhaust flame dampening. To remedy some of these problems the manufacturer replaced the nose turret with a single gun and the top turret with a lower drag four-gun unit. The company made other changes as well that increased both maximum and cruising speed by 10 percent in the Halifax B. Mk.II

Series IA. Nevertheless after the fall of 1943 Bomber Command relegated the Halifax to easier targets because of increasing losses. The Halifax III mounted more powerful engines, had better performance, and went into combat in February 1944. The British built almost sixty-two hundred Halifaxes, and they delivered one-quarter of Bomber Command's total tonnage.

When all is said, these bombers were not up to Bomber Command's task. The Manchester (see below) was considered a disaster, and both the Stirling and Halifax were disappointments. Fortunately the British were able to field much more capable aircraft, bombers that could effectively carry out the RAF's plans.

## The Avro Lancaster

The RAF's outstanding heavy bomber of World War II was the Avro Lancaster, which some will argue was the best heavy bomber of the war. (Others would award that accolade to the Boeing B-29.) Renowned aviation historian William Green writes that it was a simple, robust aircraft, ideally suited for mass production, a major factor in its success. He asserts that a great plane must have a "touch of genius which transcends the good" and "luck to be in the right place at the right time." Further, it must have above average flying qualities: "reliability, ruggedness, and fighting ability"—and skilled crews. "All these things," he concludes, "the Lancaster had in good measure."[15]

Handley Page modified a twin-engine design into the four-engine Halifax, which went into combat in March 1941. The bomber had deficiencies of defensive armament and a tendency to spin. Although Harris disliked the Halifax, it delivered one-quarter of Bomber Command's bomb tonnage. (San Diego Air and Space Museum)

Avro's Lancaster was Britain's outstanding and principal World War II heavy bomber. It was noted for having the lowest RAF heavy bomber loss rate, its large bomb load, and its ability to carry the heaviest bombs employed in the war. (National Museum of the USAF)

The Lancaster story begins in 1936, when the standard RAF night bomber was the twin-engine biplane Heyford and there was but one squadron of monoplane bombers (Hendons) in Bomber Command. That year the air service drew up specifications for a four-engine heavy bomber and a twin-engine medium bomber to be larger than the Wellington, which at this point was considered a heavy bomber. A.V. Roe proposed a design that was powered by two "new and unorthodox" liquid-cooled engines, which first flew in July 1939 and was named the Manchester. It entered combat in February 1941 and proved a distinct failure due to engine reliability and power problems, poor handling at heavy weights, and inadequate armament. It was vulnerable, on average surviving but seventeen operations, compared with twenty-seven operations in the aircraft it replaced (Hampden), and compiling the highest loss rate of all Bomber Command bombers in the war. As a result the RAF removed it from combat service in June 1942.

To remedy the bomber's problems, Avro installed four engines on the basic airframe. The Lancaster made its maiden flight in January 1941 and successful flight tests led to an immediate production decision. It was fitted with four power turrets (nose, tail, dorsal, and ventral), all mounting twin .303-caliber machine guns except the tail position with four .303s. (The ventral turret was soon removed.) The aircraft was built for production underscored by the decision to use a fixed rear wheel instead of the more complicated, expensive, and less drag producing retractable one. One of the outstanding characteristics of the Lancaster was its large bomb lift capacity and ability to carry giant bombs, including the heaviest employed in World War II, the 11-ton "Grand Slam." Unlike most combat aircraft built in large numbers, the Lancaster was little changed during the war. Most were fitted with H2S radar and other electronic aids, a few mounted .50-caliber machine guns, and some had bulged bomb doors in order to carry heavier bombs.

The Lancaster flew its first combat mission in March 1942 and became Bomber Command's most capable and primary aircraft. By March 1945 Bomber Command consisted of four hundred Halifaxes sharing bombing duties with the twelve hundred Lancasters. The latter delivered more than 600,000 tons of bombs on 156,000 sorties, two-thirds of the bombs dropped by Bomber Command after it entered service. The Lancaster's average delivery load was ninety-two hundred pounds compared with sixty-nine hundred pounds for the Halifax and fifty-six hundred pounds for the Stirling. Although thirty-eight hundred Lancasters were lost during the war, they had the lowest RAF heavy bomber loss rate.[16] It should be noted, however, that the bomber had only one escape hatch, which resulted in proportionally higher crew fatalities than those suffered on other bombers.[17] A total of seventy-four hundred Lancasters were built in British and Canadian plants and powered by both British- and American-built Rolls Royce engines.

## The De Havilland Mosquito

Less than two months prior to the Lancaster's first flight, the De Havilland Mosquito made its initial flight. It combined a small, clean wooden airframe with two Rolls Royce engines, which gave it flying performance superior to any bomber until the advent of jets. Perhaps its outstanding virtue was its versatility, for the Mosquito proved to be not only an excellent bomber but excelled in many other roles as well. Its sleek appearance once again reinforces the old flying cliché, ungrammatical as it may be, that "if it doesn't look good it won't fly good." The Mosquito looked good, and could it fly! And its unorthodox origin, outside the RAF system, helps explain its rapid development, only twenty-two months between design and combat.

De Havilland, well known for its World War I military aircraft, had concentrated on civil aircraft after the war. In 1938 the company investigated the military applications of its four-engine, wooden, fast transport, the Albatross. The design team emphasized speed above all else, regarding armament as an impediment, and favored wood construction. Lack of armament would lighten the aircraft and allow a more streamlined airframe, while wood construction would ease competition for metal and simplify repairs. De Havilland settled on a small two-engine design manned by a two-man crew. The government showed little interest in the proposal until the war's start, and even in mid-December 1939 the commander of Bomber Command stated he had "no use" for an unarmed bomber. Nevertheless two weeks later the government and De Havilland agreed on the concept of a wooden, unarmed, two-man bomber that could carry a 1,000-pound bomb fifteen hundred miles with fighter performance. The design was to emphasize producability, simplicity, and low cost. The government awarded De Havilland a contract for fifty aircraft in March 1940, deleted it from production after Dunkirk, but restored it within weeks. The GAF dealt the program a setback in October when it destroyed a warehouse housing Mosquito materials, but not the prototypes or other vital elements of the program.[18]

The De Havilland Mosquito differed radically from other bombers of the war with its wooden construction, two-man crew, and lack of defensive armament. It had exceptional flying performance that enabled it to outrun most German fighters. Almost eight thousand were built in three countries, and it served not only as a bomber but also in fighter, intruder, transport, and reconnaissance roles. (U.S. Naval Institute Photo Archive)

The Mosquito first took to the air in November 1940, a remarkable eleven months after the start of design work. Testing revealed an aircraft with fighter agility and speed approaching four hundred miles per hour—an unheard-of performance for a bomber and faster than the Spitfires then in service—and the ability to perform aerobatic maneuvers with one engine feathered. The bomber version went into action in May 1942. Initially it could carry four 250-pound bombs internally, and later an extension of the bomb bay and bulged bomb doors accommodated a single 4,000-pound bomb. With this load the aircraft could clock 300 mph at sea level and 380 mph at seventeen thousand feet. The Mosquito's high speed, equivalent to the fastest piston-powered German fighter, meant that the Mosquito could only be overhauled if the interceptor had prior warning, proper position, and superior altitude. As with most new aircraft, the Mosquito encountered problems, including tail shimmy, a tendency to swing on takeoff, a leaky cockpit, and, throughout its service, difficulties with glue.

The Mosquito proved extremely versatile, serving in such varied roles as a night fighter, fighter bomber, courier, transport, antisubmarine, antishipping, reconnaissance, and night intruder, escort fighter, and "nuisance" bomber, and it was especially valuable in the Pathfinder Force. In all almost seventy-eight hundred Mosquitoes were built. Despite feverish efforts the Germans never found a solution for the Mosquito. Not surprisingly the De Havilland bomber registered the lowest loss rate per sortie of any Bomber Command bomber.

## Bomber Command in Action

Bomber Command's limited abilities constrained its activities during 1942, but the unit flew some notable missions. In late March British bombers attacked the "fairly important, but not vital" medium-sized city of Lubeck. The 234 aircraft delivered just over three hundred tons of munitions and destroyed half of the city at the cost of thirteen bombers in what Harris considered an "unqualified success." It showed what British bombers could do in good weather and in bright moonlight conditions.

A month later Rostock was the target on four consecutive nights, the first two ineffective, the third considered an outstanding success, and the last a masterpiece. These raids destroyed 70 percent of the old city and were widely publicized. In the words of the official account, "Thus, by the end of April 1942, Bomber Command under the vigorous leadership of Air Marshal Harris had shown, not only to Britain's allies but also to her enemies, the tremendous potential power of the long-range heavy bomber force."[19] That may exaggerate the Rostock mission and is probably more appropriate for the thousand bomber raid on Cologne two months later.

This attack was a daring move by Harris because he had to employ large numbers of crews in operational training units to muster the 1,046 bombers that he dispatched. Almost 900 bombers flew through foul weather to find conditions over the target clear (Hamburg was the first choice, but weather intervened), where they dropped fifteen hundred tons of bombs, two-thirds of them incendiaries. They did this in an hour and a half, a remarkable effort and improvement over previous attacks. The bombing damaged six hundred acres, much more damage than inflicted on the previous fourteen hundred sorties against the city, and almost equaling the total (seven hundred acres) damage inflicted on German cities thus far in the war. The bombing destroyed about one month's worth of production. Bomber Command lost 3.8 percent (forty bombers) of the aircraft dispatched compared to an average loss rate of 4.6 percent during the previous year's attacks on western Germany. Although the Cologne mission was a major Bomber Command victory, it was a showpiece that could not be easily replicated. In June Bomber Command launched 956 bombers against Essen and 904 against Bremen, both unsuccessful attacks. The next largest raid of the year was against Dusseldorf in July, and that force of 630 was not exceeded until May 1943. Postwar studies reveal that the 1942 bombing had little impact on the German economy. Nevertheless the attack on Cologne indicated the power and potential of Bomber Command, sparked British morale, encouraged the Soviets, and impressed the Americans.

## Atypical Bomber Command Operations

Bomber Command burnished its record for daring, courage, and efficiency with a number of low-level daylight attacks during the war. Most were small in scale and tactical, some were costly, and many were unsuccessful. Only a few merit attention as strategic strikes. The most famous Bomber Command low-level attack, probably the outstanding British operation of the war, and among the stellar bombing attacks of all time, surely one of the best known, was the May 16, 1943, attack on the Mohne, Eder, and Sorpe dams. The official Bomber Command history calls it "the most precise bombing attack ever delivered and a feat of arms which has never been excelled."[20] Prior to the war the British considered such an attack because these dams stored water vital for German production in the Ruhr and their destruction would seriously disrupt the German economy. Yet the plan had its detractors. Harris opposed the operation, seeing it as another example of "panacea-mongers" diverting his force to use an untried weapon against promising but unrealistic targets. Nevertheless Bomber Command selected its best crews, provided them with intensive training, invented a new weapon, and devised new tactics for this endeavor.

A large explosive, detonated in a specific location is required to destroy a dam. Barnes Wallis, the inventor of the geodetic design, developed a large bouncing (or skipping) bomb designed to hit the dam's up water side and then roll down the face and explode beneath the water's surface, which would have a tamping effect on the explosion. To accurately deliver the 9,300-pound weapon, the crew had to fly an exact speed, straight and level at very low altitude (sixty feet above the reservoir's surface), and release the munition at 400 to 450 yards from the target . . . at night. Bomber Command timed the attack for mid-May to capitalize on the best moon conditions, the highest reservoir level, and the longer nights before summer.

The RAF employed modified Lancasters and organized the 617 Squadron specifically for this mission that drew on the most experienced and skilled airmen in Bomber Command. The crews not only faced the normal hazards of low-level night flying over hostile territory but also encountered fog. The attack cost eight of the nineteen bombers and fifty-six men (fifty-three killed), but the results were outstanding. The assault breached the Mohne and Eder dams and damaged the Sorpe dam. The British knew prior to the attack that the Sorpe was more important than the Eder, but it also was more difficult to destroy for both tactical and technical reasons. The resulting devastation was considerable, killing thirteen hundred, damaging fifty bridges, and causing the loss of a week or more of production in the Ruhr, as well as draining off workers and materials to repair the damage and increase the defenses of the dam system. However, damage was limited to a small agrarian area for a brief time. It was too ambitious because the airmen lacked the force to destroy the other dams that could have caused widespread and serious economic damage. Therefore the dam missions were more spectacular than significant, daring but not decisive, costly without real effect, bringing to mind one of the observations of the charge of the light brigade at Balaklava during the Crimean War: "It is magnificent, but it is not war."

Bomber Command employed Lancasters on other precision bombing missions. On September 15, 1943, 617 Squadron dispatched eight Lancasters to break the walls of the Dortmund-Ems canal with 12,000-pound bombs dropped from 150 feet. The bank held; five bombers failed to return. In a similar fashion Lancasters failed to drop the Antheor viaduct on September 16, 1943 (one of six low-flying bombers was lost) and on November 11, 1943, when ten Lancasters attacked (and all survived). A third attack on the viaduct on February 12, 1944, met tough defenses and also failed to take out the structure, although one 12,000-pound bomb hit within forty-five feet of the target. In contrast, the night attack of twelve Lancasters of the elite squadron against an aircraft engine factory at Linoges on February 8, 1944, was a rousing success. The factory was destroyed; there was neither collateral damage nor losses.

These low-level precision-bombing missions, although they sometimes achieved spectacular tactical results, yielded questionable strategic results. The RAF launched its most skilled, most experienced, and probably bravest crews on daring missions, which they sometimes accomplished but mostly at a frightful cost. The strategic air war would be fought, and won, by other means.

## Bomber Command Swings into High Gear

It was not until 1943, more than three years after the war's onset, that Bomber Command began to make itself felt. New aircraft, improved equipment, and increasing numbers gave the unit greater capabilities, which would only improve as the war continued. This increased capability was first seen in the campaign against the Ruhr, the German industrial heartland. The official history described it as not just one successful attack but "a whole series of consistent and pulverizing blows among which the failures were much rarer than the successes."[21]

The Ruhr lays in western Germany, along the Rhine River, less than thirty miles from the Dutch border and thus easier to reach than other German targets. This geographic advantage was nullified, if not cancelled, by industrial haze, the concentration of cities, and most of all, tough German defenses. The assault began in early March 1943 and ended four months later. (To be clear, while the bulk of the attacks during this period were against cities in the Ruhr, Bomber Command hit other targets as far east and south as Stettin, Pilsen, and Munich.) The British employed two major tactics. The first used the main force, sometimes led by Mosquitoes marking the target on major attacks. The second method employed small numbers of Mosquitoes, no more than a baker's dozen, hitting mainly diversionary targets. During this period there were forty-three major attacks on Germany that sustained 4.7 percent losses of the sorties flown. The RAF concluded that this toll was sustainable but realized that losses greater than 7 percent were not and that losses above 5 percent would lower effectiveness by eroding aircrew experience and morale. Despite this stiff attrition Bomber Command increased its daily average crew strength from 593 in February to 787 in August.

The Mosquito was key to the success over the Ruhr. It proved essentially invulnerable to German defenses and, fitted with Oboe, provided target marking that improved bombing accuracy. The initial attack of the campaign on March 5 against Essen illustrates the aircraft's importance and the basic pattern of Bomber Command attacks during the remainder of the war. The assault began when Oboe-guided Mosquitoes dropped red target indicator bombs on the aiming point backed up by heavy bombers of the Pathfinder Force, which dropped green target indicator bombs. The main force of three overlapping waves of bombers, first Halifaxes, then Wellingtons and Stirlings, and finally Lancasters, dropped 1,015 tons of bombs, two-thirds of them incendiaries, with one-third of the high explosives fused for long delay. The outstanding feature of the tactic was that visual observation was not involved. Raid and postraid photographs indicated that of the 422 aircraft dispatched, 153 bombers released their bombs within three miles of the aiming point, an accuracy never before achieved. The British lost fourteen aircraft and laid waste to 160 acres and demolished or damaged three-quarters of an additional 450 acres. Clearly this was a successful attack that revealed the bombers' power and a foreshadowing of what would follow.

Bomber Command concentrated its attacks on German city centers because they were easier to find and more flammable due to their density and older construction. It should be noted, however, that for the most part these were residential areas; industrial areas were located elsewhere. Thus Bomber Command directly attacked German citizens, certainly their housing, and secondarily and indirectly, German industry. The campaign against the Ruhr demonstrated Bomber Command's ability to concentrate its bomb delivery with the aid of Pathfinders using Oboe. Other attacks during this period of more distant targets using radar aiming were not as successful.[22]

The most memorable operation during this time frame took place outside the Ruhr and was one of the most destructive attacks of the air war. One factor in the British success was the introduction of a new electronics countermeasure, Window, which interfered with German radar. This was significant because the British estimated that by March 1943, 70 percent of their bomber losses were to night fighters and two-thirds of the remainder to radar-directed guns. After withholding Window from action for sixteen months, the British, facing mounting bomber losses (along with the declining GAF bomber force), revised their policy, believing they would reduce losses by one-third. Window would have its combat debut against Hamburg.

Hamburg was Germany's second largest city with a population of almost two million and a major industrial center where one-third of Germany's submarines were constructed. It was located about fifty miles from the North Sea, bisected by the Elbe River, and thus an excellent, if not ideal, radar return, which compensated for the fact it was outside of Oboe range. In addition it required only a short penetration of German air space. Bomber Command launched a force of 791 bombers (347 Lancasters) on the night of July 24, 1943, led by radar equipped aircraft that not only marked the route and target but also radioed their calculation of the winds to the main force. While 728 of the crews claimed to have released their bomb loads within three miles of the aiming point, the actual number was 306. Thus accuracy was not exceptional, but the low losses—twelve bombers—were, as Window completely disrupted the German defenses.

Six RAF Mosquitoes raided the city the next two nights, and American Eighth Air Force bombers followed up with daylight raids on July 25 and 26. The AAF launched 244 bombers, of which 154 dropped 322 tons of bombs aimed at ship and submarine construction yards. The Americans lost 17 bombers and claimed 42 GAF fighters destroyed.

All of these actions paled to what Bomber Command did on the night of July 27. Again the markers were aimed by radar and placed with exceptional accuracy, allowing 325 of 787 crews launched to get their bombs within three miles of the aiming point. The high explosives kept the fire fighters away from the fires and knocked roofs off buildings, making them more vulnerable to incendiaries. This concentration of bombs, along with the high temperatures and low humidity led to a firestorm that quickly got out of control. The rising hot air drew in oxygen and fanned the flames, created extremely high temperatures, and allowed the fires to spread with incredible speed. (In addition, there was a lack of water within the city and German fire equipment was on the other side of the river fighting the fires from the previous attacks.) The destruction and death were staggering. Some forty thousand died, and by nightfall 1.2 million had evacuated the city. The Germans downed seventeen bombers.

But Hamburg's ordeal was not over. Mosquitoes raided Hamburg on the night of the twenty-eighth, and Bomber Command hit the city with a third major attack on the night of the twenty-ninth, this time launching 777 bombers, of which 238 got their bombs within three miles of the aiming point. The attackers lost 30 bombers, 4.3 percent of the 699 crews that claimed to have bombed the target. The RAF hit the city again on the night of August 2, dispatching 740 bombers, 30 of which failed to return. Bad weather en route caused a number of crews to turn back short of the target, while clouds over the target, the absence of moonlight, and rain reduced visibility to zero, rendering the bombing ineffective. The attack was an anticlimax.

The Hamburg attacks were an impressive victory for Bomber Command. It dispatched 3,095 bombers on the four major attacks and dropped almost nine thousand tons of munitions, about half incendiaries. The unit lost 86 bombers (2.8 percent of those launched), and another 174 were damaged. In this way the bombers suffered lower casualties than they had in the campaign against the Ruhr, much to the credit of Window. The damage inflicted was unparalleled up to this point in the war. Photo reconnaissance on August 1 revealed that the bombing heavily damaged sixty-two hundred acres of residential area (74 percent of the city's residential districts) and damaged shipyards and power and transportation services. The destruction left nine hundred thousand people homeless; they fled the city with tales of horror and terror. "These firestorms," as one report made clear, "[went] beyond all human imagination . . . developing in a short time into a fire typhoon such as was never before witnessed, against which every human resistance was quite useless."[23]

This was the first massive death toll from an air attack, dwarfing Guernica, Warsaw, Rotterdam, and London. While the most quoted figure for the numbers who died is about sixty thousand, the best estimate is about forty-five thousand deaths. The attacks shook the German decision makers and prompted Albert Speer to fear the loss of the war if the Allies could do the same to six more German cities. Bomber Command tried to duplicate its Hamburg success but could not until the Dresden attack in February 1945.

Another notable attack during this time frame took place on the night of 17 August 1943 against the German V-weapons research and testing facility at Peenemünde. This raid varied from the norm as it was intended to destroy specific buildings in the facility. One innovation during this attack was the use of a "master bomber," a commander who orbited the target and directed the assault, a tactic first used in the dams attack. The bombers also employed a new marker bomb for the first time, a "red spot fire," a 250-pound bomb that ignited at three thousand feet and continued to burn on the ground for ten minutes. The attack caused considerable damage. Window and diversions initially confused the defenders, but they were able to down forty bombers and damage another thirty-two in the full moonlight.

Between the end of the Ruhr campaign (marked by the Hamburg bombing) and the beginning of the Battle of Berlin (November 18, 1943), Bomber Command made thirty-three major thrusts against Germany. These attacks were distinguished from the earlier ones by the increased numbers of bombers (from five hundred to six hundred), with a much higher proportion of Lancasters, and increased bomb loads. On 17,021 sorties the British suffered 695 losses, or 4.1 percent. The loss rate declined despite the deeper penetrations, due to the use of Window, the dispersion of targets, and weaker German defenses.

## The Tide Shifts

With these successes under its belt, Bomber Command turned its attention to the enemy's capital. Berlin was the largest city in Germany, the third largest in the world, with a prewar population of more than 4 million (by March 1944 reduced to 2.9 million) and an area of 883 square miles. It was the political as well as industrial center of the country. Arthur Harris summed up his view in November 1943, writing, "We can wreck Berlin from end to end if the U.S.A.A.F. will come in on it. It will cost between us 400–500 aircraft. It will cost Germany the war."[24] At this point he listed nineteen German cities as "virtually destroyed" and another nineteen as "seriously damaged." Bomber Command had destroyed almost twelve thousand acres of urban area, about one-quarter of what it had attacked in the thirty-nine cities. Harris stated that "there can be little doubt that the necessary conditions [to cause the enemy to capitulate] would be brought about by the destruction of between 40% and 50% of the principal German towns."[25] Harris had high hopes, but these were to be dashed over German targets, and contrary to his assertion that Germany would be forced out of the war, it was Bomber Command that was defeated early in 1944.

Berlin was a difficult target. Besides its stout defenses, it was shielded by its geography (a round trip of 1,150 miles rendered Oboe ineffective), the notorious European winter weather, and its sprawling nature (which largely neutered radar navigation and bomb aiming). And despite Harris's hopes for American assistance in the campaign against Berlin, the AAF would not engage the German capital until March 1944.

Aided by poor weather, in late 1943, Bomber Command was able to get to Berlin and other German targets with success and bearable losses, in November and December losing respectively 3.1 and 4.1 percent of those dispatched on night attacks. The new year brought an increasing butcher bill. In January 1944 Bomber Command listed 314 bombers missing, the most for any month of the war, and 5 percent of those dispatched. The loss rate for six Berlin raids was 6.1 percent and 7.2 percent for three attacks on other German cities. The overall loss rate declined slightly in February to 4.7 percent, although against Leipzig on the twentieth, Bomber Command lost 9.5 percent (seventy-eight aircraft).

Although the loss rate declined in March to 3.1 percent, in that month the battle was lost, at least for the moment. By the end of that month Bomber Command shifted its aim away from Berlin to targets in southern Germany, dispersing attacks, and using more diversionary efforts. The reason for this was simply that the defenses had overwhelmed the offensive. The March 24, 1944, attack on Berlin cost seventy-two bombers missing (9.1 percent), primarily due to unforeseen winds that dispersed the bomber stream, gave the night fighters problems, but presented the flak gunners with many more opportunities. (The British estimated that more than fifty bombers fell to German AAA.) And then Bomber Command suffered its worst defeat of the war.

On March 30, 1944, Bomber Command launched 795 aircraft against Nuremberg. Everything went wrong. The routing can be criticized for a straight leg of almost 250 miles, while poor weather conditions over the North Sea prevented any extensive diversionary missions. This was essentially a head-on assault. The weather made matters worse: clear skies on the inbound route and conditions that produced contrails, presenting ideal conditions for the interceptors. There was no master bomber. The result was, in the words of the official historian, "Bomber Command suffered the ill consequences of unusually bad luck and uncharacteristically bad and unimaginative operational planning."[26] The enemy defenses and unpredicted winds scattered the bombers so that only a few reached their target and these inflicted little damage. "Vigour, courage, resolve and endeavour" could not overcome poor planning, unfavorable weather, and tough German defenses. The Germans claimed 132 bombers destroyed, when in fact they downed 95, mostly by night fighters.[27] Bomber Command lost as many men on this one raid as Fighter Command had in the entire Battle of Britain.

Bomber Command had been defeated. In the twenty-six months between February 1942 and March 1944 the overall loss rate was 3.8 percent; however, between November 18, 1943, and March 31, 1944, Bomber Command lost 1,050 bombers, 5.2 percent of those dispatched. These losses diluted aircrew experience and undercut morale. In addition the damage inflicted was not on the scale achieved during the earlier Ruhr and Hamburg campaigns. The official history notes that in contrast to these earlier campaigns, where new technologies and tactics were introduced, Oboe and Mosquito marking in the Ruhr and Window and radar at Hamburg, the Berlin campaign saw no major innovations. Bomber Command did make one tactical improvement by compressing its bombers over the target. While in the Cologne bombing of May 1942 the unit got twelve bombers over the target a minute, by mid-November this rate had doubled.

Bomber Command had destroyed at least 2,250 acres of Berlin, killed about ten thousand Germans, and lost more than five hundred aircraft, yet it had failed to knock Germany out of the war. It had flown ninety-one hundred sorties on sixteen major raids and dropped thirty thousand tons of bombs, with a loss rate of 5.4 percent that increased to 6.5 percent when unrepairable bombers were included. This war was far from over, for by the end of March 1944 Bomber Command had dropped only one-quarter of its total wartime tonnage. Bomber Command was spared for the moment as bombsights shifted to easier targets in France when the overall Allied priority became preparation for the Normandy invasion. The Luftwaffe had won the battle for air superiority at night. Just as the Americans had been taught months earlier over Schweinfurt, over Nuremberg the British learned the cost of not having air superiority.

British heavy bombers were easy prey for German fighters. They were less maneuverable, lighter armed than the fighters, and over time their armor had been removed. Because they were unable to outfight the interceptors, they flew alone at night, shielded by the dark and electronic countermeasures; however, they could be detected visually by the telltale trail of their engines and sometimes contrails, and from their electronic emissions. In brief the heavy bombers were "outpaced, outmanoeuvred by the German night fighters and in a generally highly inflammable and explosive condition, these black monsters presented an ideal target to any fighter pilot who could find them."[28] By early 1944 British bombers could no longer evade the fighters at night.

## The Rise and Fall of German Night Defenses

Prior to the war the Germans believed that flak alone would suffice for air defense, a belief reinforced by German AAA performance in the Spanish Civil War. World War II proved somewhat different. In June 1939 the Luftwaffe organized its first night fighter unit and equipped it with twin-engine Me 110s and Ju 88Cs. Initially the Germans employed ground radio control and searchlights, but then in October 1940 they established a more sophisticated system. The GAF organized zones (or boxes) west of the Ruhr covered by radar known as the Kammhuber Line, named after General Josef Kammhuber, commander of the night fighters. By the end of 1942 the defensive line was thirty kilometers wide and extended from the northern tip of Denmark to the southeast of Paris, some nine hundred kilometers. By the beginning of 1943 the number of night fighters increased from 400 to 500, then to 775 by mid-1943. Their impact increased, accounting in 1942 for just under half of the night bombers destroyed, rising in 1943 to almost two-thirds. Consequently the Luftwaffe night defense became increasingly more effective, downing 2 percent of the sorties dispatched by Bomber Command in 1940, 2.6 percent in 1941, and 3.9 percent in 1942. In the first half of 1943 the missing rate was 3.8 percent.

Another means the GAF used to combat the British bombers was to form an intruder unit in mid-1940 consisting of a small number of Ju 88s and Do 17s. These attacked the bombers over Britain where they were most vulnerable, as they departed their bases burdened by heavy fuel and bomb loads and when they returned with limited fuel and possible damage, in a sky cluttered with aircraft that could not be easily identified. Fortunately for Bomber Command the Germans did not embrace this tactic, never getting more than 20 aircraft serviceable at one time, and canceling the operation in fall 1941. Hitler wanted bombers downed over the Fatherland where Germans could see the result. Only a few small intruder operations were flown later, and in all the unit downed perhaps one hundred British aircraft. This was a major opportunity missed by the defenders of the Reich.

Thus the GAF defenses worked well, increasingly so, as long as Bomber Command allowed crews to select their timing and routes along a broad front. In mid-1943 Bomber Command introduced two tactical innovations. The first directed a concentrated stream of bombers through a specific area as rapidly as possible to swamp the boxes. Electronic countermeasures further undermined the German defensive system, which impeded both ground and airborne radar and GCI. The missing rate declined in the last half of 1943 to 3.4 percent.

The Luftwaffe responded with revised tactics and improved equipment. Tactically the Germans began in July 1943 to use free-lance single-engine day fighters guided by ground pyrotechnic signals and searchlights in a system known as *Wilde Sau* (Wild Boar) and steadily increased their numbers to one hundred by early 1944. Accidents were high because these aircraft had short endurance, and rudimentary navigation equipment, were piloted by some rejects from line units, and were operated in all kinds of weather at night. The Germans also widened the night fighter belt and developed the system so that more than two fighters could be simultaneously guided in the same box. A better solution was *Zahme Sau* (Tame Boar), twin-engine fighters that orbited a beacon and then plunged into the bomber stream, indicated by the Pathfinders' route markers and the bombers' radar track and signal's emissions. The system moved from GCI control to a running commentary that broadcast the direction of the bomber stream (and target) to the airborne defenders. This capitalized on the concentration of British bombers in the bomber stream and focused action over the target area. (The flak gunners fired below a ceiling above which the fighters operated.) To engage the bomber stream as early as possible, by February 1944 some fighters met the bombers half way across the North Sea.

The GAF also introduced new equipment. First, it equipped night fighters with detection devices thereby making them less dependent on ground assistance. After trying an infrared searchlight in the summer of 1940, the Germans began operations with airborne fighter radar (*Lichtenstein*) in early 1942. But there was resistance to its use by the more accomplished night fighter pilots in part because the antenna array cut air speed by twenty-five miles per hour. This was downsized, and by 1943 95 percent of the GAF's night fighters were so equipped. A newer radar, SN-2, operating on a different frequency that was less susceptible to jamming, became operational in October 1943, and equipped the majority of the night fighters by early 1944. The Luftwaffe also fitted night fighters with devices to home in on emissions from British IFF, radar, and aircraft detectors. One of the most significant improvements was to arm the night fighters with heavier forward firing cannon (up to 30 mm) and twin upward firing cannon. The latter, code named *Schrage Musik* (slanting, or jazz music), achieved its first kill in May 1943 and went into service in the fall of 1943. It allowed the night fighters to approach the bombers in their blind spot, below and behind the bomber using the earth's darker background to mask their presence from the bomber's tail gunner (few British bombers had ventral gunners). It was very effective. On the other hand, the GAF were unable to upgrade to more capable aircraft. The Me 210 turned out badly, and a more capable night fighter, the He 219, was shunned, delayed, and only appeared in small numbers, which forced the Luftwaffe to rely on the older and less capable Ju 88G.

Bomber losses increased from below 4 percent of those dispatched for the period July through November 1943 to over that figure for the following three months. And while the monthly loss rate in March 1944 declined to 3.1 percent, Bomber Command suffered two clear defeats against Berlin (March 24, seventy-two bombers, 8.8 percent) and Nuremberg (March 30, ninety-four bombers, 11.8 percent). Even with the change of targets brought on by the Normandy invasion and the decline in monthly losses (less than 2.5 percent after March 31), the RAF continued to suffer high losses in attacks on Germany, 8.4 percent on July 28 for example.

Despite these successes early in the year, the Luftwaffe lost both day and night air superiority during 1944. In the end it was factors outside of Bomber Command that allowed it to successfully operate at night. One factor was the attrition of night fighter crews in combat operations and flying accidents in night and bad weather. The night fighters suffered additional and heavy losses in daylight battles in 1944 against AAF formations of heavily armed bombers and especially against escort fighters. Other factors included the loss of Western Europe, which stripped Germany of its forward early warning radars, buffer, and airfields, as well as fuel shortages that curtailed training as early as 1942 and night fighter operations by mid-1944.[29] The night defender's defeat was not due to the lack of courage, skill, or innovation, and they cannot be blamed for the loss of Western Europe. To the contrary they were poorly served by the decision makers, who only belatedly saw the importance of night fighters, rejected potent intruder tactics, neglected better aircraft, and sent the night fighters into the day battle where they were cut to pieces.

## An Interlude in Strategic Bombing

Allied airmen received a mild reprieve beginning in early 1944 and extending through the fall of 1944 with the shifting of targets from Germany to those in the occupied countries of Western Europe. While in 1943 half of the total British and American bomb tonnage fell on Germany, for the first nine months of 1944 that figure fell to a quarter, and for the period April through August, it fell to a fifth. As already noted, some of this involved bombing the launching sites of V-weapons, while the other major alternative target set was connected with the cross-Channel assault. For two and a half months prior to D day and three and a half months after that, Allied air power focused on hitting continental targets to enable a successful invasion and then permit a breakout from the beach head. These targets were in France and the Low Countries, required a shorter penetration of enemy air space, and faced lighter opposition than those flown against German targets. There were of course losses, but not on the scale endured on the deep raids into the German heartland. The strategic air war did not resume in full force until October.

This shift in target geography helps explain the decreased Bomber Command losses from 4.1 percent missing per sortie dispatched in the first quarter of 1944 to 2.2 percent in the second quarter and 1.8 percent in the third. Part of this change was due to the shorter penetration of enemy territory and the dispersed nature of the targets, which deprived the defender of the "target rich environment" of the bomber stream. At the same time the defeat of the Luftwaffe day fighters allowed relatively cheap day operations, even though the British crews lacked the training and experience of the Americans and their aircraft lacked the firepower and altitude capabilities of the AAF bombers. During the first five years of the war, Bomber Command's day missing rate was 5.1 percent, compared with 0.4 percent after June 1, 1944. Night losses also declined. Bomber Command losses fell to a mere 0.84 percent in the last quarter of 1944 and 1.2 percent in 1945. During this period (March 26 to September 16) Bomber Command delivered one-third of its total tonnage.

## The Oil Campaign

Bomber Command knew that the Germans were short of oil and had assigned it the number one target priority, a position it retained until July 1941. Other target sets and operational difficulties led to it being downgraded. The early years of the war saw the German oil position improve as the Nazis obtained Polish and Rumanian oil and increased synthetic oil production. After achieving a peak stockpile of oil in the summer of 1940, German consumption, production, and stocks remain essentially in balance until the fall of 1943, after which stocks rose to a peak (since 1941) by April 1944.

Although the strategic bombing between April and September 1944 period focused on supporting the ground forces, the AAF began a concerted effort against oil beginning in April, an effort reluctantly, but effectively followed by Bomber Command starting with an attack on June 12. The AAF exceeded RAF tonnage dropped on oil targets in eight of the last eleven months of the war and dropped 58 percent of the Allied tonnage on this target set. It should be noted, however, that RAF bombers delivered heavier bombs that caused much greater damage. Harris regarded oil as another panacea target and resisted orders from his superiors to concentrate on this objective and continued to blast cities. Meanwhile the tight German oil position was made worse by Soviet advances that by the end of August 1944 overran Polish and Rumanian oil fields.

Bomber Command's efforts against oil targets in June did not go well when it sent out 832 bombers on three nights and lost 93 aircraft, 11.2 percent. Not only were losses high, results were limited. The British did better during the last quarter of the year, suffering only 1.1 percent losses against oil targets.

The Germans worked vigorously to protect and restore their oil production. In May 1944 the Germans quickly responded to the bombing by shifting flak units to synthetic oil facilities, and by the end of September they had massed 70 percent of their fighters to defend the oil industry. They built concrete blast walls, decoy plants, and shelters and employed smoke screens. The main effort, however, was to deploy men and equipment in order to rapidly repair the facilities. The Allies also had to contend with the onset of poor bombing weather in the fall and winter that hindered their efforts. Despite these problems, the bombing, combined with the capture of German occupied oil fields, was very effective in starving the German war machine.

One of the reasons for the great success of the oil campaign was accurate intelligence, the best the airmen had for any of their targets. It can not be overemphasized that one of the perennial problems of strategic bombing is determining the key targets and their locations and assessing how badly they have been damaged by attack and thus when they should be reattacked. Through intercepted German radio traffic decoded by Ultra, the airmen enjoyed an advantage unique in the bombing campaign: precise, timely knowledge of the extent of the damage inflicted, that indicated when and which targets required reattack. (With this exception, the airmen received little aid from Ultra as key information on the air war was not transmitted by radio.)

The impact of the bombing was dramatic. Between May and August 1944 German aviation gas stocks fell by half, and half again by November. Motor gasoline saw a similar decline while diesel fuel fell at a lesser rate. Despite drastic conservation measures the Germans had only one-third the stockpiles of fuel in January 1945 as they had one year earlier.

The months between March and October 1944 brought a marked improvement in Bomber Command's capabilities. All measures of power increased from March to October 1944: the daily average of bombers grew from 1,000 to almost 1,500, monthly sorties rose from 8,200 to 15,400, and monthly bombing tonnage increased from 27,700 to 61,200 tons. British bombers achieved better accuracy due to decreased losses (less distraction on bomb runs and increased experience) and from navigational aids moved forward with the advance of the ground troops that extended Britain's range into Germany. New marking tactics employing a very low-level visual delivery, the use of the "master bomber," and lower bombing altitudes were other factors.

At the same time targets had changed. Whereas Bomber Command had been pounding German cities with area bombing prior to the Normandy invasion, now it aimed at three other target systems: oil, transportation, and a "final concentrated and catastrophic blow against morale."[30] There was also a greater proportion of bomber effort against targets in Germany. But Harris again went his own way, for despite directives, Bomber Command devoted two-thirds of its effort in October against area targets as Harris stubbornly sent "his" bombers against German cities. As in the case of ball bearings (see chapter 5), Harris fought his own war regardless of his orders. And while his superiors considered replacing him, one author states that the British air chief, Charles Portal, had sufficient cause to fire Harris three times in 1944, at this stage in the war that was not a viable option. The final phase of the war saw the ever more powerful Anglo-American bombers pound Germany relentlessly as the Luftwaffe became increasingly impotent. Bomber Command dropped two-fifths of its total tonnage after September 17, 1944.

## Dresden

For some time the planners had been considering a massive attack on German civilian targets. In response to the German V-1 attacks, the Allies considered a major Anglo-American attack on Berlin, and as early as August 1944 the British chiefs of staff discussed hitting morale to knock Germany out of the war. Close examination of an operation called Thunderclap concluded it was unlikely to succeed at the time, but that the option should remain open. In October the airmen considered a major daylight strategic and tactical air effort against the Ruhr aiming at both material and morale. This plan was also dropped but then in late January 1945 a similar plan against German morale surfaced, in this instance to capitalize on the great westward flow of refugees by dropping twenty-five thousand tons of bombs in four days on Berlin. Such an attack was also seen as having a positive military and political impact on the Russians. On his part, Harris suggested that the operation be extended to similar blows against Chemnitz, Leipzig, and Dresden. This concept found favor at the highest RAF level, although there were some reservations because of the distance from British bases, the need to consult with the Soviets, and the belief that oil was a better target. Despite some of his earlier statements and his later position, Winston Churchill applied pressure to bomb these German cities. These attacks were now framed around hampering German troop movements. British and American airmen decided in late January that oil should continue as the top priority but that Berlin, Leipzig, and Dresden would be hit when their destruction would cause great confusion among civilian evacuees and thus hamper German military movement. The change in priorities was not discussed by the Combined Chiefs of Staff (CCS) nor mentioned to the Soviets at the February 1945 Yalta Conference. However, the Russians did ask that the Germans be prevented from moving reinforcements to the eastern front and suggested that bombers "paralyze" Berlin and Leipzig. As discussed below, the AAF hit Berlin on February 3, 1945. Ten days later Dresden was destroyed.

Dresden had a prewar population of six hundred thousand, now swollen by crowds of refugees. Bomber Command launched more than eight hundred bombers against the city on February 13, followed by three hundred AAF day bombers on the fourteenth, more than two hundred on the fifteenth, and four hundred on March 2. The British night attack was the most destructive of these strikes and has become, along with the two Japanese atomic bombings, the most infamous bombing attacks in history. This status resulted from the wide-scale destruction (over thirteen square miles), a high death toll (between twenty-five thousand and forty thousand), along with the city's beauty ("Florence on the Elbe"), stoked by an Allied press account that cleared the military censors, and fed by wartime Nazi and postwar Communist propaganda that produced a legend of Allied war crimes. (The myth includes such distortions as a death toll of two hundred thousand and allegations of AAF fighters strafing civilians.) In fact Dresden was a legitimate military target because it contained war industries and was a communications hub. The substantial death and destruction in Dresden resulted from a lack of air defenses, little previous damage, inadequate air raid protection, and weather, all of which combined to create a "perfect storm." This was what Bomber Command attempted to do on every raid but accomplished only here and at Hamburg in 1943. (The fickleness of the situation is seen in the Allied attack on Chemnitz days later by more than seven hundred bombers, which dropped over thirteen hundred tons yet caused little damage at the cost of twenty-three bombers compared with eight lost against Dresden.)

The result was a major uproar that resounded in the American and British press and in the House of Commons. Churchill composed a message for the British chiefs of staff that the "bombing of German cities simply for the sake of increasing the terror . . . should be reviewed," for it was not in British interests to occupy a ruined land. "The destruction of Dresden," Churchill wrote, "remains a serious query against the conduct of Allied bombing." The prime minister continued: "I feel the need for more precise concentration upon military objectives, such as oil and communications behind the immediate battle-zone, rather than on mere acts of terror and wanton destruction, however impressive."[31] The RAF hotly resented and rejected this critical note and it was withdrawn within days, replaced by a less strident one. In it Churchill deleted the quoted material cited above and called for a review of area bombing considering Britain's long term interests.

## Communications, Transportation, and Coal

German communications were vital to both the German economy and military. In planning for the cross-Channel invasion these became entangled. An attack on communications would serve tactical and strategic ends, and both the tactical and strategic air forces could engage in the campaign. This effort went forward despite the resistance of the British and American bomber commanders, the AAF's Carl Spaatz preferring to bomb oil, and the RAF's Arthur Harris cities. Yet bombing of communications was an outstanding success as German reinforcements to the battlefront were delayed and battered helping the ground forces make good their assault.

The strategic question was more difficult. The Germans had done much to thwart the bombing effort against their economy by concealing, dispersing, hardening, and building their factories underground, as well as directly defending them. However, the Germans still had to transport parts and raw materials from one facility to the next and move the finished product to the troops. To do this the Germans had a vast, redundant rail and canal system that was well organized, well maintained, and rapidly repaired. Allied decision makers were divided on which target set was best, oil or communications, so both were hit. German oil dried up and the railway system collapsed under Allied bombing in August–September 1944.

One key was to cut off coal mined in the Ruhr from the rest of Germany. Bomber Command attacks reduced canal traffic to a trickle by fall 1944 and rail movement to a lesser extent. The result seriously reduced coal coming out of the Ruhr and ore going into it, consequently steel production plummeted and coal required for rail transport fell short, and by the end of March 1945 the Ruhr was separated from the rest of Germany. (However, this was only a week before ground forces accomplished the same goal.) The communications bombing reduced production and hindered weapons from reaching the troops.

Bomber Command waged a long, lonely, and lethal campaign against Nazi Germany. It was expensive in treasure and costly in lives, with forty-seven thousand aircrew fatalities on combat operations and a total of seventy-nine thousand casualties to all causes. In addition the Allied bombing flattened many European cities and killed hundreds of thousands of civilians. The campaign lasted from almost the very beginning to the very end of the war. Between Dunkirk and D day it was the only offensive operation waged by Britain in northern Europe. The results were ambiguous and remain controversial to this day. As this bombing is so intertwined with the American bombing, I will hold off further comment for the moment.

Chapter 5

# U.S. Strategic Bombing in Europe

## DAY BOMBING

The initial U.S. reaction to the rise of aggressive dictatorships in 1930s was to enact legislation to prevent the United States from being "dragged" into another conflict. The great powers of the world began to rearm during the early and mid-1930s, an action America did not follow until January 1938. When the European war exploded in September 1939, the United States declared neutrality. The nation wanted to stay out of the war yet favored the Allies and sold them military equipment. The Japanese attack on Pearl Harbor and the German declaration of war thrust the United States directly into the war.

The United States was ill prepared for war, although the airmen did have a plan. In response to President Franklin Roosevelt's request for overall production requirements, in one week in the summer of 1941 six graduates of the Air Corps Tactical School wrote AWPD/1 (Air War Plans Division), which called for sixty thousand combat aircraft for the AAF (the Air Corps became the Army Air Forces in June 1941) and 2.2 million men. When the Japanese attacked Pearl Harbor, however, the AAF had fewer than three hundred heavy bombers in fourteen heavy bombardment groups of the seventy tactical groups it had activated. AWPD/1 went beyond production requests and was a remarkably accurate projection of how the war would be fought. For example, it forecast that because of the time required for training and production, an all-out offensive could not be launched before April 1944. It specified 154 key precision bombing targets in four systems (electric power, transportation, oil, and morale) and highlighted the Luftwaffe as an intermediate objective that had to be neutralized to allow successful bombing operations. The plan also suggested immediate efforts to develop a long-range escort fighter. The airmen held out the prospect that "if the air offensive is successful, a land offensive may not be necessary."[1]

Earlier in March 1941 the United States and Britain met and agreed to a number of points that would determine how the Western Allies would fight the war. The America-Britain-Canada agreement called for a "Germany first" policy and a "sustained air offensive" to weaken the Germans before an eventual invasion. America went to war with strategic bombing as a clear part of its victory plan.

## The Fortress and Liberator

Unlike the British, who used a number of medium and heavy bomber types in their strategic bombing offensive against Germany, the AAF employed just two. The more famous of these, probably the best known of all of America's strategic bombers, was the Boeing B-17 Flying Fortress. Certainly it deserves special recognition as the first modern heavy bomber of the 1930s and for its record in World War II. The Fort was well regarded as a reliable, rugged, forgiving, and honest aircraft that did its job well with young and inexperienced flyers and got most back from danger.

The Fortress emerged from a 1934 competition for a multiengine bomber that could carry a one-ton bomb load a minimum of one thousand miles (possibly twenty-two hundred) at a top speed of

The *All American* (B-17F) survived a midair collision with a German fighter over North Africa and made it home, but after landing its aft section collapsed. The B-17 was renowned for its ruggedness and ability to bring its crew home and thus was highly regarded by its crews. (USAF via Boeing)

200–250 mph. The Boeing entry first flew in July 1935. (The Fort's quality is evident in that it was still in front-line service over Europe a decade later alongside the younger Lancaster and Liberator.) The four-engine Boeing bomber was what the airmen were looking for, an aircraft that could fly high, fast, and far; fight off aerial defenders with its heavy armament; and deliver bombs on target. The story becomes dramatic and tragic, however, for after setting a cross-country nonstop speed record from Seattle to Dayton, the prototype was destroyed in a takeoff accident (due to pilot error) in October 1935. Instead of a hoped-for buy of sixty-five aircraft, the Air Corps got only a baker's dozen, while the Army bought a larger number of the more conservative, cheaper, and much less capable twin-engine B-18.[2] Meanwhile the War Department pushed close support aircraft and restricted experimental and development funds for fiscal years 1939 and 1940 to close support aircraft. Thus between October 1935 and July 1939, the Air Corps procured only 14 heavy bombers, 13 B-17s and 1 B-15, despite requests for 206 of the former and 11 of the latter.

Over the years Boeing modified and improved the Flying Fortress. The first model to see combat was the B-17C, a model that replaced all but one gun with .50-caliber machine guns and added self-sealing tanks and armor. Nevertheless the B-17C proved wanting in RAF operations over Europe beginning in July 1941. However, these were very useful combat trials pointing out a number of the bomber's deficiencies. The B-17D, basically the same aircraft, did no better against the Japanese early in the Pacific war. The "E" model addressed many of these shortcomings. The most visible change was a redesigned empennage with a dorsal fin that began amidships and flared into the large signature "big tail," which permitted the addition of a tail gun position and increased stability, especially at high altitudes. Boeing also added powered top and belly turrets armed with twin .50-caliber machine guns, a gun mounted in the radio compartment (firing upward and aft), one gun on each side in the open waist positions, and two .30-caliber machines guns in the nose. Although the "E" was seven tons heavier than the first Fort, it was capable of reaching 317 mph and was the first model used by the AAF in the European war beginning in August 1942. The "F" model went into European combat in January 1943 and looked and was very much the same as the "E," although later bombers in that series had increased fuel capacity. The last big change came in the "G" model, which entered combat in late 1943, the last and most numerous (eighty-seven hundred built) of the Forts. Boeing mounted a chin turret with twin .50-caliber machine guns to counter the very effective GAF head-on attacks, improved the tail armament, and staggered and enclosed the waist positions. The AAF had a peak of 4,600 Forts in August 1944, and by the end of August 1945 the AAF had accepted 12,700 B-17s. The Forts dropped almost 60 percent of the bombs delivered by American heavy bombers in Europe.

The B-17's teammate was the Consolidated B-24 Liberator. It has been overshadowed by the more photogenic, better publicized, more loved, and well

known Fortress, and thus "the boxcar with wings" has been largely underappreciated. It deserves better, for more B-24s were built than any other American aircraft in history (18,200) and it proved more versatile than the B-17, serving in bomber, maritime reconnaissance, and transport roles. At its peak in September 1944 just over six thousand Libs were in AAF colors, and at their maximum overseas strength the AAF had thirty-three B-17 groups (September 1944) and forty-six B-24 groups (June 1944) in service. In addition twenty-five hundred Liberators served with British forces.

In late 1938 the Air Corps asked Consolidated to consider building the B-17, to which the company responded by suggesting a new and better bomber, for which it got a contract the next March. The designers built the XB-24 around a high-aspect-ratio wing, which had performance advantages (especially in range) over existing types, and a large fuselage. The aircraft also featured tricycle landing gear, which was just coming into use, roller-shutter bomb bay doors,

The Boeing B-17 was an old design compared to most World War II bombers, having first flown in July 1935. Seen here is the "G" model that went into action in late 1943 and was the ultimate and the most produced Flying Fortress, distinguished by its chin turret. B-17s dropped 60 percent of the bombs dropped by U.S. heavy bombers in Europe. (U.S. Naval Institute Archive)

Compared to the B-17, the Consolidated B-24 was built in larger numbers, was much more versatile, and had some performance advantages. But it did not receive as much publicity or praise because it lacked the Fort's ruggedness, looks, and easy handling. Its most famous operation was the August 1943 low-level Ploesti raid seen here. (U.S. Naval Institute Photo Archive)

which cut drag, and a distinctive twin tail configuration. The slab side, boxcar-like, broad fuselage could accommodate up to eight thousand pounds of bombs and endowed the aircraft with great versatility but did not enhance its appearance. The bomber made its initial flight in December 1939, a remarkably short time after its inception. The aircraft's major virtues were range and versatility.

The Americans fitted their first bomber versions (B-24B and C) with self-sealing tanks and dorsal and tail turrets, both armed with twin .50s, and replaced the mechanically supercharged engines with turbo supercharged ones that led to the elliptical cowling shape that gave the Liberator its unique look. Maximum speed increased to 310 mph. The B-24D was the first large-scale production version with thirty-seven hundred built. In short order the AAF modified the "D" to carry ten .50s (the last versions including a retractable "ball" turret) and a maximum bomb load of 12,800 pounds. Thirteen B-24Ds made the first AAF heavy bomber attack in Europe with a strike on the Ploesti oil fields in June 1942. None were downed, although only seven landed as planned in Iraq, while two made it to Syria, one crash landed, and four were interned in Turkey. The Libs flew their first combat mission from Britain in October 1942. Their numbers grew so that for most of the war it made up one-third of the Eighth Air Force's (the AAF's strategic bombing unit in northern Europe) bombers and outnumbered the Forts in bomber units in the Mediterranean and Pacific theaters. The B-24G and successive models were armed with a nose turret mounting twin .50s. More "J" models were built (sixty-seven hundred) than any other version.

## The Mighty Eighth

In early 1942 the AAF established the Eighth Air Force as a balanced unit to support the European invasion. In short order, however, it was redirected toward strategic bombardment operations, although it employed medium bombers until October 1943. The Americans were confident, overconfident as it turned out, despite British experience, skepticism, reservations, and opposition. "I know that our crews, equipment, and training will be far superior to theirs [the British] and our results consequently better," Ira Eaker (commander of the Eighth's bombers) wrote Hap Arnold, the AAF chief prior to the first AAF strike. He added, "We can do day bombing without prohibitive losses."[3] The story of American strategic bombing against Germany is how, despite reverses and at times heavy losses, the Eighth adapted, went on to victory, and made a significant contribution to the defeat of Germany.

AAF operations out of Britain took some time to begin and only slowly built up in size and effectiveness. It can not be overemphasized how much the Americans relied on the British for a wide range of assistance, extending from airfields and supplies to intelligence and food. Initial operations consisted of heavily escorted short-range raids of targets in occupied Western Europe beginning in August 1942, when a dozen B-17s hit a French marshaling yard at Rouen. Eaker flew on the attack, as did AAF leaders on occasion.[4] GAF fighters engaged the bombers and their Spitfire escort and damaged two bombers. The airmen considered the bombing good, although only one-third of the bombs fell within two thousand feet of the aiming point.

Mission number 14 on October 9 against Lille was notable as being the first comprised of more than 100 bombers, the first with B-24s, and the first fierce air battle. The results were not heartening, as only 69 of the 108 bombers dispatched bombed the primary target and accuracy was poor. The photo interpreters could identify only half of the bombs dropped and concluded that a mere 3 percent of these hit within fifteen hundred feet of the aiming point.

In addition, a number of the bombs were duds and others caused French casualties. The discouraging performance was blamed on inexperience—one-quarter of the crews were flying their first combat mission—and the weather. The air battle resulted in the loss of four bombers and damage to forty others.

The bomber gunners claimed fifty-six GAF fighters destroyed and the escorting fighters another five. These claims were immediately suspect because the number was a sizable portion of the estimated Luftwaffe strength in the West. American reassessments reduced this figure to twenty-one destroyed while German sources indicate a loss of only two fighters. The issues of high victory claims and low admissions of losses were problems throughout the war and plague researchers to this day. The AAF attempted to accurately gauge these claims, tightening up the criteria for awarding credits, ever mindful of the impact on intelligence estimates and aircrew morale. A reassessment in early January 1943 reduced the previous overall figure of 223 GAF fighters destroyed to 89. Nevertheless, German records indicate that even this figure was too high.

After their first fourteen missions the Americans were confident of their ability to conduct strategic bombing operations with both accuracy (Eaker believed getting 40 percent of bombs within fifteen hundred feet of the target) and minimal losses. Their optimism was fueled by the fact that AAF bombing accuracy was superior to what had been seen thus far in the theater. The Eighth's commander also thought he could attack anywhere in Germany with acceptable losses (4 percent) if he had a force of three hundred bombers over the target. The airmen did recognize there were problems, the most serious of which were inadequate training and the weather.

These difficulties were compounded by three outside events. The newly opened North Africa theater (Operation Torch) drained off bombers and fighters and, even more significant, the Eighth's longest range fighters (P-38s) and the most experienced bomber crews. A second problem was that the Eighth was not getting adequate reinforcements because heavy bomber units were being sent to the North African and Pacific theaters. The number of heavy bombers fully operational and crews available in Britain did not reach the two hundred mark until April 1943, four hundred in August, and seven hundred in November. The third difficulty emerged when antisubmarine operations were granted top priority in late October 1942, a position held until June 1943. This campaign siphoned off bombing units to fly maritime reconnaissance and redirected the aim of the bombers to submarine pens on the French coast, even though their 11.5-foot roofs and 8-foot walls were impervious to American bombs. Consequently, while the bombing badly damaged the surrounding area and killed French civilians, it only inconvenienced the Germans, who moved all of the operations into the sub pens. Grand Admiral Karl Doenitz commented in May 1943 that the French "towns of St. Nazaire and Lorient have been rubbed out as main submarine bases. No dog nor cat is left in these towns. Nothing but the submarine shelters remain."[5] Likewise attacks on the submarine industry had little impact on U boat production. The most significant contribution of air power to the Battle of the Atlantic was the role of maritime reconnaissance, especially long-range aircraft covering the mid-Atlantic.[6]

It can be argued that greater concentration on the sea war, particularly maritime reconnaissance and less on strategic bombing, would have yielded better results, sooner, with less cost. Likewise it can be argued that the Allies would have profited more from greater emphasis on tactical (close air support and interdiction) than on strategic bombing. For that matter it can be argued that the U.S. ground offensive would have benefited from more landing craft, better tanks, and higher quality leadership from the resources absorbed by the strategic air war. But these hypotheticals must be weighed against what the strategic air war accomplished, as will be described.

Meanwhile in late 1942 there were uncertainties at the highest level concerning the conduct of the war, as well as the conduct of strategic bombing. The RAF continued to press the Americans to join them in night bombing, a move strenuously opposed by the AAF although it held a weak hand, for after a year of war, the Eighth's numbers were small and it had not yet bombed Germany. At the Casablanca Conference in early 1943, Eaker successfully made the case for the U.S. concept directly to Churchill; the British and American leaders agreed to a British night and an American daylight bombardment campaign against Germany, "around the clock bombing" (Eaker's phrase). The culmination was a document issued by the Combined Chiefs of Staff on January 21, 1943, that called for "the progressive destruction and dislocation of the German military, industrial and economic system, and the undermining of the morale of the German people to a point where their capacity for armed resistance is fatally weakened."[7] The leaders listed the target priorities as German submarine construction yards, aircraft industry, transportation, oil, and other war industrial targets. It is important to emphasize that bombing was not intended to win the war alone but to be a partner in the overall victory. One other significant accomplishment of the Casablanca Conference was the publicly stated policy of unconditional surrender. This was now officially a total war between the two coalitions, which would grant the airmen great latitude in their application of strategic bombing.

In early 1943 Eaker adapted an operational plan based on half a year's combat experience. It stated that the unit believed it needed eight hundred aircraft to be able to dispatch three hundred per mission, the minimum able to fend off GAF fighters. As this force built up and trained it intended to operate under heavy fighter cover, a period it anticipated would extend until the end of July 1943, when it should have nine hundred bombers. Only two deep penetrations were planned, against the Ploesti oil facilities and the bearings plants in Schweinfurt. Between July and October the Eighth would focus three-quarters of its attention on fighter factories and repair installations and the remainder on submarine yards. After October the force would grow to about twelve hundred bombers and extend its scope of targets; it would then build to a peak of twenty-seven hundred bombers by March 1944 as it prepared for the Normandy invasion. The CCS approved this plan in May 1943.

The Eighth marked a milestone in its bombing campaign on January 27, 1943, when it dispatched ninety-one bombers on its first strike against targets in Germany. To this point the attacks in Western Europe had gone fairly well. In 1942 the Eighth had lost or written off thirty-four bombers, 4.1 percent of effective sorties.[8] But the loss rate was increasing, and with deeper penetrations into Germany, stiffer resistance was expected. There were other problems as well. Maintenance hindered operations. During 1942 only three-fifths of the dispatched bombers attacked targets, while during the entire war that rate was four-fifths. Although the B-17 was not a new bomber, it encountered a slew of problems with superchargers, generators, and, most of all, propeller governors. These were exacerbated by the inexperienced air and ground crews. As Eaker noted in January 1943, there was a difference between prewar regulars and wartime personnel made up of "sturdy amateur[s] with all the 'guts' in the world."[9]

In the first half of 1943 the Eighth dramatically built up its strength, made deeper penetrations into Germany, and suffered increased losses. The shorter missions were escorted by the range-limited Spitfires until May, when the longer range P-47s joined the fray. On April 17 AAF aircraft hit the aircraft factory at Bremen and lost sixteen bombers, 15 percent of those that attacked the target. By the end of July the Eighth was dispatching over three hundred bombers and suffering even heavier losses; whereas it has lost 5.9 percent during the first half of the year, on the last six missions of July it lost 9.9 percent.[10]

In May 1943 the Allied decision makers made a number of key decisions. First, they scheduled the cross-Channel Normandy invasion for May 1, 1944. Second, operations in the Mediterranean would be confined to knocking Italy out of the war. They also directed that the Ploesti oil fields be hit as soon as possible. The Combined Bomber Offensive (CBO) plan listed fighters as the intermediate objective; submarines, the remaining aircraft industry, ball bearings, and oil as primary objectives; and rubber and military transport vehicles as secondary objectives.

## The Luftwaffe Adapts to American Daylight Tactics

To bolster their fighter numbers the Germans increased fighter production and redeployed units to homeland defense. These measures increased single-engine fighters in the West from 500 in January 1943 to 810 in July and 964 in October 1943. Beginning in March 1943 twin-engine fighters joined the day battle, providing greater numbers, heavier firepower, and longer endurance than single-engine fighters. While American crews reported contact with over 100 GAF fighters on one mission in March and one in April, they observed that number three times in May and four times in June. In late July the Luftwaffe was getting over 200 fighters up to oppose the AAF bombers, with 225–275 observed on July 30, 1943.

The heavily armed American bombers flying in formations at high altitude presented a different and more difficult problem to the defenders than the British heavy bombers that flew alone at lower altitudes at night. The tightly flown American formations unleashed a hail of gunfire that could deter, disrupt, and destroy attacking fighters. In addition AAF bombers proved more rugged than British bombers, as it took an average of twenty to twenty-five 20-mm hits to down a B-17. The German response was to increase cannon armament and employ air-to-air rockets to attack from behind and outside the range of the bombers' .50s (eight hundred to one thousand yards). One of the rockets could down a bomber and a barrage could disrupt a bomber formation.[11] In contrast German air-to-air bombing, first used in February 1943, proved ineffective.

The Luftwaffe also organized its fighters under centralized control to mass them against the American formations. In the second quarter of 1943 they began to equip airfields in Western Europe to rapidly rearm and refuel shifting aircraft to allow multiple sorties. The Germans learned by November 1942 that head-on attacks were the most effective tactic against the AAF bombers, especially on the bomb run, where open bomb doors masked the forward fire of the belly guns and the bombardier–nose gunner was preoccupied with bombing. Additional advantages were that head-on attacks did not require computation of deflection, the bomber's crew and engines were more vulnerable from the front, and the bomber's nose armament proved inadequate. Fighter effectiveness increased. Whereas seven bombers were lost in one hundred combats with fighters in 1942, that ratio rose to ten by the end of 1943, with a peak in March–April 1944 of eighteen bombers downed per one hundred combats.

The defenses were assisted by other means. Flak was less a problem than fighters until mid-1944, after which time the number of bombers lost to enemy fighters was less than those downed by flak or flak and fighters. The Germans also used smoke screens to hinder AAF bombing and reconnaissance, smoke decreasing bombing accuracy by a quarter.

## American Tactical and Equipment Modifications

The AAF also made changes from its peacetime tactics to deal with wartime realities. The Eighth began combat operations with six aircraft of a squadron flying in formations a few miles apart but were forced by German defenses to compress the formation to increase defensive firepower, leading to formations of eighteen then thirty-six aircrafts that flew at about twenty thousand feet. By April 1943 the Eighth had increased the numbers in a single formation (combat wing) to fifty-four. This proved unwieldy, and in January 1944 the Eighth experimented with a thirty-six-bomber formation, a size that became standard for the remainder of the war.

In January 1943 some formations began to drop their bombs on the signal of the lead bombardier, a change that brought both defensive and offensive advantages. The Eighth could not employ each bombsight and bombardier for bombing because this would loosen the formation on the bomb run and weaken the bombers' defensive capabilities. And as the nose position was too cramped to permit both the nose gun and bombsight to be manned simultaneously, freeing up the gunner on the bomb run increased forward firepower. Group bombing also had offensive advantages. Operations analysis concluded that of bomb-aiming methods tried by the Eighth, the group-leader technique was twice as effective as the next best method and therefore it became the standard practice by the second quarter of 1943.[12] By this time the airmen were also using automatic flight control equipment (AFCE), which linked the bombsight directly to the aircraft's autopilot, allowing the bombardier to make rapid and smooth corrections in the critical last moments before bomb release. The AAF had to overcome initial problems of maintenance and acceptance of this device, but outstanding results led to its standard use by May 1943. Although accuracy figures are especially questionable because only a percentage of the bombs dropped could be identified, the Eighth believed that visual bombing accuracy in the last quarter of 1943 was double that of the first three-quarters of the year.

The Eighth employed zigzag routing to avoid flak concentrations, routing and diversionary formations to deceive the fighters, as well as fighter sweeps and medium bomber attacks to harass the defenders. The key weapon, fighter escort, was limited by the range of the fighters. For the first year escort consisted mostly of short-range Spitfires because the untried but longer range P-38 Lockheed Lightnings has been deployed to the African theater.

Another defensive weapon was electronic countermeasures. The AAF used Chaff for the first time in October 1943. It was effective against German radars, but there were problems with production because each mission required about fifty to sixty tons of aluminum. Initially Chaff was dumped by hand out of the bombers, but then in October 1944 the Eighth sent bombers and fighters ahead of the formation to drop the device in order to shield the formations. Later the airmen installed machines to automatically dispense Chaff, and in May 1945 it was dropped in bomb casings. In October 1943 the AAF first used Carpet, an electronic jammer that reduced flak damage by one-fifth. The first versions were barrage jammers that blocked a set frequency, requiring at least half the bombers in the formation to be so equipped. In the last half of 1944 the Eighth introduced a version which swept for radar emissions and when one was detected automatically began jamming. The Eighth established an ECM squadron consisting of a dozen B-17s (later transitioning to B-24s) within the RAF's ECM group, which went into action in June 1944. It assisted Bomber Command until November, when it began operating on daylight missions, flying orbits over the North Sea to prevent the Germans from monitoring AAF radio traffic. By February 1945 all the Eighth's bombers deployed Chaff, a number of ECM aircraft flew ahead of the formation equipment with Carpet, some with ECM operators.

The most serious AAF problem proved to be German fighters. The Americans continued unescorted operations despite heavy losses, not so much because of stubbornness but because of limited fighter range resulting from the universally held belief that it was impossible to field an aircraft with both fighter performance and bomber range. The unsatisfactory performance of the Me 110 during the Battle of Britain confirmed this view. One American's solution was a "convoy escort," a large, heavily armed aircraft that flew with the formation. In the 1930s the Air Corps had experimented with the twin-engine XFM-1, but it had proved unacceptable. During the war the AAF modified B-17s and B-24s to an escort role by adding armament and deleting bombs. The XB-40 was a B-17F fitted with two additional turrets, double the guns positioned in the waist, additional armor and ammunition, all of which increased weight four to five tons and reduced flying performance. A dozen of these aircraft flew nine combat missions in May through July 1943. The AAF cancelled the project in July mainly because of the aircraft's dissimilar performance relative to the "normal" bombers after they had unloaded their bombs.[13] The aircraft's chin turret, however, was a success and mounted on later AAF bombers, the B-17Gs and B-24Hs.

## August Glory and Gore

Two of the most famous American bombing missions of all time took place in August 1943. Both were aimed at critical targets, featured daring and innovative tactics, demonstrated that the bomber could get through, met fierce resistance, and suffered heavy losses. They wrote a glorious page in AAF history, yet their impact on the war was much less than anticipated and their cost was unacceptable. These missions and the ones over the next few months cast increasing doubt on AAF strategic bombing operations.

The first of these missions was against the refineries at Ploesti, Rumania, which Allied intelligence believed supplied 60 percent of Germany's crude oil. It was one of the most daring AAF heavy bombing missions, perhaps the most thoroughly planned and prepared, yet it was undone by the friction of war. The Russians had already hit Ploesti oil targets a few ineffective times, as had Americans in an attack in June 1942. The AAF considered a high-level medium-sized attack from Syrian bases but chose instead to launch a large, minimum-altitude attack from North Africa, in stark contrast to the AAF high-altitude doctrine. Such tactics promised better bombing accuracy and the element of surprise. The airmen in Europe disliked the concept for good reason: Low-altitude operations were much more dangerous. The AAF would employ 177 B-24s (selected because of their superior range) with two additional fuel tanks fitted in the bomb bay and a low-level bombsight substituted for the Norden. Low-level operations encouraged the airmen to remove the ball turret and prompted them to equip some bombers with fixed forward-firing machine guns and some with dual .50s in the waist positions.

Just after sunrise on August 1, 1943, the AAF launched almost 180 bombers from Libyan bases led by the commander of 9th Bomber Command, Brigadier General Uzal Ent, along with the commanders of each of the five participating B-24 groups.[14] The missions got off to a bad start as one Lib crashed shortly after takeoff and a second went down half way across the Mediterranean. The Liberator formations strung out and were further scattered by bad weather. (The airmen did not use their radios to regroup in order to maintain radio silence and preserve secrecy; unfortunately German radar and signals intelligence had already detected the attack). As a result two groups reached the target area before the remaining three. The greatest misfortune was the attacker's misidentification of the turning point, which carried two of the five groups away from the target and led Ent to order his formation to attack targets of opportunity, only six bombers of which bombed Ploesti. The other lost group flew into the target area on the reciprocal heading of the plan and bombed, but its bombers hit targets assigned to other units, which meant they later passed through the smoke and exploding bombs as well as Libs on the opposite course. The bombers flew at very low altitudes (one hundred to three hundred feet), met intense antiaircraft fire from twice the number of guns expected, encountered some fighters, and had to dodge chimneys and balloon cables. Consequently the disrupted American bombers, many badly battered, were unable to fly the planned recovery route in formation and instead returned in dribs and drabs, many to the nearest airfield within range, twenty-one diverting into other Allied bases (Cyprus, Malta, and Sicily). The Ninth Air Force commander expected losses as high as 50 percent and was not far off as between forty-nine and fifty-three bombers were destroyed, two were written off as junk, and seven were interned in Turkey—approximately one-third of those dispatched and almost half of those that bombed Ploesti.

Results were about half of expectations: Three of the targeted refineries were untouched and the other six reduced production by 46 percent. The Germans quickly responded by putting idle capacity into production and repairing the damage. More critical, there were no follow up attacks until the spring of 1944. The high cost of low-level bombing was once again demonstrated. The AAF awarded five Medals of Honor (three posthumously), the most for any AAF/USAF mission. Any operation that costs one-third of the force, does not inflict decisive damage, can not be followed up, and merits the award of five of the nation's highest decoration deserves sharp criticism. Probably the raid's most important positive result was that it forced the Germans to divert resources to defend this vital target.

The other mission two and a half weeks later and on the first anniversary of AAF operations from Britain was against two key targets, bearing plants at Schweinfurt and a fighter factory at Regensburg. As early as 1941 the British had considered attacking bearings, which appeared to be an ideal target because they were indispensable in all mechanized military equipment and production was concentrated in seven plants that turned out 80 percent of German bearings. The plants at Schweinfurt were believed to produce half of the total Axis bearing production and were listed for attack in the CBO plan. In early August British and American planners generated a plan that called for an AAF day raid followed by an RAF night assault, but bad weather foiled this effort. Therefore the Eighth proceeded alone, combining the Schweinfurt mission with an attack on another high-priority target, an aircraft factory at Regensburg. The plan called for one unit (4th Wing) to hit the more distant Regensburg target and, in a first, fly on to Africa. The 4th would be followed by the 1st Wing, which would strike three factories in Schweinfurt and then return to Britain. The planners expected that the Regensburg mission would attract the bulk of the defenders inbound to the target and then surprise the defenders when it exited to the south. Meanwhile the Schweinfurt force would slip toward its target while the GAF was distracted. The mission would include fighter escort for the Regensburg force prior to the target and escort for the returning Schweinfurt bombers. There would also be bomber and fighter-bomber attacks on two marshaling yards and four airfields. Timing was critical because of the limits of fighter escort range and the distance to the targets.

Weather again upset the airmen's plans. Fifteen minutes prior to the 4th Wing's taxi time, the Eighth's acting commander, Brig. Gen. Frederick Anderson, was told that his bases were weathered in. Anderson delayed the 4th's movement as long as possible (ninety minutes) to enable it to get to Africa before dusk. The 4th got off, but the weather at the 1st's bases remained too poor for take-off.[15] Anderson decided to launch the 4th separately because of the importance of the target, the fickle weather, and the fear that the recent raids would encourage the Luftwaffe to reinforce its Regensburg defenses. He also decided to send the 1st Wing out three and a half hours later to allow the weather to clear and the escorting fighters to rearm and refuel for a second sortie.

Col. Curtis LeMay, commander of the 4th Wing, led 146 B-17s through stiff resistance to Regensburg and later commented that "our fighter escort had black crosses on their wings."[16] His bombers achieved excellent results due to a lull in the defenses at the target, the lower than usual bombing altitudes, and the relatively long bomb run. (One unit, finding the target obscured by smoke, made a complete turn and a second bomb run.) The plant lost eight to ten weeks' production. The Forts continued on to Africa.

The Schweinfurt raid of 230 Forts, led by its commander, Brig. Gen. Robert Williams, met even tougher resistance.[17] One reason was that fighters the Germans shifted westward for the anticipated return of the 4th were in excellent position to meet the 1st's effort. The unit did not bomb as well as the Regensburg force due in part to the defenses, bombing three thousand feet higher than usual, and congestion on the bomb run. Nevertheless, production in Schweinfurt in September and October was cut by one-third.

The August 17 attacks are remembered and significant because of the fierce German defenses. There were no new German weapons or tactics, just more GAF fighters, about five hundred sorties, almost twice the previous one-day high. Initially the Americans claimed the destruction of 280 GAF fighters with over one million rounds of .50s expended. These claims were scaled down to 148, while the Germans put their losses for the day at 40 (one source put it at 27). The Eighth lost 60 bombers (20 percent of those bombing); in addition two crews flew damaged bombers into Swiss internment and 4 damaged aircraft were written off.

The heavy losses shook the American airmen. The Eighth's highest one-day losses thus far had been twenty-six bombers, and the unit had gone seven months and flown thirteen hundred sorties before losing its first sixty bombers. The twin missions showed the German defenders at their best and how weather could foil American plans. It also demonstrated that shuttle missions were not practical because of inadequate base facilities in Africa.

Despite these losses the AAF persisted and continued to be savaged by GAF defenses and thwarted by the weather. On a September 6 mission against ball bearing plants in Stuttgart, weather allowed only 46 of 338 bombers to (poorly) bomb the target. The Eighth encountered fewer than two hundred fighters, yet the German defense along with the range and weather resulted in forty-five bombers lost in action, five interned in Switzerland, and three written off as wrecks.

The air war rose to a climax in mid-October, when four missions over a week demonstrated that the Luftwaffe had made the skies over German too dangerous for unescorted bombers. On the eighth the bombers attacked shipyards at Bremen and Vegesack and an aircraft factory at Bremen. The bombers encountered 300 GAF sorties and, despite P-47 escort, lost thirty bombers (plus two more salvaged) on 357 effective sorties. The next day the bombers flew a more ambitious mission. It is notable as being the deepest penetration yet for the Americans and for the outstanding bombing at Marienburg, where the B-17s employed lower than normal bombing altitudes and a longer bomb run to get half of their bombs within the factory area, destroying two-thirds of the factory. The deep raid caught the Germans by surprise, yet the attackers suffered 28 bombers lost, 3 interned in Sweden, and 3 written off of the 352 aircraft that bombed. The October 10 mission against Munster endured similar losses, 29 bombers missing and 3 salvaged. (Of 14 bombers dispatched by the unlucky 100th "Bloody Hundredth" Bomb Group, only 2 returned home, 1 with two engines out and another crash landing.) The bomber crews claimed to have downed 182 GAF fighters and escorting fighter pilots claimed another 21; German records admit losing 25 fighters in combat that day.

Schweinfurt was the target on October 14, a target the leaders knew would provoke heavy resistance and produce substantial losses. This is clear in the message that General Anderson had read at the crew briefings:

> This air operation today is the most important air operation yet conducted in this war, the target must be destroyed. It is of vital importance to the enemy. Your friends and comrades that have been lost and that will be lost today are depending on you. This sacrifice must not be in vain. Good luck, good shooting, and good bombing.[18]

The Eighth launched 321 bombers, all but 29 of which were B-17s. The weather was similar to that of August 17, but the better trained pilots were able to assembly without grave difficulties. The Germans changed their tactics and engaged the escorting P-47s early on. The Thunderbolts flew 145 effective sorties and claimed to have destroyed a dozen GAF fighters for the loss of one of their own. (Two others crashed upon return and were written off.) Once the escort turned for home the Luftwaffe jumped on the Forts. In a three-hour battle the defenders employed 90 to 100 twin-engine fighters and 300 to 500 single-engine fighters. In all the Eighth lost 60 bombers, with another 4 written off as salvage and 1 landing in Switzerland. The bomber crews initially claimed 288 fighters destroyed, then reduced that number to 184 and eventually 99, while the fighter pilots claimed 13 victories, numbers that sharply contrast with German records that show 38 lost on that day. Bombing accuracy of the 226 Forts that dropped on the primary was considered good as four of the sixteen groups got half or more of their bombs within one thousand feet of the target.

The cost was high, but the airmen were elated. The day after the mission Eaker wrote Arnold, "I class it pretty much as the last final struggles of a monster in his death throes. There is not the slightest question but that we now have our teeth in the Hun Air Force's neck."[19] On his part Arnold told reporters, "Now we have got Schweinfurt."[20] The bombing knocked out 60 percent of Schweinfurt's production in what proved to be the most damaging of the sixteen raids against the bearings plants. Although this shook the German leaders, the attacks did not yield the expected benefits. German stockpiles, redesign of equipment, imports from Sweden,[21] reorganization of production, and the dispersal and protection of the plants muted the much hoped for impact, as did a four-month lapse before the next raid on Schweinfurt. The Germans claimed that no equipment was delayed because of a lack of bearings. It appears the bearing attacks were not worth the cost.

More significant was the demonstration by German defenders that daylight, unescorted, good-weather bombing could not be conducted with acceptable losses. On the four missions between October 8 and 14 the Eighth lost 148 bombers, along with 12 written off and 4 interned, a third of the unit's daily effective strength for the month. October 1943 saw the highest percentage of losses the Eighth would sustain throughout the war, 10.4 percent of effective sorties listed as lost, missing, and salvaged. Along the route to Schweinfurt, scattered with the twisted metal of downed bombers and broken bodies of bomber crews, the unescorted bomber theory died. The AAF strategic bombing offensive had been repulsed; the American prewar theory had been found wanting.

## American Adjustments

The reality of combat forced the AAF to make a number of changes in response to European weather and German defenses. The result was a force structured and using tactics that were different than what the prewar planners had envisioned. The ability to make these changes, relatively rapidly, was a significant accomplishment for the AAF and critical for its eventual aerial victory and contribution to the Allied triumph.

Four major factors help account for the turnaround in the air war and the AAF's victory over the GAF. The first was sheer numbers. The Eighth increased its effective bomber force from 80 in January 1943 to a peak of 2,480 in March 1945, with a similar trend for fighters from 66 in April 1943 to 878 in March 1945 (peaking at 1,031 in November 1944). At the same time the AAF's strength in the Mediterranean theater increased from 300 bombers and 285 fighters in November 1943 to 1,551 bombers and 741 fighters in May 1945. The U. S. Ninth (tactical) Air Force and the RAF added to these numbers. To control and coordinate this mammoth American air force the AAF established the Fifteenth Air Force in November 1943 to operate in the Mediterranean theater alongside the Eighth based in Britain under a headquarters formed in January 1944, the United States Strategic Air Forces in Europe (USSTAF). The Fifteenth flew its first mission against Germany in October and moved to Italian bases in December. To be clear American and British bombers were never unified under one commander.

A second factor was the adjustment to the weather and geography as operating in Germany proved far more challenging for the American airmen than what they had encountered during their training over the southern United States. The high latitudes of Europe limited winter daylight to ten hours, and the weather rendered one-quarter of the days during the year unsuitable for visual operations. This varied by season and time of day: From October to March as many as half the days were unsuitable for visual bombing. In addition weather forecasting was less reliable than it is today. During the war an average of 10 percent of the missions were cancelled due to weather, a figure that declined from 1943 (23 percent) to 1945 (5 percent).[22] Part of this improvement was due to increased information as the airmen employed ships and aircraft to collect weather information over the Atlantic as well as aircraft to observe weather in the assembly areas (late 1943) and en route and in the target areas (summer 1944). Bad weather abilities of the pilots also improved.

To cope with the visibility issue the Americans made a major change to their bombing procedures in the last half of 1943 that permitted nonvisual bombing. In 1942 the Eighth requested two British devices, Gee and H2S, but it was not until February 1943 that Arnold pushed nonvisual equipment as an urgent need, although this presented both a manufacturing and training challenge. The AAF trained and equipped select crews as Pathfinders, but unlike their RAF namesakes, who used ground and sky markers to guide in the main force and flew ahead of the main bombing force, the Eighth's Pathfinders led the formations and signaled bomb release. The timing of this innovation worked out fairly well for the Americans, for just as the summer weather was ending and German defenses were inflicting heavier losses, the Eighth flew its first combat mission employing radar aiming in late September 1943. The Eighth also used Oboe-equipped Pathfinders for the first time a month later. Understandably the initial bombing results left much to be desired. Although bad weather operations were hazardous for all combatants, German losses in weather were greater than American as GAF pilots were now receiving less training than their foes. Nonvisual bombing was adopted because the Americans believed it was better to bomb less accurately than not to bomb. While the AAF planned to bomb visually and is known for this, 40 percent of the bombs dropped by U.S. heavy bombers in Europe were aimed by non-visual methods. To be clear, AAF accuracy with this method was far less than with visual bombing, and less than that achieved by the RAF. Nonvisual bombing techniques allowed the AAF to bomb more often and indirectly enact attrition on the defenders; however, they were both inaccurate and indiscriminate, raising efficiency and morality issues.

The third and most important factor in the turnaround in the air war proved to be the addition of the long-range fighter escort. The American airmen had considered this not only technologically impossible, a view held by other airmen as well, but theoretically unnecessary. Two points deserve emphasis: The provision of long-range escort was the key to the Allied aerial victory and the Americans did what both the Germans and British failed to do. The Eighth introduced three fighters in the strategic air war against Germany, the P-47 in early 1943, the P-38 in October, and the P-51 in December.[23]

The Lockheed P-38 Lightning was fast, could outturn GAF fighters, and had a superior combat radius. However, it was initially hindered by engine cooling and cockpit heating and defrosting problems. It also lacked the diving ability and ruggedness of the P-47, had an inferior rate of roll, and, with its twin booms and engines, was easily identified, unlike its teammates, which could be mistaken for German fighters. It was not as well regarded as a fighter in the European war, as were the P-47 and P-51, and instead gained its fame in the Pacific theater.

The P-47 Republic Thunderbolt was large, heavy, and very tough. It was heavily armed with eight .50s, fast in a dive, and available early in 1943, when it was needed. It did well in the air-to-air combat but was overshadowed by the more photogenic and better performing P-51. Overlooked is the fact that prior to April 1944, by which time the AAF had won the air superiority battle, the Republic fighter claimed three-fifths of GAF aircraft downed by the two fighters.

The North American Mustang design was prompted by the British, neglected by the Americans, and developed outside the normal AAF channels, similar to the RAF's Mosquito. (That two of the war's outstanding aircraft were so developed is certainly a comment on the system.) It languished with its inferior U.S. power plant until the Allies mated its magnificent airframe with the equally outstanding Rolls Royce Merlin engine. The resulting fighter could outperform the P-47, P-38, Spitfire, and, most important, any conventional Luftwaffe fighter, making it the best piston-powered fighter of the war. Its most important attribute was its range, yet it was belatedly deployed as an escort fighter due to the high command's failure to appreciate both the problem and the aircraft. The AAF did not see the need and urgency until mid-1943, and it was not until June that Arnold firmly demanded long-range escort fighters and that the Eighth ordered large-scale production of drop tanks. As late as the fall of 1943 Eaker allowed all P-51s to be assigned to the AAF ground support air force, and it was not until January 1944 that all incoming Mustang units were allocated to the Eighth. Eventually that unit would transition all but one of its fifteen fighter groups into that aircraft.[24]

Drop tanks were another technology that made a big difference in extending fighter range. While initially technical and safety concerns inhibited their combat use, when employed, their impact on combat was dramatic. The Mustang's range superiority was clear from the start, demonstrating a combat radius of 475 miles without tanks (compared to the P-47's clean radius of 230 miles and the P-38's only slightly greater radius) that by March 1944 was extended to 850 miles with two 108-gallon wing tanks.

Finally, two high-level decisions, one American and one German, were significant in the battle for air superiority. One of the first changes made by the new Eighth Air Force commander, Jimmy Doolittle, in January 1945 was to revoke the existing escort policy, which tied the fighters closely to their charges. This violated standard AAF policy and was unpopular with bomber crews, who liked seeing friendly fighters close at hand, but it gave the American fighter pilots the initiative, which they eagerly and effectively exploited. Pilot aggressiveness was further encouraged by granting victory credits for aircraft destroyed on the ground and adopting a policy allowing returning escort fighters to strafe enemy targets, especially airfields. Low-level tactics were costly, and some question the wisdom of using expensive machines and skilled pilots against parked aircraft and much cheaper flak.[25] It bears repeating that the GAF lost air superiority due to a lack of experienced pilots, not a lack of aircraft. A month earlier Hermann Goering, the Luftwaffe chief, made the other key decision when he ordered GAF fighters to ignore the U.S. fighters and concentrate on battling the bombers. This handed the initiative to the American fighter pilots, who took advantage of this gift.

## The Day Battle Climax

The AAF's primary mission was to win air superiority in preparation for the cross-Channel invasion. Arnold made this clear in his 1944 New Year's message to the Eighth and Fifteenth Air Force, concluding, "Therefore, my personal message to you—this is a *MUST*—is to, '*Destroy the Enemy Air Force wherever you find them, in the air, on the ground and in the factories*.'"[26] By the beginning of 1944 the AAF was ready to do this because its strategic bombing force was bigger, more versatile, better equipped, and using more appropriate tactics than had been the case only months earlier.

The Eighth licked its wounds after the October setbacks and over the next three months limited its operations to targets in western Germany and the Western European occupied countries. The Eighth launched it next major mission consisting of 593 heavy bombers on January 11, 1944, against five aircraft factories in central Germany. Because of the weather, only a third of the force launched bombed its primary objectives, with good to excellent results, while the other bombers hit targets of opportunity. The GAF countered with five hundred fighters that swept through the small P-51 escort in the target area and savaged the spread-out bombers. The Americans believed that the Luftwaffe defenses had improved with the defenders staying in contact longer and adjusting to the bombers' reaction, using rockets against the tight formation or pouncing on the bombers that dispersed to avoid the rockets. Sixty bombers went down and another two were written off for salvage. The bombers claimed 228 victories and the escort fighters another 30 victories for 8 losses, while the Germans admit to losing 53 fighters destroyed and 31 junked. Action subsided until the end of February, when one of the milestone air battles of the war took place.

Called "Big Week," the bombing flurry was dominated by poor weather, fierce battles, and heavy losses. On February 19 USSTAF weathermen forecast a clearing trend over Europe, prompting the airmen to plan a major operation, a series of Anglo-American bomber attacks on German aircraft factories. It began that night when Bomber Command attacked a Me 109 plant in Leipzig and ended with a Bomber Command attack on Schweinfurt on the night of February 25, which incidentally was the only coordinated effort by the three air forces. In the operation Bomber Command flew five missions, the Eighth five, and the Fifteenth four.

The operation was highly touted at the time. Arnold wrote that Big Week "may well be classed by future historians as marking a decisive battle of history, one as decisive and of greater importance than Gettysburg."[27] Since then airmen and historians have elevated it to almost mythical status as the decisive battle of the air war. However, it is unknown to most of the public, for it was neither an Allied disaster similar to the Ploesti, Schweinfurt, or Nuremburg actions nor as devastating as the Hamburg or Dresden raids. Instead it was a concentrated series of attacks that delivered an effective body blow to the GAF. The three strategic air forces pounded mostly German aircraft factories with nearly twenty thousand tons of bombs on sixty-three hundred sorties; the RAF delivered just over half the tonnage, while the AAF contributed three-fifths of the sorties. Allied losses were high, 378 bombers and 42 fighters, although not as bad as feared, for the planners believed that as many as 200 bombers might be lost in one day. The Luftwaffe was hard hit by the battle. The AAF crews claimed 449 fighters destroyed in the air battle, while the Germans admit to the loss of 355 fighters. Regardless of the numbers, the change in GAF defensive tactics during the following weeks indicates the impact of Big Week: The Luftwaffe no longer rose to oppose every AAF raid. On the other hand, bomb damage (as usual) was overestimated. (Postwar investigation revealed that German single-engine fighter production in the first half of 1944 was twice Allied intelligence estimates and that the bombing destroyed less aircraft production than the July and October 1943 raids.) However, the bombing encouraged the Germans to disperse their aircraft industry, which decreased production not only during the move but also from the loss of efficiency from the dispersed smaller units and made Germany more dependent on its transportation system. The air war now took a decisive turn.

Poor weather prevented an immediate follow up to Big Week. With the aim of providing air superiority for the upcoming cross-Channel invasion, the airmen sought to lure the Luftwaffe into the air. One way adopted in March was to stop flying diversions and route the attacking bombers directly toward their targets. Another was to pick out targets the Germans had to defend, such as Berlin. Weather disrupted the Eighth's first major raid on Berlin on March 6 and protected the primary targets from all but 51 of the 658 attacking bombers, forcing the remainder to scatter their bomb load throughout the German capital. The defenders launched 460 fighter sorties, which accounted for the bulk of the sixty-nine bombers downed, four interned in Sweden, and three written off. The bombers claimed ninety-seven GAF fighters and the fighter escort (eight hundred dispatched) another seventy-eight, compared with German records that indicate the loss of sixty-four fighters. The escort lost eleven fighters with two junked. In contrast the mission against Berlin on March 8 saw fewer bomber losses (thirty-seven) and claims (sixty-three), as did an attack in heavy weather on the ninth that met little German opposition, with twenty GAF fighters observed, resulting in only eight missing bombers and one enemy fighter claimed by the bombers and none by the escort. However, on March 18 the Eighth lost forty-three bombers and another four that were scrapped.

By the end of March it was evident that the AAF owned the skies over Germany. In the first quarter of the year the Germans lost more than twelve hundred fighter pilots on all fronts, about half of the GAF fighter pilot strength during that period. Allied losses were high, but they could endure the attrition; the GAF could not. Beginning in March the Luftwaffe changed its defensive strategy from maximum resistance to every raid to selective resistance that concentrated on escortless or scattered bomber formations. By this time the Anglo-American fighter force had greater numbers as well as an edge in both aircraft and average pilot quality. The GAF had lots of aircraft and in months would field jets that were superior to Allied fighters. Their downfall was the quality of their pilots. While the Germans had a small core of very experienced and very successful pilots, the mass of their fighter pilots were new, inexperienced, and extremely vulnerable.[28] The attrition of the war coupled with fuel shortages forced a dilution of the flying training program and consequential erosion in pilot quality. By the summer of 1944 few GAF fighter units had pilots, aside from their commanders, with more than six months experience.

While Luftwaffe defenses declined, they still were able to inflict stinging losses on the American bombers. In April there were five missions on which the Eighth lost more than thirty bombers. On the eleventh the bombers attacked fighter factories in central and eastern Germany and in Poland, dropping over two thousand tons of bombs with the loss of sixty-four bombers, the internment of nine, and the junking of five. The mission of April 29 was even worse, in that fewer bombers took essentially the same losses: sixty-three missing, one interned, and two scrapped. Four missions in May saw more than thirty lost, the highest on the twelfth, with forty-six missing and nine written off on an attack on oil targets. On that day the GAF reacted with a strong force of 515 sorties against the 886 bombers and

950 escorting fighters, resulting in bomber claims of 115 fighters and escort claims of 66 in the air, while German records indicate the loss of 60 fighters in the air.

## Strategic Operations to War's End

After the first quarter of 1944 strategic operations took second priority to support of the cross-Channel invasion and the campaign against the German V-weapons. The AAF flew strategic missions when support of the troops was not essential, allowing the AAF to target oil (on visual missions) and the RAF to hit cities. Flak became more significant as German fighter resistance was minimal, with only occasional pitched battles. German fighter defenses proved sparse in October with no claims by the bomber gunners between October 8 and November 1, a month notable for the onset of poor bombing weather. When the GAF came up to fight, it suffered disproportionate losses. On November 2 the Germans launched some 500 fighters, of which 36 were claimed by the bombers and 134 by AAF fighters. German records indicate the loss of 133 aircraft on this day; in exchange 40 AAF bombers went down (16 attributed to fighters) along with 13 fighters (4 more were junked). A giant fighter battle erupted on November 27, when the Germans misread a fighter sweep for a heavy bomber attack and launched 600 to 750 fighters, the most that challenged an AAF mission during the war. The 460 AAF fighter pilots claimed 98 victories while suffering the loss of 11 fighters and 2 scrapped. Although the Eighth was sending out over one thousand bombers on missions, only three times during the month did it suffer more than thirty losses.

The Germans launched the surprise Ardennes offensive (the Battle of the Bulge) in mid-December under the cover of poor weather as a shield against Allied air power in mid-December. For weeks thereafter Allied air forces supported the ground forces. The Eighth Air Force flew its largest mission of the war on December 24, 1944, when the unit dispatched an astounding 2,046 bombers and 853 fighters, and of these, 1,884 bombers dropped over five thousand tons of bombs. The cost was 12 bombers downed and 23 written off along with, respectively, 10 and 2 fighters. The bombers claimed 18 and the fighters 74 enemy aircraft destroyed.

In the closing months of the war, oil became the AAF's target of choice. The most important of these targets was the plant at Leuna, the largest synthetic oil plant of its kind in Germany and important in producing chemicals. Understandably it was also the best protected plant, with heavy flak guns increasing from sixty-five or so prior to the first AAF bombing attack in May 1944 to ten times that number by November. The Germans also effectively deployed smoke and fighters that, along with the weather, greatly hindered accuracy during the twenty AAF and three RAF attacks. The Eighth got 29 percent of its bombs within the target area (1.2 square miles) on five visual attacks, 12 percent on three attacks using radar and visual methods, and 5 percent on the remaining attacks using only radar.[29] The defenders enacted a high toll, 128 AAF bombers. Postwar investigation of oil targets revealed that only a small proportion of the bombs created damage. Overall, 84 percent of the Allies' bombs dropped on oil targets impacted outside the factory area, another 3 percent hit decoys, 7 percent fell in open areas, 2 percent were duds,[30] and a mere 3.5 percent hit production areas or pipelines. The American airmen erred in using bombs that were too light and were nullified by blast walls, whereas the heavier RAF bombs caused much greater damage. Aiming points were also poor, but with the miserable accuracy this was of little import. Nevertheless, Leuna production tumbled and during the bombing amounted to only one-tenth of its preattack capacity.

The most significant mission following the Dresden attacks of mid-February was Operation Clarion, a maximum Allied air effort of both strategic and tactical forces on German transportation, specifically marshaling yards, railway stations, barges, and bridges. It had originated as an attack on German morale, specifically visual attacks on small German communications centers. A number of top AAF leaders objected to this plan on moral grounds, Eaker writing on January 1, 1945, that "we should never allow the history of this war to convict us [of] throwing the strategic bombers at the man in the street. I think there is a better way we can do our share to defeat the enemy."[31] (He went on to state that if the AAF was to attack civilians, it should wait until morale was lower and the weather was better.)

The mission contrasted with the Eighth's previous operations by assaulting from half the normal bombing altitude (to ten thousand feet, with some units bombing from as low as six thousand feet), in small formations (as few as thirty bombers), and against a large number of dispersed targets. Almost half of the fourteen hundred effective sorties hit the primary targets, and 96 of the 124 squadrons bombing used visual aiming and got almost three-fifths of their bombs within one thousand feet of the aiming points, twice the percentage within five hundred feet as compared with the rest of the winter bombing. Despite these more vulnerable tactics, the Eighth's losses were low: seven bombers and thirteen fighters lost and one fighter junked. Clarion continued the next day, but weather interfered so that only 470 bombers hit the primary targets and 720 hit targets of opportunity. Losses again were low, two bombers lost and five written off, along with eight fighters downed and two scrapped. The operation demonstrated the weakness of GAF defenses and power of the Allied air forces but did not strike an especially hard blow. German morale held up, as did the transportation system.

The bombing continued as the war ground to a close. On April 7, 1945, the Eighth launched more than thirteen hundred bombers against targets in central and northern Germany and lost seventeen bombers and five fighters, writing off five bombers and two fighters. The mission is notable because a number of GAF fighters attempted to ram the bombers and did down seven and damage three others in this manner. Germans losses were high, with the bombers claiming the destruction of sixty-two fighters and the escort sixty-seven. The remaining missions of April are best remembered for the high number of claims of GAF aircraft destroyed on the ground: on April 10, 335; on April 16, 727; and on April

17, 293. The Eighth's last strategic bombing mission of the war was on April 25 against targets in Czechoslovakia and Austria. All of the 550 bombers that bombed unloaded on their primary targets; six Forts were lost and three were written off. The AAF fighters flew almost five hundred sorties and claimed one bomber, although another fighter engaged a twin-engine aircraft that was later identified as a Soviet bomber. Two fighters were lost and another two were scrapped.

## AAF and RAF Bombing Accuracy

The subject of bombing accuracy is a difficult one. Photos interpreters could not account for all bombs dropped and excluded gross errors, resulting in figures that were optimistic. Despite these serious limitations we can draw some tentative general conclusions. Accuracy of visual bombing improved as the war progressed, although this was not a straight line trend as larger formations attacking the same target, higher bombing altitudes, weather, smoke, and stiff German defenses degraded accuracy. Aside from the defenses, the biggest problems were the weather and season. The AAF heavy bombers aimed 40 percent of their bombs by nonvisual means, two-thirds of this by radar. Under the best circumstances nonvisual bombing (radar in 5/10s visibility or Gee-H) had a CEP of one mile, while visual bombing under the best circumstances (good to fair visibility) had a CEP of about one-third mile; however, radar bombing in the worst conditions averaged a CEP of about four miles, and visual bombing in poor visibility averaged four-fifths of a mile. Only 14 to 20 percent of the Eighth's bombs fell within one thousand feet of the aiming point.

While this was daylight bombing, it certainly was not precision bombing. Blind bombing, a term that some of the airmen detested, allowed the AAF to operate on days that would not permit visual, precision bombing. Thus the pressure on the Germans was maintained. While the Allies had strict restrictions on bombing in the occupied countries, these did not apply to Germany. As a result operations in the first quarter of the year increased from eighteen days in 1943 to fifty-two in 1944 and sixty-seven in 1945. The common view was that it was better to bomb, even if less accurately, than to stay on the ground.

A comparison of the AAF and RAF bombing offensives is difficult to evaluate because it depends on a number of variables. Bomber Command's accuracy also improved from April 1943, when it got 30 percent of its bombs within three miles of the aiming point, to April 1945, when it got 90 percent within that distance. AAF visual bombing was more accurate than RAF night bombing, but the RAF bombing was more accurate than AAF nonvisual bombing.[32]

## Other AAF Innovations

Although shuttle missions between Britain and the Mediterranean proved impractical, similar missions to the Soviet Union went ahead. Russian bases would make bombing of Eastern Europe much easier, extend the bombers' radius of action, spread German defenses, and, perhaps most important, might lead to the use of Siberian bases in the Pacific war. Difficult negotiations and primitive infrastructure hindered the operation aptly code named Frantic. On June 2, 1944, Gen. Ira Eaker led a force from Italy consisting of 130 B-17s and 69 P-51s that hit railroad targets in Hungary on the eastward thrust and an airfield on the return four days later. On June 21 the Eighth launched more than 1,000 Forts, mainly against Berlin, including 147 bound for Russia that bombed a synthetic oil plant along the way. The Germans responded that night with about 150 twin-engine bombers that hit Poltava, one of the three AAF bases in Russia. The GAF lost two bombers in their attack, which hit all seventy-three B-17s on the field, destroying forty-seven of them along with three C-47s, one P-38, and vast amounts of stores. The Russians beefed up their defenses and the missions continued, but without great success considering the resources invested. The imaginative concept was undermined by the tardiness of its execution, reluctant Soviet cooperation, and the rapid Russian advance.

Shuttle bombing looked good on paper and did put distant targets at risk. However, two factors proved more important. First, the key factor in accurate bombing was the weather over the target, not over the bases. Second, effective operations depended on the maintenance and supply structure. The two shuttle attempts, the Regensburg raid and Frantic, convincingly demonstrated the problems.

## The Luftwaffe's Last Gasp

Timing is critical in many aspects of life and often decisive in history. The massive buildup of AAF bombers and fighters and the introduction of American long-range escort fighters came in time to ensure Allied air superiority for the Normandy invasion. Meanwhile the concurrent increased output of German fighters was rendered ineffective by the decreased quality of Luftwaffe fighter pilots and the improved quality of Allied fighter pilots flying mounting quantities of superior propeller-powered fighters. The declining impact of the GAF fighter force obscures the improvement of German aircraft in the closing months of the war. The Eighth noted heavier fighter armament and that the number of bomber losses increased relative to the number of fighters encountered.

The biggest challenge to Allied air superiority in the closing months of the war was German rocket and jet-powered fighters. On August 5, 1944, three P-51s were quickly downed by three rocket-powered Me 163s. The delta-winged fighter's exceptional speed (nearly 600 mph compared with the P-51's top speed of about 450 mph) was counterbalanced by its limited endurance, only 2.5 to 8 minutes of powered flight. The Germans built fewer than four hundred and it proved to be a less serious threat than jet-powered fighters.

The twin-jet Me 262, although not as fast (540 mph) as the rocket fighter, was a safer aircraft to fly, had greater endurance, and mounted heavier armament (four 30-mm guns). Yet efforts to standardize on the jet in 1943 were nixed by the bureaucracy and Hitler. The demands of the war and Allied bombing delayed the jet's introduction into the war and limited its deployment. Of fourteen hundred built, no more than a few hundred saw operations.

The Germans also pushed a bare-bones jet fighter, the He 162, that could reach a top speed of 522 mph. The Germans planned to build the wooden and metal aircraft in great numbers and operate them off of meadows and highways flown by hastily trained pilots. The plane first flew in December 1944, a mere seventy days after its initiation. Heinkel built about one hundred, although none saw action. The Germans had a clear advantage over the Allies with jet and rocket propulsion as the British jet-powered Meteor had an inferior top speed (493 mph) and only saw combat against the V-1s. The first true American jet fighter, the P-80, was even further behind in production.

The British and American airmen beat the jet threat the old fashioned way, with superior numbers. The Allies successfully attacked the jets where they were most vulnerable, on takeoff and landings. Maintenance problems (the jet engines had to be overhauled after only a few hours of operation) and well-trained Allied pilots were also keys to the failure of the jets to regain air superiority. The Eighth Air Force claimed that 740 of its bombers and 1,200 of its fighters encountered jet and rocket fighters, of which 52 bombers and 10 fighters were lost, and in exchange 183 Me 262s and 6 Me 163s were destroyed. A more recent source claims that the Me 262s destroyed 446 Allied aircraft at the cost of 190 Me 262s.

Meanwhile German flak improved and accounted for the bulk of American aircraft losses during the war, especially after June 1944.[33] GAF defenses were extensive, in February 1945 (months after their peak) consisting of 680 128-mm, 1,840 105-mm, and 9,300 88-mm guns. Almost one million men manned the heavy guns at their maximum strength, although less than a third were flak troops, the rest boys, prisoners of war, factory workers, and limited-service troops. During the course of the war numerous improvements increased flak effectiveness. Radar control was one of the most important, although the Germans trailed the British in its use, and as late as August 1944 were still using more than fifty-five hundred sound detectors. The Luftwaffe introduced dual fuses (contact and timed) into combat in 1944, which increased the 88-mm hit rate by five times, the 105-mm rate by three, and the 128-mm rate by two. Controlled fragmentation (groove projectiles) increased projectile effectiveness. Nevertheless, flak proved disappointing to the defenders. Prior to the war the Germans estimated that on average forty-seven shells would down an aircraft while in fact it took more than eight thousand 88-mm, six thousand 105-mm, and three thousand 128-mm rounds, without the dual fuse. Flak was unable to halt Allied bombing or inflict prohibitive losses and damage, but it did create substantial losses and decrease bombing accuracy.

The Germans attempted, and failed, to field two key antiaircraft innovations. The Americans introduced proximity fuses in the Pacific war in 1943 and against the V-1s in the European theater in mid-1944, a device that increased flak effectiveness by a factor of five. A postwar American study concluded that proximity fuses would have increased German flak efficiency by a factor of 3.4, making B-17 operations much more costly and B-24 operations impractical. The Germans also tested a number of flak rockets, later known as surface-to-air missiles

(SAMs), but none saw combat. Proximity fuses fitted to rockets would have decimated the AAF's heavy bomber formations.

## The Results and Lessons of the Strategic Air War against Germany

The Americans began a detailed evaluation of the bombing of Germany during the war that continued after the war and extended to Japan. The United States Strategic Bombing Survey (USSBS) employed hundreds of personnel to produce two hundred studies on the European war and one hundred on the Pacific war. The British undertook a similar project that was much smaller and relied on the USSBS for basic data. The investigators studied the bomb damage, researched enemy records, and interviewed captured officials. The USSBS remains the most extensive source on the subject.

The overall conclusion of USSBS was that "Allied air power was decisive in the war in western Europe."[34] This bold statement was much to the likening of the airmen and has come to mean to many that air power was the most important factor in winning the war. On reflection, however, credit hinges on the definition of "decisive." Certainly air power was important to Allied victory, but it was just one of several factors that won the war. In the European theater the principal agent of victory was the Soviet army, which severely drained the German army. (Two-thirds of German military fatalities were on the Russian front.) The Allies employed equivalent technology and overwhelming numbers of men and machines to defeat the overstretched Germans. In hindsight it appears that the connection of air power with cutting edge technology, the romantic aura of aviation, and the spotlight of wartime publicity gave the airmen more credit than their actions merited. In any case the AAF's record in the war along with the potential of the newly developed atomic bomb bolstered the promise of strategic bombing in the future that led to the establishment of the U.S. Air Force (USAF) in 1947.

What did the strategic bombing accomplish? The most important result was its contribution to the defeat of the Luftwaffe and winning air superiority. The Luftwaffe was significant in the German victories in the first years of the war. But GAF advantages eroded as the British and Soviets adapted to the air war and the Germans found themselves spread thin across Europe and North Africa. The strategic air war was instrumental in forcing the Germans to transition from an offensive air force to a defensive one. (In 1942 60 percent of the GAF's manpower and equipment was geared for offensive warfare, in 1944 82 percent was focused on homeland defense.) By January 1944 two-thirds of German fighters were located on the western front (as distinct from the Russian and Mediterranean fronts) along with three-quarters of German heavy flak. The AAF began a concerted effort against German aviation plants in July 1943, yet this only amounted to 3.5 percent of the total tonnage it dropped during the war. German aircraft production increased during the war, peaking around March 1944 with acceptances peaking in July. Without the bombing, which caused direct destruction as well as loss of production due to the dispersal effort, output would have been much higher. This understates the impact of this bombing because the Luftwaffe rose to protect the aviation industry and in so doing suffered heavy losses. To be clear, the Germans had a lot of aircraft in the closing months of the war and jets superior in performance to Allied aircraft. The GAF was defeated by the attrition of its fighter pilots fighting a multifront war and the decline of the quality of its pilot training graduates mainly due to a lack of fuel. German timing was faulty as the great flow of aircraft came after the decisive air battle had been fought and lost.

A second major accomplishment of strategic bombing was to cut off German oil. Because the Germans were short of oil, they cut the length of their flying training program by early 1943, so that as the war continued British and American pilots had increasingly more flying training hours than the Germans. Oil shortages also drastically restricted the mobility of the German army, forcing German units to employ animal power to move equipment.[35] By mid-1944 fuel shortages restricted combat operations. (During the December 1944 Ardennes offensive the Germans planned on capturing Allied fuel.) A further consequence of the diminishing German oil position was its detrimental impact on the chemical industry, crippling ammunition, fertilizer, and rubber production. For these reasons it has been argued that oil was of first importance in the defeat of Germany. An important note: One of the reasons for the successful oil campaign was excellent intelligence. Allied use of Ultra prompted the reattack of oil targets shortly after they were repaired, the only such effective use of the broken code in the strategic air war.

The bombing imposed heavy costs on Germany—not only the direct cost through the loss of production but also indirect costs as resources were siphoned off by the bombing. Air power destroyed or heavily damaged one-fifth of German dwelling units. The USSBS estimated that 3.5 million workers were engaged in rescue and clean up operations, reconstruction, and replacement of civilian goods. One-third of German heavy guns, one-fifth of heavy gun ammunition, half of electronics, and one-third of optics were employed in air defense. Two million workers were engaged in producing air defense weapons, and another one million manned this equipment. The Germans also committed great resources to build underground factories to shield their vital industries, an effort that in December 1944 required almost half of the labor force involved in industrial construction, resources diverted from building fortifications.

Two other key target systems were transport and coal. Transportation was necessary to move coal, the chief power source throughout the country. The dispersion of German industry as a consequence of the bombing made transportation more important, difficult, and expensive. The Germans had a large, well-organized transportation system that finally broke under the battering of Allied bombs.

The bombing reduced German production in a number of ways. In the aircraft industry, for example, this was due to destruction of factories and machinery as well as death and disruption of the work force and the dispersion of the industry. The USSBS concluded that there was a 9 percent loss of overall production in 1943 and 17 percent loss the next year, while another source asserts that the Germans lost one-fifth of their production in the last sixteen months of the war. Bombing critics often note that German armament production peaked in July 1944. However, this fact should be evaluated against another: 72 percent of the bombs fell on Germany after July 1, 1944. Nevertheless, Germany did not suffer any significant equipment shortages because of the bombing.

In hindsight the bombing could have been better conducted. The USSBS concluded that it was a major error not to have bombed electrical and chemical targets (especially tetraethyl, the basis of high-test gasoline), not to have focused on aircraft engines rather than airframes, and not to have attacked transportation earlier. Others have observed that the destruction of abrasive grinding wheels would have severely hobbled the German economy. There is also speculation that Germany could have been defeated sooner if the synthetic oil plants had been bombed earlier and instead of the attacks on bearings.

The Allied bombing effort against Germany was large and costly. British and American aircraft dropped some 1.4 million tons of bombs on Germany. (In stark contrast the Germans unloaded 78,000 tons on Britain and the Soviets some 6,700 tons on mainly Finnish and Hungarian cities.) Americans and British losses were high. The Eighth and Fifteenth Air Forces lost eighty-seven hundred (downed, missing, and written off) bombers, forty-two hundred (downed, missing, and written off) fighters, and suffered seventy-three thousand casualties; Bomber Command lost ninety-two hundred bombers and sustained sixty-four thousand operational casualties.

There were also opportunity costs, the alternate use these men and machines could have made. The commitment to air power (one-quarter to one-third of U.S. production went into aircraft, 40 percent of this into strategic bombers)[36] was a major reason the United States limited its ground forces to ninety divisions in contrast to the much higher percentages fielded by the other major powers. One author has calculated that the resources involved in the strategic bombing forces were equivalent to twenty-five armored divisions and an adequate number of infantry replacements. (Even with the limited ground forces, the United States was running short of infantry replacements in late 1944, prompting combing out of training, ground, and air units.) Or what if the strategic bombing effort had been converted into tactical aviation?

The strategic bombing enacted a heavy toll on Europe. The USSBS estimates German civilian deaths at 305,000, which is undoubtedly low, while an East German account asserts the number is 410,000, a source that had no reason to underestimate the truth. In addition tens of thousands of civilians in the occupied countries were killed by Allied bombs, 58,000 or more in France alone. Major European cities and their treasures were ravished by the bombs as well. London suffered the destruction of six hundred acres, an area of destruction exceeded in twenty-seven German cities.

Where does this leave us? Air power was certainly important. It lessened the casualties of Allied ground forces by direct action against the German army and prevented the GAF from attacking friendly forces. But as Noble Frankland, one of the premier historians of Bomber Command, makes clear, there was no revolution in warfare. In contrast to the air power theorists, World War II was attritional warfare extended to the skies. One air power historian concludes his treatment of this subject by writing that strategic bombing "vindicated the treasure expended on it. If in the final analysis it accomplished its ends more by brute force than by elegant precision, the fault lay in the unrealistic assumptions of prewar doctrine as to wartime accuracy, the vagaries of European weather, and the limitations of radar technology."[37] Strategic bombing made a major yet expensive contribution to the Allied victory in Europe. It also was to have a major impact on the Pacific war.

Chapter 6

# Razing Japan

## THE ZENITH OF WORLD WAR II STRATEGIC BOMBING

In many respects the bombing of Japan was in sharp contrast to the American bombing of Germany. It began later in the war, was of a shorter duration, and employed the most advanced bomber of the conflict. The Pacific campaign was waged over much great distances, mostly over water, and encountered even greater weather problems than those found in Europe. On the other hand Japanese defenses proved weaker than the German's, and thus most of the American bomber losses were not caused by the defenders. These differences led to tactics that varied from the prewar American strategic bombing theory and those initially employed over Europe. It ended with the use of the atomic bomb, what appeared to be the ultimate strategic bombardment weapon. The Japanese campaign proved more efficient and destructive than the German one, but was it decisive?

### The Boeing B-29 Superfortress

The B-29 came late to the war, one of only two AAF aircraft that first flew during the war and yet served in combat.[1] Understandably it was hurriedly pushed into production and service. This pressure, its cutting edge performance, and numerous advanced features ensured it would have problems, and it seemingly had more than its fair share. Curtis LeMay, who solidified his reputation employing the bomber, later commented that the "B-29 had as many bugs as the entomological department of the Smithsonian Institution. Fast as they got the bugs licked, news ones crawled out from under the cowling."[2]

Even after developing the outstanding B-17, American airmen pushed for bombers with greater performance, especially longer range. In January 1940 the airmen sent out a request to aircraft manufacturers for a bomber with a 5,333-mile range, 400-mph top speed, and a one-ton bomb load. Boeing won a contract to build three experimental bombers (and a fourth airframe for static tests) with the first to be delivered by March 1941. In the event the Boeing design failed, the airmen also awarded Consolidated a contract for the B-32 (see below).

The design the airmen finally bought in 1940 was a considerable change from the B-17. The bomber featured tricycle landing gear, a high-aspect-ratio wing, and giant flaps to lower takeoff and landing speeds. Although the Superfortress was larger and twice the weight of the Fort, its performance was markedly superior because it had the same drag as the B-17 and 83 percent more engine power. Defensive armament was also greater, with four turrets (two atop and two beneath the bomber), each with two .50-caliber machine guns and a tail position mounting twin .50s and a single 20-mm cannon.

The Superfort introduced three significant innovations into strategic bombers. The first was to add a flight engineer, who despite the protests of pilots who were concerned about not having full instrumentation at their position, greatly aided in flying this large and complex aircraft. A second major improvement was cabin pressurization, which allowed the crew to operate at altitude without oxygen masks. The third innovation was the gunnery system: four turrets remotely controlled by gunners positioned in sighting blisters. Besides improving streamlining this system removed the gunner from the noise and vibration of the firing and allowed the master gunner to switch control of the turrets among the gunners, which meant that if a gunner was disabled, firepower from that turret was not lost. The experience over Europe with head-on attacks encouraged the airmen to upgrade forward defense by doubling the number of guns in the top forward turret. This proved to be less an advantage than hoped for, but fortunately the B-29s over Japan faced less fighter opposition than did the B-17s and B-24s over Germany and the system did an adequate job.[3]

Engines were the bomber's most serious problem and would dog it throughout its service. The Wright R 3350 was the most powerful engine available in 1940, put into production before its problems were solved, pushing these difficulties into the field. Rapidly expanding production added quality-control and schedule issues. The major technical issue involved inadequate engine cooling, partially a result of engine design and a tight engine cowling. The airmen employed a number of efforts to combat this problem, with eventual success; however, engine failures were common, responsible for half of all aborts and aircraft losses and damage in 1944. The B-29 engine averaged only 170 hours of service compared with 400 hours for the B-17 engine, which contributed to an abort rate three times that of the B-17. Engine failure was blamed for 28 percent of the accidents in the first five hundred B-29s, compared with 4 percent of the accidents in the first five hundred B-17s.

The Superfort's engine problems frequently became engine fires. Between February 1943 and July 1945 the AAF attributed one-fifth of B-29 accidents to engine fires, four times the rate of fires in the B-17s and B-24s. To make matters worse, the aircraft's fire extinguisher system proved ineffective. Little wonder the crews were wary of the B-29 and (at least in training) would abort missions for reasons that would be disregarded in other aircraft. This led to the bizarre suggestion that the bomber be configured with engines that could be jettisoned.

These problems did not dissuade the AAF, which aggressively pushed the B-29. One indication of the pressure to field the Superfortress is that the AAF ordered 1,650 before the aircraft's first flight in September 1942. Flight testing was hindered by engine problems, bad weather, a lack of facilities, and the crash of one of the experimental bombers. In early 1944 Arnold, frustrated by the continuing problems, exclaimed, "Is that airplane so damned rotten [that] they have to put it in cellophane?"[4] The B-29's accident rate improved as the war continued but was considerably higher than the other AAF heavy bombers during the war.[5]

Building the B-29 was a complicated undertaking because the bomber used the most powerful engines, the most sophisticated radar, and the most complex fire-control system of the day. Boeing had to work out problems with the aircraft, incorporate numerous modifications, and gear up for mass production, which required the expansion of its facilities and reliance on other manufacturers. American workers and manufacturers rose to the challenge and turned this superior bomber out in amazing numbers, thirty-four hundred by mid-August 1945. In all the program cost $3.7 billion dollars, the most expensive weapons program of the war, including the atomic bomb project.

## A Unique Bombing Situation

The airmen realized that the bombing of Japan would differ from the bombing situation in Europe. Observers noted that Japanese cities were very vulnerable to fire because of their density, the light construction of their buildings, and the marginal firefighting capabilities of the Japanese, certainly compared with the Germans. This was made clear by the 1923 Tokyo earthquake, which killed some 110,000 people and destroyed one-fifth of the city's buildings. Billy Mitchell, novelists, and the Japanese wrote of the vulnerability of the country to fire, and some wrote of how a low tonnage of bombs would cause vast destruction and panic.

This Japanese vulnerability was offset, however, by a number of factors. First, Japan and the Pacific were seen as secondary to the European theater, where Germany was perceived as a more dangerous threat. Second, it was a major problem to bridge the vast Pacific Ocean, secure bases, and deliver bombs to Japan. Third, there was much less intelligence available on Japan than on Germany. Finally, and perhaps most important, the airmen believed in precision bombing.

Prior to the war the airmen devoted little to no attention to fire bombing and did not develop an effective incendiary weapon. Yet in 1934 an airman's student paper at the army's Command and General Staff School proposed that instead of matching Japan's army head on, "the tremendous striking power of an air force [should be] directed at the paper cities of congested Japan."[6] Three years later the chief of the Air Corps, Oscar Westover, noted that Japan knew of the power of fires following earthquakes and that incendiary bombing of her "tinderbox cities" would equal many such quakes. To be clear, these were the exceptions, for Air Corps Tactical School lectures in the late 1930s emphasized the power of precision attacks on Japanese targets. In brief, American airmen recognized Japanese vulnerability to fire bombing, nevertheless they planned on conducting precision bombing.

## Changing Views on How to Fight in the Pacific Theater

When war erupted in Europe, airmen thinking about the Pacific problem thought in terms of the American strategic bombers as a deterrent. In September 1939 Lt. Col. Carl Spaatz sent a paper to Hap Arnold that proposed dispatching heavy bomber groups to the Philippines which he believed would restrain Japan from interfering with U.S. national interests. But the United States had few bombers and few bases, and its focus centered on Europe. In early 1940 Claire Chennault, a fighter plane advocate now serving the struggling Chinese, wrote Arnold about the potential of incendiaries against Japanese cities. The Air Corps chief rejected this idea, championing the virtues of precision bombing and insisting that U.S. national policy was against the use of incendiaries. This did not discourage Chennault, who recommended a force of five hundred bombers crewed by Americans to "burn out the industrial heart of the Empire with fire-bomb attacks on the teeming bamboo ant heaps of Honshu and Kyushu."[7] While the American airmen rejected Chennault's proposals, President Roosevelt was more receptive and ordered a top-level investigation of the concept, an action in sharp contrast to his 1939 appeal to all combatants to spare civilians from bombing. The scheme received a mixed reception and was dropped. Chennault did not give up on the concept, however, writing Roosevelt again in May 1941, once more getting a warm response from the president, and again a negative reaction from the War Department.

This brought the U.S. Navy into the picture. Roosevelt ordered the Navy to study the idea of carrier attacks, presumably with incendiaries, on Japan. This was well outside the sea service's capabilities because it did not even have incendiary bombs until August 1941 and did not get adequate stocks of them until summer 1942. The Navy had earlier considered flying boat attacks on Japan, but by February 1940 it realized its PBYs were inadequate for such operations.

Deterrence became the American military policy in the Pacific by necessity because it was considered a secondary theater and because of inadequate American military forces. In the spring of 1940 the United States left the fleet in Hawaii as a deterrent, instead of returning it to the West Coast as had been the pattern. This move did not deter the Japanese but encouraged them to attack Pearl Harbor. To be clear, the root cause of the American-Japanese conflict was the continuing Japanese war in China. After Japanese actions led to a complete American trade embargo, they were faced with either ending the war with China without victory or seizing the resources they required. The Japanese bet on a wider war and the lack of American will for victory.

Meanwhile the United States ramped up the pressure and preparations against Japan. In July 1941 the president approved a plan to send American twin-engine bombers and crews to China to conduct incendiary bombing of Japan. Meanwhile the United States reversed its long-held position that the Philippine archipelago was indefensible and initiated actions to reinforce the garrison. The change in policy may have resulted from the optimistic and persuasive requests of the top officer in the Philippine army, Douglas MacArthur, or from a mistaken view of the B-17's capabilities. Another possibility is that the decision makers were grasping at straws and believed they had to do something, anything.

The result was a military debacle. The defenders of the islands were reinforced, but with too little, too late. The first B-17s arrived at Clark Field in mid-September, with the projected bomber force of about 165, essentially all of America's Flying Fortresses, scheduled to be in position in February or March 1942. It was a major bluff or gamble because the B-17s could reach only southern Japan and bases, bombs, and air defenses in the Philippines were inadequate.

In mid-November army chief George Marshall held a secret press briefing in which he stated that the United States was on the brink of war with Japan. He went on to assert America's intention to fight a merciless war that would include immediate use of B-17s "to set the paper cities of Japan on fire."[8] Civilians would not be spared; this would be all-out war. He stated that negotiations were in progress to obtain bomber bases in both China and Siberia and that bombing alone could defeat Japan. The press, undoubtedly encouraged by Marshall's interview, ran stories of American intentions to bomb Japan from bases in Alaska, Guam, the Philippines, and Siberia. In fact American bombers lacked the capacity to do very much, and in any event, seventeen of the thirty-five Forts on Clark Field were destroyed on the first day of the war. No B-17s bombed Japan.

The incendiary bombs require discussion before moving on. The American airmen had little incentive to develop incendiary bombs because of the emphasis on precision bombing, although incendiaries were a very effective munition. The M-47 incendiary bomb, which the United States put into service in 1940, weighed seventy-three pounds and was judged to have twelve times the effect against combustible building and one and a half times the effect against noncombustible or fire-resistant buildings as a 500-pound high-explosive bomb. Prompted by a request from Arnold in September 1941, the Americans went on to develop an even more effective fire bomb. The AAF first tested this 6-pound device in early 1942, and in tests in mid-1943 against replica German and Japanese villages, it demonstrated not only that it was the most effective fire bomb but also the vulnerability of Japanese urban areas. While there were some questions about the validity of these tests—the climate in Utah is much drier than that in Japan—additional tests at Eglin Field, Florida, in April 1944 confirmed the fire bomb's effectiveness.

Prewar air plans for strategic bombing (AWPD/1 of September 1941) only briefly mentioned the Pacific theater. The following plan (AWPD/42 in 1942) gave it more attention, but these plans mirrored those drawn up against Germany. It was not until February 1943 that the planners looked at fire raids on Japanese cities as a possibility, although there were questions over the legitimacy of such attacks. By June 1944 a top targeting committee established the target priority: aircraft, coke, oil, electronics, bearings, urban areas, and shipping. In June 1944 (when the B-29 bombing began) the targeting committee established a subcommittee to study how to burn out the six major Japanese cities on Honshu. It reported that if 70 percent of the six were destroyed, it would take out 20 percent of Japan's war industry and inflict 560,000 casualties. The subcommittee admitted that such attacks would be spread across the entire economy and not on a specific industry, would not immediately impact on the combat areas, and would have only a questionable effect on civilian morale. The planning that followed pushed up the priority of cities. In September the planners noted how Japan had been effectively cut off from resources in Southeast Asia, the prospect of an early invasion, the success of the antishipping campaign, and the possibility that Japan could be blockaded by mines planted by B-29s. It further commented that bombing of cities should await the arrival of sufficient forces. These staff papers continued to underestimate the tons required to burn out Japanese cities at the same time raising humanitarian concerns.

## B-29 Planning and Problems

The strategic bombing campaign against Japan faced many hurdles. The first was that of geography. Whereas the distance between the British bases and Berlin was about six hundred miles, the distances in the Pacific theater were much greater. From Honolulu to Tokyo is thirty-nine hundred miles, from Dutch Harbor in the Aleutians it is twenty-seven hundred miles. The distance between Manila and the Japanese capital is nineteen hundred miles, and from Guam sixteen hundred miles. So for all of its superlative performance, the B-29 could not begin strategic air attacks until the Allies closed with the enemy.

A second factor was bases. A number of potential basing areas were considered. Weather and logistics weighed heavily against the Aleutians, while Soviet noncooperation ruled out Siberia. The only other options were to either use China or seize territories within range of Japan. In either case these bases would have to be built, unlike the situation in Europe, where most of the American bombers moved into existing airfields that already had the necessary infrastructure. Then the airmen would have to supply and support the most sophisticated aircraft of World War II at the end of a supply chain that literally would extend half way around the world.

At the Quebec conference in August 1943, the airmen proposed basing twenty B-29 groups on airfields to be built in China and supported from India by four thousand converted B-24 transports. But a review of the plan found that there would not be adequate transport available to support such an enterprise until Germany was defeated. At about this time Gen. Joseph Stilwell, commander of the American theater in the area, came up with a less ambitious plan that would ease some of these logistical concerns. It would move the bombers to permanent bases near Calcutta and stage them out of advanced bases in China. Another idea that surfaced in September 1943 was to move up the planned invasion of the Marianas from early 1946 to mid-1944 and use it as a staging base for B-29s stationed in the Marshall and Caroline islands. Such a scheme would allow bombing of Japan by March 1945. This was packaged with the idea of also using Aleutian and Siberian bomber bases along with fighter escort bases in the Bonins. In December 1943 the U.S. Joint Chiefs of Staff (JCS) approved a plan that initially would use the advanced bases in China supported by bases in Calcutta to begin operations in March and tentatively approved basing in the Marianas, scheduled for seizure in October, with operations beginning by the end of the year. This was the plan that was followed, with the March 1944 modification that advanced the Marianas assault up to mid-June 1944.

## Moving into Battle

The AAF planned to begin operations with the XXth Bomber Command, consisting of two bomb wings that would operate out of China. Although the first B-29 bomb wing, the 58th, had as its nucleus twenty-five highly experienced pilots and twenty-five highly experienced navigators and drew many high-time instructor pilots, only 10 percent of the unit's officers had previous overseas experience. Training was a major problem, hampered by a dearth of B-29s due to the aircraft's developmental difficulties, the B-29's low in-commission rate, a shortage of mechanics, poor weather, and especially the slow flow of engines. Little wonder that training fell behind schedule with serious gaps in high-altitude and long-range operations, the core of the B-29's proposed tactics. In training only 10 percent of the gunners fired the guns as the wing had received only two B-29s with the central fire-control system and no radar training was conducted as there were neither radar equipped aircraft nor radar operators. The situation with the ground crews was about the same. Consequently the unit shipped out having completed less than half of its prescribed training. This is typical of a wartime situation, the ideal versus the practical, although Curtis LeMay later noted that although B-29 training was poor, it was better than what was given to those who served in the European war.

The bombers posed additional difficulties, for the Superforts that arrived at the four midwestern training fields in the middle of February were hardly ready for deployment, much less for combat. Numerous modifications had to be made, radar had to be installed, and engines had to be changed. The air move was slipped half a month as the airmen launched an all-out effort to prepare the bombers for combat. The obstacles were many: few hangars, a shortage of parts, an inadequate number of experienced personnel, winter weather, threats of a strike, and friction with the civilian workers.

The AAF moved the support personnel and materiel by sea and the B-29s by an 11,500-statute-miles air route to India. The first B-29 touched down near Calcutta on April 2, and by the end of the month 130 had arrived. The bases they found left much to be desired. One crewmember's first impression of India was that "there were no barracks, no paved streets, nothing but insects, heat and dirt."[9] Food was poor, recreational facilities almost nonexistent, sanitation minimal, and disease prevalent. The bases lacked concrete taxiways, tools, and engine stands. Supply of parts was slow. But if things were bad in India, they were even more primitive in China. There bases were built by half a million Chinese breaking up rock by hand and pulling huge rollers (requiring five hundred to six hundred Chinese workers) to tamp down the one-and-a-half-mile long, six-foot-thick runways. The construction took three months, the first B-29 landing on April 24.

Supply of the advanced bases in China was extremely difficult. In the twelve hundred miles between the Indian and Chinese bases stood the Himalayan mountains and some of the most difficult terrain on the planet. This, combined with a lack of weather forecasting facilities, emergency airfields, and navigation aids, made flying the "Hump" a hazard that sometimes exceeded combat, so that the airlift became one of the epic stories of the war. The AAF used transports, combat B-29s, and twenty B-29s converted to transports and in September brought in B-24s that had been converted to transports (C-109s). About 90 percent of the cargo was gas. The Japanese made a few ineffective efforts to impede the airlift, attacking Chinese bases twice and the Indian bases once, and achieving only a few air-to-air interceptions. Nevertheless losses were high, as through October 1944 twenty-three aircraft went down to noncombat causes. Supply constraints restricted operations, so while the wing believed it could fly 432 sorties a month, its peak effort from the Chinese bases was 310 sorties in October. In the end the strategic air offensive from China looked better on a map than on the ground and was more an exercise in politics than in military force.

## Operations from India and China

The tremendous effort yielded little military gain. The first combat mission was a daylight strike against railroad yards in Bangkok on June 5, 1944, that was marred by mechanical problems. Of the unit's 112 bombers available for action, 14 failed to get airborne, 1 crashed shortly after takeoff, 13 others returned early, and 4 were lost to difficulties with the fuel transfer system. Only 4.5 tons of the 353 tons of bombs dropped on the primary target by seventy-seven B-29s fell in the target area. Meanwhile Arnold pushed for a hundred-plane mission against Japan to coincide with the invasion of Saipan. On June 13 the XXth dispatched ninety-two bombers to China, a dozen of which landed short of their destination and one was lost. On the fifteenth sixty-eight B-29s lifted off on a night attack of a steel plant in southern Japan. It cost the AAF five bombers to get forty-seven to the primary. The payoff was slim; intelligence officers could account for only 28 percent of the bombs released and found the closest struck thirty-seven hundred feet from the aiming point. So bombs fell on Japan, which was a psychological blow, but they were not of any military value.

The missions that followed were similar, great effort achieving limited results. Logistics restricted the number of operations from the Chinese bases, while the B-29's range and Chinese bases confined the strikes to targets in southern Japan. In fact the B-29s flew only nine raids against Japanese targets from China. One of these was flown the same day (August 10) as the AAF attacked a key oil facility in Sumatra. This was impressive, at least on paper, because the two targets, Palembang and Nagasaki, were three thousand miles apart. The oil attack was notable as perhaps the longest nonstop bombing mission of the war, thirty-seven hundred statute miles round trip. To achieve this distance the bomb load was reduced to one ton and the operation was flown at night, below twenty thousand feet, by individual aircraft. The mission staged out of Ceylon, and again the force made little to no military impact with the thirty-eight tons of bombs and mines dropped by thirty-nine bombers. Although some aircraft flew as long as eighteen hours, only one was lost. The mission against Nagasaki consisted of sixty-one bombers attacking in the daylight and ten (delayed by a takeoff crash that blocked the runway for hours) at night. The daylight attack is notable as it was the first major aerial clash between the B-29s and the Japanese defenders. The Japanese got fifty to sixty fighters aloft, and they made almost 150 attacks, primarily head on, that knocked down three Superforts, one by gunfire and two from a collision the Americans believed was deliberate. Flak destroyed another two Boeing bombers. The bombers' guns worked well, and the gunners claimed seventeen Japanese fighters destroyed. In the end it cost the Americans thirteen bombers, five to enemy action, to deliver 112 tons of bombs.

The majority of the unit's missions were against targets that read like a tourist's guide of East and Southeast Asia. Probably the most notable of these attacks was the operation against the Hankow dock area on December 18, the outfit's first daylight incendiary attack. Of the 101 bombers that left India for the staging bases, 84 bombed the primary target with 511 tons of incendiaries. In contrast to the other raids, which were inconclusive, this one burned out about 180 acres in fires that lasted three days. A XXth Bomber Command report noted the "tremendous potential destructive capability of B-29 aircraft against suitable incendiary targets."[10]

During the operations from India and China serviceability improved, although it remained a major problem. The airmen had to deal with primitive base conditions, parts shortages, and inexperience. The B-29's engines remained the major difficulty, responsible for 45 percent of mechanical failures for the period July through November 1944. Radar was another problem area, malfunctioning on 60 percent of the sorties in November and between July and November causing 9 percent of the airborne aborts. Strikingly, operational causes accounted for 70 percent of the bombers lost by the XXth Bomber Command.

The object was to get bombs on selected, vital targets, but those hit from China and India were not the critical, bottleneck targets envisioned by the airmen at the ACTS. In addition, bombing accuracy was poor. A study in December 1944 estimated that in visual conditions from thirty-one thousand feet, the Superforts could get 5 percent of their bombs within one thousand feet of the aiming point, increasing to 30 percent from twenty thousand feet. Radar bombing achieved a CEP of two miles, which meant that less than 1 percent of bombs fell within one thousand feet of the target. One analyst in December 1944 wrote that radar bombing "can produce but insignificant strategic damage and that radar bombing can best be described as wholly indiscriminate in respect to accuracy."[11]

The one bright note in these operations was that Japanese defenses were weaker than those the AAF faced over Europe. The Japanese had fewer aircraft, less modern radar, and only some of their fighters could operate satisfactorily above twenty thousand feet, whereas the B-29s were designed to fly and fight at thirty thousand feet. Likewise the Japanese had fewer and less capable flak than their German allies. This, combined with the fact that the B-29 was a better bomber than those employed against the Germans, meant lower combat losses. On its first twenty-five missions the unit encountered 2,200 fighters, but only 11 percent were able to attack in a coordinated manner. The Japanese downed 11 B-29s with cannon and machines guns, 2 with air-to-air bombs, and 5 in midair collisions. (The airmen concluded that at least 2 of the latter were intentional.) In exchange the gunners claimed the destruction of 130 fighters.

Arnold fired the XXth Bomber Command leader, Kenneth Wolfe, after less than a month of combat operations. The always impatient AAF commander was intolerant of anything but results, and the unique command arrangement that gave control to Washington led to micromanagement.[12] It can be argued that Wolfe was unfairly treated in view of the conditions in China and India and the problems with both the bomber and logistics. And it should not be forgotten that Arnold appointed Wolfe to this position in the first place, although his prowess was with aircraft development not with running a combat unit. Wolfe's successor, the hard driving Curtis LeMay, later stated that Wolfe faced "an utterly impossible situation" and had done a good job. Even LeMay could do little better.

LeMay arrived at his new post at the end of August 1944, when the XXth Bomber Command had flown seven missions, only one of which had achieved decent results despite high casualties, twenty-nine B-29s lost, more than 8 percent of those that had bombed the primary target. LeMay, probably the best American World War II bomber leader, could not make the XXth a success. He did instill enthusiasm, push training, and change formations, but most of all, he gained the respect and confidence of his men. One officer described LeMay as "a tough, hard-working, no-nonsense general who said relatively little, but got big results. . . . He was a hard, but fair Commander and he fought for his men."[13]

The unit's operations from China and India were costly, with twenty-nine B-29s lost to enemy action, fifty-one others on operations, and another forty-five lost to noncombat causes. And the claim that these early missions helped the development of the later B-29 operations is wishful thinking as there was no time to assimilate these lessons, for the successor unit was training and operating nearly in parallel with the XXth. For the XXIst Bomber Command was well into combat, flying its seventeenth mission (and ninth over Japan) before the older unit flew its twenty-fifth mission. In the end this Indo-Chinese effort got the B-29 into combat as quickly as possible, placated the president, who wanted to aid the Chinese, kept it under central (AAF) control, and trained LeMay in the particulars of the B-29 and the Pacific strategic air war.

When the JCS pushed up the date for the invasion of the Marianas, the AAF leaders, who recognized the limitations of the Chinese and Indian bases, revised the plan, which would send a second bomb wing to the XXth. The latter's limited results and drain on tight supplies in the theater were other factors in the January 1945 JCS order to move the unit in April to the Marianas. The 58th Bomb Wing began operations, under the XXIst Bomber Command, in early May.

## Operations from the Marianas

The path of the second B-29 unit to war was somewhat easier than its predecessor. The 73rd Bomb Wing sequenced through the same bases and encountered some of the same problems as did the 58th. However, it received more and better training, averaging more than three times the flying time in Superforts as did the first unit. The first B-29, *Joltin' Josie—The Pacific Pioneer,* landed on Saipan on October 12, piloted by the commander of XXIst Bomber Command, Gen. Haywood Hansell.

Hansell had a rich background in bombers, having taught at the Air Corps Tactical School, where he played a part in the development of the American strategic bombing theory. He had both planning and command experience, having been a key participant in writing both AWPD/1 and AWPD/42 and having commanded a bomb wing in the Eighth Air Force for half a year before going to Headquarters AAF in mid-1943.

The B-29 bases on the Marianas were built on the three major islands: Guam, Saipan, and Tinian. Base construction was hindered by the tough and sustained Japanese resistance, poor weather, rough terrain, and differing naval priorities. The resulting facilities for the aircraft were less than planned and those for the personnel were primitive at first. There were Japanese stragglers in the jungle, but thirteen Japanese air attacks proved more serious, destroying eleven B-29s, damaging forty-nine others, and inflicting more than 240 casualties.

The AAF established the B-29 target list for the bombing of Japan based on the European experience. The aircraft industry was selected as the "overriding intermediate objective," industry second, and shipping third. The planners recommended that the six largest cities not be hit extensively until an adequate force was available to destroy them in a month and that the B-29s engage in the mining campaign. Intelligence in the Pacific bombing campaign proved to be even more of a problem than the one in European due to the secretiveness of the Japanese and because, unlike the situation in Europe, the Americans lacked the useful, if not essential, aid from the British and underground. On November 1, 1944, a B-29 photo plane, aptly named the *Tokyo Rose,* found clear weather over the Japanese capital, a rare occurrence, and orbited for one hour at thirty-two thousand feet taking pictures. This was the first American aircraft to fly over Tokyo since Doolittle's April

1942 raid, and it demonstrated beginner's luck, for thirty of forty-nine reconnaissance missions during the rest of the year encountered weather problems.

The bombing of Japan differed considerably from the bombing of Germany. Distances were much greater, about two and half times that of the average flown against Germany. Most of this was over water, which made navigation more difficult, as did less weather information than was available in the European campaign. The third major obstacle was that the Superfort was not fully developed and was operating at weights 7.5 tons over its design limit. In addition, the crews were marginally trained and had minimal flying experience. High noncombat losses ensued.

The Americans scheduled the first strike on Japan from the island bases as a joint AAF-Navy operation for November 10, 1944, but Navy problems nixed this plan. To compound Hansell's problems, foul weather lashed the islands, forcing the airmen to cancel the mission on November 17 and attempts over the next five days. It was not until November 24 that the bombers of the XXIst Bomber Command headed north for the land of the rising sun.

The 73rd's commander led 110 Superforts on the first mission. The pattern the AAF followed during these early attacks was to fly at two thousand feet until about 250 miles from Japan, where they climbed to their bombing altitude of twenty-seven thousand to thirty-three thousand feet, where they usually met high winds and clouds. Thus only twenty-four B-29s unloaded their bombs on the primary target while forty-eight dropped on secondary targets, all with poor results. The Japanese defenders launched about 125 fighters to meet the Superforts and made over 180 attacks, but they downed only one bomber, an unintentional collision according to the Americans. Another bomber ran out of fuel and ditched twenty-five miles short of the runway. The bad news was the problems with winds and clouds, poor bombing results, and fuel problems; the good news was the weak Japanese defenses, which were to characterize the high-level daylight bombing of Japan.

Hansell continued to attack Japanese targets using high-altitude, precision, daylight tactics. During his tenure the XXIst flew ten such missions against Japan as well as two experimental incendiary raids. The most successful of the precision attacks was on January 19, 1945, against Japan's third largest aircraft manufacturer. The 73rd launched eighty bombers, sixty-two of which bombed the primary target from a lower than normal altitude (twenty-seven thousand feet) through unusually clear weather, permitting better than average accuracy: 21 percent of the bombs impacted within one thousand feet of the aiming point. The bombing knocked the engine factory out of production for six months and the airframe plant out for four.

Hansell had additional problems besides the weather and distance. Serviceability was poor, and airborne aborts were frequent. Engines were the major difficulty, although it appeared that about every other system added to the maintenance woes. Operations at high altitude caused engine strain, greater consumption of fuel, and freezing of equipment. Inexperienced ground crews had to contend with all kinds of shortages along with coral dust.

The XXIst's commander took a number of measures to deal with these problems. He pushed for additional training of aircrews and the formation of lead crews. In early December Hansell began to send three sorties a day on "weather strike missions" to harass the Japanese, gather weather information, and conduct photo and radar reconnaissance. Another program reduced the bomber's weight by decreasing fuel loading by 3,600 pounds, enabling one bomb bay tank to be removed which netted a further weight savings of 1,500 pounds, and deleting 1,900 pounds of armor. In January the unit began removing the 20-mm tail gun, which had proved both unnecessary and ineffective.

These efforts could neither conceal the 73rd's modest results nor immediately improve them. Arnold decided in December to replace Hansell with LeMay. Writers have favored Hansell on his firing, for he was both a sympathetic character and later very accessible to historians. A case can certainly be made on his behalf. For like Wolfe he probably should not have gotten the job in the first place as his forte was planning, not command. It also should be emphasized that the XXIst was improving despite unprecedented problems. Arnold was unsympathetic and well known for his intolerance of slow or limited results. On the other hand Hansell was inflexible in adhering to the prewar bombing doctrine, LeMay outranked Hansell, the Indo-Chinese operation was about to shut down, and LeMay had the well-deserved reputation for getting things done and was probably the best combat bomber commander in the AAF. LeMay was not pleased with the XXIst, writing after a few weeks that the staff he found was a worthless lot of misfits. He thought little better of the rest of the outfit, an initial opinion he also had of the XXth Bomber Command.

For all of his unhappiness, at first he made few changes. Through the middle of March the unit flew eight missions against Japan, six precision and two that tested incendiary tactics. These were flown slightly below the altitudes of the missions under Hansell and were even less successful; on three of these not one bomb fell on the primary target. There were signs of improvement, however, as mechanical problems decreased and bomb loads increased.

Two of these early missions under LeMay are noteworthy. The Americans considered the air battle of January 27 the most savage of the strategic campaign against Japan. The Japanese hit the B-29s with 984 attacks and downed nine bombers, a loss rate of 14 percent of the effective sorties, while the bomber gunners claimed sixty fighters destroyed. Two weeks later eighty-four Superforts bombed the Nakajima plants at Ota in clear weather and achieved good results that the intelligence officers estimated destroyed three months of single-engine fighter production and one month of twin-engine fighter production. A later study called this mission "undoubtedly the most effective single attack by the XXI Bomber Command during the first four months of operations."[14]

LeMay employed and continued some of the same techniques begun under Hansell to improve the performance of the XXIst, such as more training, organizing lead crews,

The Boeing B-29 was the first operational bomber to mount remote-controlled gun turrets and have a pressurized cabin. However, it had a higher accident rate than the B-17 and B-24 and suffered more losses to mechanical problems than to Japanese defenses. Here seen raining incendiary bombs on Japan, the Superfort went on to serve in the Korean War and Cold War. (USAF)

and introducing production line maintenance (centralized at the group level). These policies along with growing ground crew experience and improving weather led to increased flying time on each bomber and a greater percentage of aircraft attacking the primary target. Another factor that helped improve the XXIst was the stream of new aircraft and crews that joined the battle.

The February 1945 seizure of Iwo Jima also improved overall performance. On February 19, 1945, the Marines invaded Iwo Jima, a small (two by two miles) volcanic island about half way between the Marianas and Japan. This action not only denied the Japanese a base for radar, fighters, and intruding bombers but, more important, gave the AAF a valuable base. The airmen were thinking mainly in terms of their European experience, advancing escort fighters within range of Japan, and they did base four fighter groups there. But the relatively weak fighter opposition and the Japanese decision to conserve their air force to combat the expected American invasion rendered this unnecessary. Therefore the fighters were released after a week and a half of escort missions to free-ranging strike sweeps, which became their major employment.[15] Although the AAF also planned to stage bombers out of the island, it flew only seven mining missions and one bombing mission in this manner. From the airman's perspective the major advantage of taking Iwo Jima was to provide succor for damaged and fuel short B-29s. The first landed there on March 4 and was followed by twenty-four hundred others. Four-fifths landed for fuel, and although not all would have been lost otherwise, the existence of an emergency field certainly bolstered aircrew morale.

LeMay completely changed the Pacific strategic air campaign in March 1945 with a radical deviation from the AAF doctrine of high-level, formation, precision, daylight bombing. As already described, there had been calls for fire bombing Japanese cities prior to the war due to their vulnerability. Operations exposed even more compelling reasons to change tactics, such as Japan's dispersed cottage industries, limited intelligence, the weather, and, most of all, the lack of results with the existing techniques. The AAF had directed test incendiary raids that were conducted under both Hansell and LeMay. These six missions flown at high altitude, both day and night, yielded inconclusive results, albeit the last on February 25 destroyed or damaged one square mile of Tokyo. LeMay deserves considerable credit for implementing the new tactics, clearly one of the boldest command decisions of the war, yet it should be emphasized that these ideas emanated from a number of individuals. In brief the radical concept consisted of flying at low altitude (below ten thousand feet), at night, in solo attacks, initially without armament, dropping only incendiaries, and aiming to burn out Japanese cities.

The first of 325 bombers took off in the late afternoon of March 9, 1945, for Tokyo. The weather was clear as the bombs began to fall early the next morning, with a record 86 percent of those launched dropping seventeen hundred tons of bombs on the city. The Twentieth Air Force lost fourteen bombers, but the Japanese lost far more. Huge fires swept Tokyo, burning out sixteen square miles, destroying a quarter of city's buildings, including twenty-two AAF targets, and rendering one million homeless. Estimates of the number of dead vary, with the most credible range between eighty thousand and one hundred thousand. The exact number will never be known, but this attack was certainly one of the deadliest air attacks of all time.

The raid on Tokyo was not a fluke; it could and would be repeated. For the XXIst Bomber Command proceeded to burn down all but one of Japan's major cities, and many more as well. In the next ten days the B-29s hit Nagoya (Japan's third largest city) twice, Osaka (second largest), and Kobe (sixth largest), in all destroying thirty-two square miles of urban area. In this short period the B-29s destroyed 41 percent of the urban area destroyed during the entire air war against Germany. The bombing stopped when the XXIst Bomber Command ran out of incendiaries and because the air and ground crews were exhausted. American casualties were low, twenty-two bombers lost, only three to enemy action, for a loss rate of 1.5 percent. The B-29 effort was now hitting its stride as more aircraft and crews reinforced the effort and more experienced ground crews and lower altitudes improved serviceability, allowing the unit to get more than three hundred bombers over Japan on each mission. Lower altitude operations also permitted the bomb loads to more than double the average of three tons in November and December.

The XXIst stood down for six days and then flew three precision raids: two low-level night attacks on engine plants that were unsuccessful and a day medium assault that achieved excellent results with 22 percent of the bombs hitting within one thousand feet of the aiming point. Then the B-29s switched to other targets the airmen considered diversions. The first was to support the invasion of Okinawa with a series of attacks of airfields in southern Japan that comprised three-quarters of the unit's efforts between April 17 and May 11 and cost twenty-two bombers.

The other change was to engage in mining operations. Little known to most, it proved to be very effective, quite cheap, and one of the B-29's most significant contributions to victory. American submarines had savaged the limited Japanese merchant fleet on which the Japanese depended for imports of raw materials and food. Mines were another major weapon in this campaign. The Allies had planted more than twelve thousand mines in areas distant from Japan, mostly using aircraft. The XXth Bomber Command had a minor role in these operations, dispatching 176 bombers that seeded almost one thousand mines on their primary targets without a loss. The situation around the home islands was even more critical for the Japanese because of their limited rail system, and here the B-29s dominated the mining operations. The Navy requested help from the AAF, requests the airmen initially resisted as a diversion from strategic bombing. But the AAF found itself in a sticky situation, for if it did not cooperate with the Navy, it might lose the mission and aircraft. Although the XXIst was neither trained nor equipped for such operations, it quickly adapted, beginning night, low-level, radar-aimed, mine-seeding missions at the end of March. The Shimonoseki Strait was the primary target because it was a bottleneck with twice the density of traffic of any other point in Japan. The airmen achieved great success measured in the tons of shipping sunk and in delays of traffic.

Meanwhile, during April, when it was not bombing airfields, the Twentieth Air Force adopted a policy of launching daylight precision attacks from medium or low altitudes when the weather was good, and when not, radar-guided attacks on urban areas. On April 7, 1945, the AAF encountered brisk Japanese fighter resistance when two daylight thrusts against aircraft factories in Tokyo and Nagoya provoked 764 fighter attacks that claimed 6 B-29s destroyed and 151 damaged. (The bomber gunners claimed the destruction of 101 Japanese fighters.) This marked the first Superfort mission escorted by fighters, which posted twenty-one victories at the cost of two friendly fighters.

By mid-April the XXIst Bomber Command had built up its incendiary stocks and thus was able to continue fire bombing Japanese cities. Three strikes in April burned out twenty-two more square miles of Japanese cities and in May the B-29s fire bombed cities five times. On the twenty-fifth the Superforts hit Tokyo for the last time, 464 bombers raining 3,260 tons on Japan's capital. This destroyed nineteen square miles, the most for one mission, with the loss of twenty-six bombers, also a record. The Americans continued to fire bomb the major cities until mid-June, by which time they had destroyed 102 square miles of Japanese urban areas: half of Tokyo, one-third of Nagoya, over half of Kobe, and one-quarter of Osaka. The only major Japanese city not targeted was Kyoto, spared for cultural and religious reasons. On June 17 the Twentieth Air Force changed its tactics.

As the campaign progressed, the B-29 force grew in size and efficiency, Japanese defense buckled, and bomber losses declined. Now the XXIst had an abundance of power and a growing shortage of targets.

The unit continued to hit the few remaining precision targets at the same time it switched its focus to smaller cities ranging in population from 80,000 to 300,000. Each of the Twentieth's four bomb wings would attack a city at night from below ten thousand feet. Destruction was large, and resistance at these smaller cities even weaker than at the major ones. For example, on three night missions in late June the bombers destroyed fifteen square miles of builtup area at the cost of four bombers, none believed due to enemy action. As the bombers' sorties and tonnage increased, the loss rate fell from 1.9 percent in May to 0.8 percent in June, 0.4 in July, and 0.3 in August. The overall B-29 loss rate for the war was 1 percent of aircraft airborne.

The Twentieth Air Force introduced a new technique when the fifth and last B-29 Bomb Wing (315th) went into action on June 26. It flew the B-29B, a Superfort that differed from her siblings in two ways. It was fitted with a different radar designed for precision radar bombing. The Eagle radar employed a fixed antenna, which gave it better resolution than the standard rotating antenna, mounted under the belly parallel to the wing and appearing as short second wing. The device had two drawbacks, however. First, it only scanned an area 60 degrees to the front, limiting its navigational use, and second, the antenna added drag, reducing speed by five miles per hour and range by one hundred miles. As the bomber would be used in night bombing the AAF removed the turrets, blisters, guns, and ammunition which reduced weight almost four tons that offset the additional drag of the antenna. Boeing added a tail turret with three radar-directed .50s, which in tests proved 2.5 times as accurate as manual sighting and achieved 3 to 4 times as many hits.

The unit went into action on June 26, 1945. In all the 315th flew fifteen night attacks against oil targets, picked not for their importance but because they had not been attacked and thus would allow the effectiveness of the B-29B to be better assessed and because they were sited on the coastline where the land-water contrast made them an outstanding radar target. The unit achieved excellent results with accuracy at least as good as that of daylight, visual bombing and on a number of occasions exceeding it. (In contrast the radar tail gun turned out to be a failure in combat, searching without locking on, locking on but not searching, and, most of all, failing to distinguish between hostile fighters and friendly bombers.) The effective destruction of Japanese oil refineries was a wonderful test of the system but did not help win the war as Japanese oil production already was in decline due to strangulation of imports.[16]

Another new tactic the XXIst employed beginning in late July was to issue warnings to Japanese civilians before the B-29 attacks. On the night of July 27 the Americans dropped leaflets that stated that four or more of the eleven listed cities would be destroyed in the next few days and cautioned that although the United States only wanted to hit military targets, "bombs have no eyes." The leaflets and later shortwave radio broadcasts urged the inhabitants of the cities to evacuate. The B-29s blasted six of the named cities on the night of July 28 without a loss. The only response the Japanese had to these continued warnings and attacks were words.

(The Twentieth lost only four bombers to enemy action after July 28.)

The remainder of the conventional bombing offensive against Japan can be briefly summarized. Incendiary bombing predominated as sorties and tonnage rose and losses fell. On August 1 the XXIst launched its largest attack of the war on what was described as "Air Force Day." The airmen got 853 bombers airborne, of which 784 hit their primary targets with 6,521 tons (all incendiary except for 242 tons of mines). One bomber was lost to noncombat causes.

## Atomic Bombs

The atomic bombing of Japan remains a controversial subject more than sixty years after the event. I am convinced that the decision to use the nuclear weapons was motivated primarily by the desire to win the war as quickly as possible, with the fewest American casualties, and especially with the hopes of avoiding what was believed would be a very costly invasion of the Japanese home islands. The problem was not to defeat the Japanese—they were clearly beaten—the problem was to get the Japanese to surrender.

The United States conducted a crash program to develop nuclear weapons, fearing that the Germans would get them first. They didn't, the Allies did. A number of factors led to the decision to drop atomic bombs on Japanese cities. America was war weary and some speculated that the war could continue for a number of years with horrendous casualties. After the great effort to develop the weapon, there was no good reason, and no advocate, for not using the bomb. The decision makers sought a way to demonstrate the bomb other than destroying a city but could not. Fanatical Japanese resistance on Iwo Jima and Okinawa, their atrocities, kamikazes, attack on Pearl Harbor, and unwillingness to surrender, as well as American racism, were other factors. The Japanese were given a chance by the Potsdam Declaration, which demanded surrender and which, if refused, promised "prompt and utter destruction." A few days later the Japanese rejected the ultimatum.

The AAF began to modify B-29s for nuclear delivery in late 1943 and the extensive Project Silverplate modifications reduced the aircraft's weight by almost four tons.[17] At the same time the nuclear bombs were designed to be carried by a Superfort, which limited their weight and size.[18] Meanwhile the AAF organized the 509th Composite Group under the command of Lt. Col. Paul Tibbets, who had flown on the initial Eighth Air Force mission in Europe, had considerable combat experience, and been involved in the B-29's development. After training in Utah the unit deployed to Tinian with fifteen Silverplate bombers.

The 509th received the core for the uranium weapon on July 26 and was ordered to deliver the first bomb on the first visual bombing day after August 3 on one of four targets: Hiroshima, Kokura, Niigata, and Nagasaki. Additional bombs would be dropped on these targets as soon as they were ready. The Americans launched six B-29s on August 6, the *Enola Gay*, piloted by Tibbets and named after his mother, carrying the bomb. The mission went off almost flawlessly with the weapon exploding

**(Right) Col. Paul Tibbets commanded the unit that in August 1945 employed two atomic bombs against Japan. He piloted the B-29 *Enola Gay*, named after his mother, which dropped the first bomb on Hiroshima. (Historical Research Agency, USAF)**

82
ENOLA
GAY

over Hiroshima with a force equivalent to 12,500 tons of TNT.

The city was caught completely by surprise for although an air raid warning had been sounded, the alert was cancelled twenty minutes later. Because of the crowded urban conditions, dry weather, Japanese construction, and lack of prior bomb damage the city was fire prone. The blast destroyed 4.7 square miles of the city; this is relatively easy to measure, while the death toll is more difficult to determine. Initial American estimates were between 64,000 and 86,000 killed, a figure that later (1947) was raised to 100,000; Japanese studies put the number at 130,000. Whatever the precise figure, the casualties were very high and a considerable proportion of the quarter of a million inhabitants.

The airmen scheduled a second nuclear attack for August 11, but a forecast of poor weather forced the airmen to advance the timetable. On August 9 Maj. Charles Sweeney flew *Bockscar* toward its primary target, Kokura. In contrast to the first mission, he ran into various problems. Sweeney made three passes on cloud-covered Kokura before diverting to his secondary target because his orders were to bomb only under visual conditions. Running short of fuel, and harassed by the weather, he began his bomb run using radar and only at the last moment did the bombardier pick up the target visually and release the bomb. Because of the geography and the miss distance, the destruction and casualties at Nagasaki were less than the first attack: 1.5 square miles destroyed and 35,000 to 70,000 killed of the 250,000 residents. Critically short of fuel, Sweeney made one circuit of the stricken city before flying directly to Okinawa. After a "hot" (high-speed) approach and landing, the bomber stopped only ten feet from the end of the runway. Sweeney then taxied off the runway, at which point two engines died from fuel starvation.

Despite the two atomic bombs, the hopeless situation, and the entry of the Soviets into the war on August 8, the Japanese did not immediately surrender. The Japanese decision makers were split over the surrender terms, and only the intervention of the emperor on August 10 resolved the issue. A coup attempt by hard-line officers followed, extinguished with few casualties. Meanwhile the bombing continued. The last B-29s lost in action went down on August 8, and combat missions ended on the fourteenth.

## The B-32

The Boeing B-29 was not the only very heavy, long-range bomber engaged in the Pacific air war. The Consolidated B-32 was an upgraded and enlarged version of the company's successful B-24, which was developed simultaneously with the B-29 and was to serve as its backup. It was larger and heaver than the Liberator and differed from it in that it had a cylindrical fuselage, rounded nose, and, more important, initially had both a pressurized cabin and remote, retractable gun turrets. It was powered by the same engines as the B-29, first flew in September 1942 (two weeks before the B-29), and exhibited a number of problems. Delays were chronic and not helped by the crash of one of the two experimental aircraft in May 1943.

The Consolidated B-32 was an improved B-24 built as a backup to the B-29. The deletion of its remote-controlled retractable turrets and pressurized cabin rendered it inferior to the B-29. Various problems derailed ambitious AAF plans for the bomber and only fifteen saw combat in the closing weeks of the war. (National Museum of the USAF)

Convair made a number of changes so that the production model was somewhat different from the prototype. Problems with both pressurization and the fire-control system led to their deletion, which cancelled two of the B-32's major advantages over the Liberator and meant that the B-32 was less advanced than the B-29. The revised aircraft armament consisted of five manually operated power turrets (nose, two upper, belly, and tail) each mounting twin .50s. In addition, directional instability pushed the designers to a single tail empennage, requiring a major redesign. One significant innovation was to fit reversible propellers on the inboard engines, which reduced landing roll, a first and its one technical advantage over the B-29. Even the bomber's name caused problems. Originally named "Terminator" by the company in January 1944, the AAF changed it to "Dominator" that August. But the latter name riled the State Department, so it was changed back to the original in 1945.

The AAF planned to deploy the bomber to the Mediterranean theater and, after service tests were completed, transition units in both the Eighth and Fifteenth Air Forces into the new bomber. But delays dogged the aircraft so that by the end of 1944 the AAF had accepted only thirteen B-32s. Mechanical problems, delays, and complaints of faulty workmanship almost led to its cancellation in late 1944. In 1942 the National Advisory Committee for Aeronautics (NACA) had predicted the bomber would fail, and many in the AAF agreed. As late as February 1945 the acting chief of the Air Staff, Brig. Gen. Patrick Timberlake, opined that the "B-32 in its present form is not an acceptable bomber."[19]

Meanwhile Gen. George Kenney of the Far East Air Forces came to the rescue. He had failed to get B-29s for his command in mid-1943 and in March 1945 volunteered to combat test the B-32. No one else seemed to want Consolidated's bomber and Arnold asked his staff, "Why are we building the B-32?"[20] A few days later Arnold authorized combat service tests for the bomber, and by late May three B-32s were in place in the Philippines, where they flew twenty sorties on eleven missions, delivering 134 tons of bombs mostly against targets on Formosa. The detachment commander noted that the B-32 carried twice the bomb load of the B-24, fifty miles per hour faster, to a much greater range, and with about the same ground and aircrew. The bomber proved stable and rugged; however, sources differ on the issue of the aircraft's maintenance. Maj. Gen. Ennis Whitehead, commander of Fifth Air Force, thought the bomber could do the job and recommended that his unit get a group of B-32s. However, the official AAF history states that the test crews were "pessimistic regarding technical defects of the B-32's."[21] Although the war was almost over, the test unit, now numbering nine bombers, moved to Okinawa and conducted a couple of missions over the South China Sea. More dramatically the unit flew four reconnaissance sorties over Japan on August 17 and 18. On both days they were intercepted by Japanese fighters, the bomber gunners claiming one destroyed on the seventeenth and two on the eighteenth, the last American victory claims of the war. The B-32s suffered minor damage from AAA and fighters and on the last mission had one man killed and two wounded. Only fifteen B-32s reached the theater, and although no B-32s were lost in combat, one was lost in testing, two in training, and two in accidents after the war's end.

The AAF cancelled the contract for 1,588 B-32s and scrapped all of the existing B-32s. Hindered by delays and technical problems, it was not pushed as was the B-29 and was barely able to get into the war, where it performed in a mediocre fashion. While it was an improvement over the B-24, it did not approach that aircraft's accomplishments or those of the B-29. As a result it has been relegated to the dustbin of history, a forgotten also-ran. As one historian has concluded, "To the end, the B-32 was unwanted and unloved."[22]

## Japanese Attacks on America

Japan initiated the war against the United States with attacks on Pearl Harbor and the Philippines, which were meant to cover its

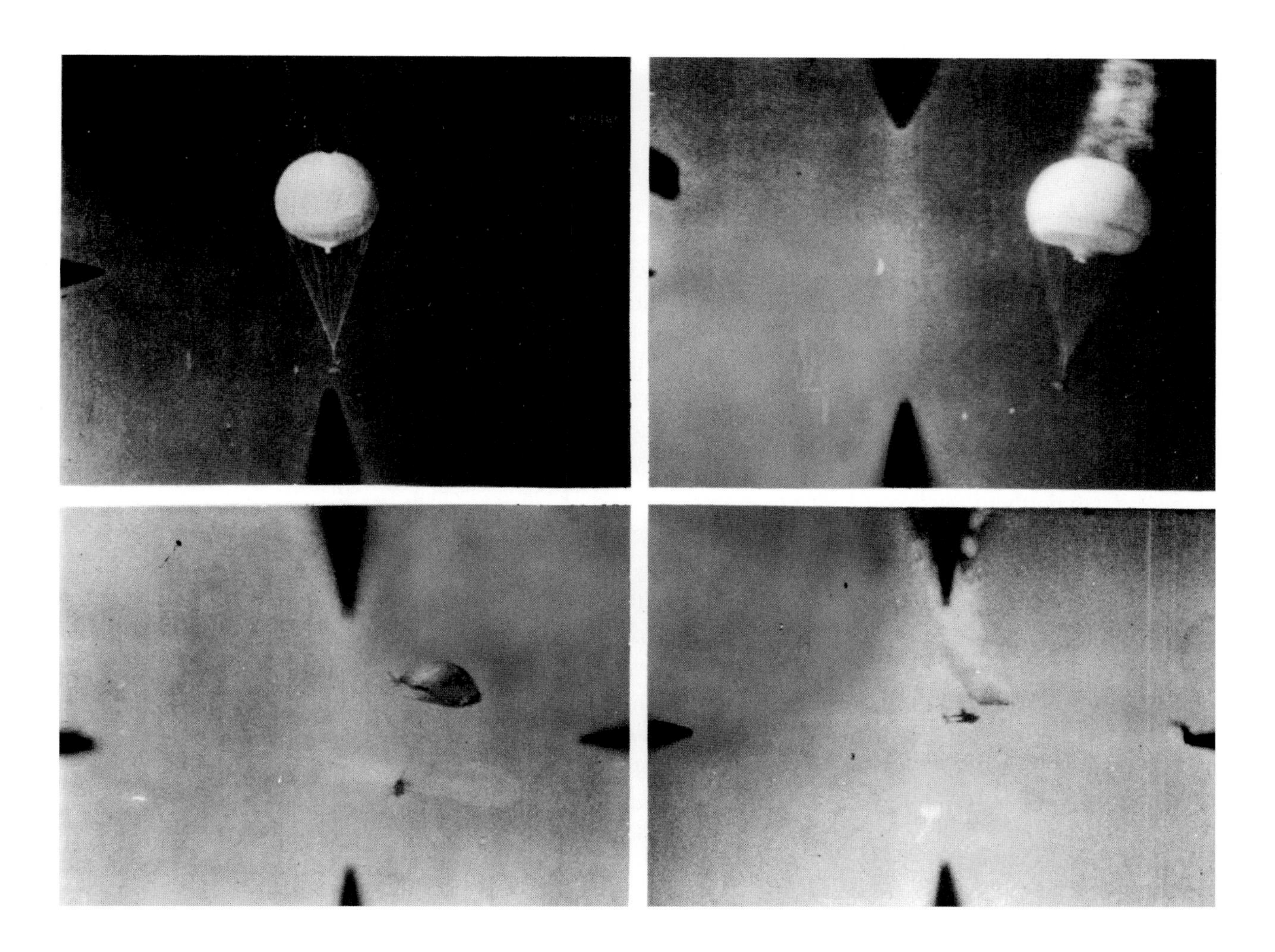

While B-29s burned out most of urban Japan, the Japanese could only respond with a few ineffective pin pricks. In their largest effort the Japanese launched over nine thousand armed balloons; however, the one thousand that reached North America caused little damage and killed only six. Gun camera film *(lower right)* shows the destruction of a balloon by a P-38. (National Museum of the USAF)

main thrust to seize the resource rich areas of Southeast Asia. The Japanese counted on a fierce defense, broad ocean areas, and weak-willed Americans to lead to a successful conclusion of the war, and it had neither the plans, technology, nor resources to attack the U.S. mainland. In contrast to their spectacular efforts and victories in the western Pacific, Japanese actions against North America were pitiful. The Japanese did occupy a few islands in the Aleutians for a number of months, shelled U.S. West Coast targets (without physical effect) from submarines on three occasions, launched two flying boat sorties that ineffectively bombed Hawaii in March 1942, and twice in September 1942 launched a floatplane (from a submarine) that unsuccessfully attempted to start forest fires in Oregon. However, their largest long-range effort underscores the differences in capabilities and effect between the two combatants: B-29s versus paper balloons.

As early as 1933 the Japanese considered using the prevailing winds to send bomb-armed balloons against the United States. They developed balloons to be launched from submarines and tested two hundred of these during 1943–44. But it was a larger, hydrogen-filled balloon that would use the jet stream at thirty thousand to thirty-eight thousand feet to cross the Pacific Ocean with twenty-five to seventy pounds of antipersonnel and incendiary bombs that saw service. The Japanese planned to launch fifteen thousand of these devices and actually launched ninety-three hundred beginning in November 1944, just as the Superforts were beginning their bombing campaign from the Marianas, a mismatch similar to that of the Pacific war. About one thousand reached North America, landing from Alaska to northern Mexico and as far east as Michigan. Censorship prevented the Japanese from learning of the fate of their weapons. They inflicted minor damage (two small brush fires and a temporary electric power loss at the Hanford, Washington, atomic plant, that delayed the atomic bomb) and few casualties (one woman and five children killed in Oregon). The contrast with the AAF bombing campaign is striking.

## Summary

The American airmen got the best bomber of the war into action quickly as they overcame developmental and operational difficulties. Paradoxically the aircraft's two major innovations, cabin pressurization and remote-controlled armament, were less important in the campaign against Japan than the bomber's range and, most of all, its versatility. It proved most effective in low-level night incendiary bombing and mine laying, tasks not considered during its design.

The B-29s dropped 156,000 tons of bombs and mines against the Japanese. In contrast to the prewar theory and operations in Europe, two-thirds of the bomb tonnage consisted of incendiaries and almost three-quarters of these missions were flown at night. The destruction was immense: 178 square miles of urban area, twice the 79 square miles of urban area destroyed in Germany, and 56 square miles of Tokyo, compared to 10 square miles of Berlin. The bombing killed more than a quarter million Japanese, probably closer to one-third of a million. The chief precision target of the B-29s was the aircraft industry on which it dropped 9 percent of its total tonnage. Damage to Japanese industry was high, but production was also hurt by the blockade and the belated effort to disperse the factories. By July 1945 Japanese production was one-third to half, oil refining was less than 15 percent, electrical power and coal consumption was half, aircraft airframe production was 40 percent, and aircraft engines were 25 percent of peak output.

The Superfort's mining effort consisted of 6 percent of the unit's tonnage, over twelve thousand mines weighing just under nine thousand tons. Fifteen bombers went down in this effort. The aerial mining accounted for 63 percent of the Japanese merchant ships destroyed during its duration from March to the end of the war. About half the mines were seeded in the Shimonoseki Straits, reducing its traffic by June to 10 percent of the premining total and to less than 2 percent by August. In combination with the submarines, which sank 60 percent of Japanese shipping, the mining effectively isolated the Japanese from imports of food and raw materials.

The American effort effectively strangled Japan's industry and people. Japan's food supply shrank because of the reduction of domestic food production (in 1944 it was one-quarter below prewar levels), the destruction of one-quarter of Japan's emergency rice supplies by the fire raids, and the drop in fish catches by more than half. The result was that the average daily per capita caloric consumption dropped from two thousand in 1941 to seventeen hundred in 1945. Had the war continued, the Japanese would have faced famine.

The B-29s were also a major factor in undermining Japanese morale as the Japanese government could not conceal their helplessness against the bombing attacks and the great damage inflicted on the country. While low morale could not overthrow an authoritarian government, it did lead to high absenteeism at the factories, rising to 12 percent in the undamaged aircraft plants and 40 percent in those damaged. And the bombing did have an impact on the thinking of the top leaders. Therefore the most important contribution that the bombing campaign made to victory over Japan were connected with mining and morale.

Any critique of the bombing of Japan has to begin with the realization that the bombing was intended to support an invasion of Japan, which was to be the war winning blow. The authoritative United States Strategic Bombing Survey states that the Japanese economy was destroyed twice: first by the blockade and then by the bombing. It held that the blockade should have been given greater attention and that shipping and railroads should have been the focus. There are some who insist that the urban bombing was wrong and that precision bombing should have been pursued. The most vocal advocate of this view, during and after the war, was Haywood Hansell. He believed precision bombing was feasible and would have been less costly in civilian lives, although it would have taken months longer.

The Pacific war cost the Twentieth Air Force 414 B-29s, 148 attributed to enemy causes, 151 to operational causes, and 115 to unknown causes.[23] The unit listed 1,090 men as dead, 1,732 as missing in action, and 362 as surviving enemy captivity, internment, or missing status. Compared with the bombing of Germany, the bombing of Japan was much more efficient in terms of duration, destruction, tons dropped, sorties flown, and bombers lost.

There are those who believe the bombing could have defeated Japan without the planned invasion. These included not only top AAF leaders (Arnold and LeMay) but also the USSBS, which stated that "even without the atomic bombing attacks, air supremacy over Japan could have exerted sufficient pressure to bring about unconditional surrender and obviate the need for invasion."[24] The official AAF history quoted two Japanese decision makers who stated that the B-29 was the basis of the Japanese surrender. A more balanced view is that the bombing devastated Japan but did not win the war. Instead its greatest impact was to reinforce the blockade and depress Japanese morale. This is ironic in view of the prewar theories and wartime operations.

## Strategic Bombing and Morality

Having discussed the strategic bombing of World War II, the mass deaths of civilians and the attacks on Hamburg, Dresden, Tokyo, Hiroshima, and Nagasaki, the issue of morality must be addressed. Strategic bombing kills civilians because noncombatants are frequently intermixed with combatants, aircrews make errors, munitions and bombing systems malfunction, and key strategic targets are usually in or near cities and populations in the enemy's homeland. In addition sometimes civilians are directly targeted to attack their morale. Does intent alter the equation? That is, does the goal of the bombing matter? Is the killing of civilians while aiming at a military or economic targets justified? Is it acceptable to aim at the population with the intent of terrorizing civilians to break or lower their morale? In any event, are all civilians innocent? Certainly babies are. But are war workers, be they men, women, or even children? Are those who support the war with their labor, taxes, voices, and votes innocent? What of the citizens of the conquering nation who benefit from the wars, or those who acquiesce in their government's actions? Are these people anymore innocent than draftees in the military?

Moralists critical of strategic bombing should consider the results of the World War I Allied blockade that starved to death 800,000, certainly mostly women, children, and old men. (Strategic bombing in World War II probably killed at least 600,000 in a war in which 50 million or so were killed.) While this does not excuse the situation, it provides some context. That is, the great evil is not strategic bombing but war itself. Leaders should consider the proportionality of their acts, that the hoped for destruction of a target is both probable and worth the death of, or risk to, nearby civilians. Numbers do count, but what is the proportion? Is the death of some innocents worth the saving of the lives of some combatants or other innocents either directly or by shortening the conflict? These have always been difficult, and perhaps impossible, questions to answer.

Chapter 7

# The Postwar Era

## THE END OF PROPELLER-POWERED BOMBERS

Strategic air warfare was significantly changed by technologies that emerged from World War II. First, the employment of the atomic bomb gave one aircraft the destruction capability of hundreds of conventionally armed bombers and appeared to make accuracy less important. Electronics was a second significant technology that affected strategic bombing. On the offensive side radar aided bomber navigation and permitted nonvisual bombing. Overall, however, electronics benefited the defense even more, with radar stripping the cloak of invisibility and surprise from bombers and aiding interception, proximity fuses increasing the efficiency of projectiles, and guidance for both surface-to-air and air-to-air missiles improving the probability of kill. A third technology was jet and rocket propulsion. German jet-powered fighters saw little combat and had no military impact in the war; however, in short order jet power markedly changed aviation. Other developments seen in World War II went into service later, including development of the V-1 and V-2 cruise and ballistic missiles. The Germans also tested, but did not field, surface-to-air missiles, which later would make high-altitude bomber operations impractical.

There were a number of other key innovations that appeared in the immediate postwar period. One of the most significant, largely unappreciated by laymen, was aerial refueling, which emerged into operational use in the late 1940s and early in the 1950s and gave essentially unlimited range to the bombers. Another development was the miniaturization of nuclear warheads, allowing the use of smaller aircraft than had previously been the norm, thus blurring the lines between "strategic" and "tactical" aircraft. The period from 1945 to 1965 saw rapid change and was a most eventful period for our story.

## Context of the Times

Between World War II and the Korean War the U.S. military was shaped by three major factors: a military policy of economy, the independence of the air arm, and the Cold War. After the long years of economic depression and war, Americans wanted to enjoy peace and prosperity. Therefore, despite disturbing events overseas, until 1950 the United States maintained a policy of economy and greatly reduced its military. The impact on the air arm was dramatic. The AAF had grown to 243 groups in 1944, and while wartime planning contemplated a peacetime strength of 105 groups, by August 1945 the planners were forced to downscale their plans to 70 groups, 25 to 26 of which would fly very heavy bombers. Lack of funding made it impossible to field this number until the Korean War changed everything, so that by 1952 the Air Force reached 70 groups.

The second major event was the creation of a Department of Defense and the establishment of an independent U.S. Air Force. The Army mainly supported this development, but the Navy did not. And now there was a three-way fight over budgets, intensified by the government's policy of economy, that would fan the flames of interservice rivalry.

The primary task of these forces was to deal with the Cold War. This face-off between an American- and a Soviet-led coalition dominated the foreign policies of the world, spawned an arms race, amplified a number of conflicts, and provided the military with its justification for forty years. World War II reshuffled a multipolar world into a bipolar world and encouraged the struggle for supremacy between two powers with contrasting ideologies, systems, and aims. (This is not to imply that there was equal responsibility for the dangerous international situation, far from it.)

## The American Air Force, 1945–1950

The American atomic advantage was far less than believed at the time. One factor that undercut the U.S. edge was the condition of the delivery systems. When the war ended in 1945 the United States had the largest and most powerful air force in the world, but the stampede to demobilize gutted the AAF. Numbers dramatically plummeted, skills declined, and morale dropped, all undermining the effectiveness of the air arm.[1] The AAF established the Strategic Air Command (SAC) in March 1946 as its strategic bombardment force, and by year's end it had 148 B-29s (30 configured for nuclear delivery), a far cry from the thousands of bombers it had deployed in the war. Even these numbers conceal the reality of the situation as only three of SAC's five groups in the United States were considered more than 60 percent effective and a year later, SAC rated two of its eleven groups as effective. Only one unit, the 509th Bomb Group, was capable of delivering the atomic bomb. The United States' nuclear advantage was a "hollow threat."[2] The situation gradually improved as the number of delivery vehicles increased, by 1948 to over 100 nuclear-capable aircraft, by 1950 to 225 bombers, and by 1951 to 364 aircraft.

The air arm also lacked atomic bombs. The United States had 9 bombs in 1946, 50 by 1948, and 450 by 1950. The bombs, rather their components, were under the control of a civilian agency (Atomic Energy Commission) stored apart from the bombers, and required time-consuming assembly. By mid-1948 two could be assembled in a day. These bombs weighed about five tons each and were difficult to load into a B-29. Consequently it would take five to six days for the AAF to launch nuclear strikes.

SAC was in poor shape not only because of restricted finances but also due to poor leadership. Its initial commander, George Kenney, had little strategic bombing experience, was engaged in the proposed UN air force, and spent considerable time in speaking engagements. He left his deputy, Clements McMullen, who also lacked strategic bombing experience, in charge. McMullen instituted a disastrous personnel policy that undercut morale and efficiency and neglected realistic combat training. In a May 1947 practice mission against New York City the airmen planned to launch essentially all its available bombers, and got one hundred to the target, with another thirty unable to get off the ground, a clear reflection of the unit's capabilities. As later events demonstrated, this exercise overestimated the command's readiness.

In early 1948 the Air Force asked Charles Lindbergh to survey SAC's combat capability. The famed aviator discussed his findings with Chief of Staff Hoyt Vandenburg and filed his report in September. It was damning. Lindbergh criticized pilot proficiency, crew teamwork, high accident rates, inexperienced personnel, the overwork, and maintenance. He noted the need for higher standards in the atomic age than those that had proven satisfactory with the mass air forces of World War II and specifically recommended changing SAC's training program, advocating more realistic training missions. Lindbergh did not spare either Kenney or McMullen. The situation demanded change, and it came in short order.

In October Kenney was posted to head Air University. Clearly this transfer from the nation's primary military unit, certainly the USAF's, to an educational assignment was neither a promotion nor a lateral move. Curtis LeMay became SAC's new commander, and has come to symbolize many things, and is probably the personification of strategic bombing.[3] LeMay was a pilot and crack navigator before the war, commanded a bomb group that pioneered new formations and bombing techniques over Germany, was promoted to head a bomb wing he led on the Regensburg portion of the famous August 1943 Schweinfurt-Regensburg mission, and went on to command B-29s, first operating out of India and China and then the Marianas, where he innovated the very destructive and effective fire raids on Japan. No AAF officer had more experience or success in strategic bombing than LeMay. His task with SAC was as difficult as it was important. LeMay emphasized training, standardization, flying safety, and his personnel, making the unit an elite force.[4] He commanded the unit for an unprecedented eight and half years and is correctly credited with making it a formidable fighting machine and dominate force in the USAF, U.S. military, and world military. SAC was the core of American deterrence for decades, operating as close to a wartime basis as possible.

LeMay pushed realistic exercises to gain a clear view of SAC's true capabilities and to impress upon the crews the need for improvement and hard work. In January 1949 SAC ran a practice mission against Dayton, Ohio, using pictures from 1938 (about the currency of most information the USAF had on Soviet targets), "bombing" from thirty thousand feet (as opposed to the customary fifteen thousand feet), and not using radar reflectors on the target, which made target identification much easier and had been the procedure in the past. Thunderstorms in the area added an unplanned and realistic touch. Engines and radars failed, pressurization and oxygen did not work, and consequently results were poor. The average accuracy (CEP) of the bombers that made it to the target was 10,100 feet, far from precision bombing and considerably worse than the 1,500 feet the staff briefed. LeMay later wrote, "You might call that just about the darkest night in American military aviation history. Not one airplane finished that mission as briefed. *Not one*."[5] A lot had to be done. Under LeMay's firm hand the command quickly improved its performance. During exercises against Oklahoma City (May 1950) and Omaha (October 1950), the accuracy was twice that recorded against Dayton.

SAC became known for doing things by the book, the SAC book. One measure of success was the command's improved flying safety record. In 1948 SAC's major accident rate was 10 percent higher than the USAF's, dropping to less than half the USAF's rate by 1952, an advantage that command retained at least up through 1976. And for all the complaints from innovative, energetic airmen and outside critics, SAC's tight and disciplined controls were appropriate to conserve resources and handle nuclear weapons.

At the same time SAC was struggling to build up and improve its force of bombers and bombs, the military wrestled with the best way to employ strategic bombardment. The Soviet Union was a much more difficult target than either Germany or Japan because of the greater distance between bases and targets, Russia's larger size, inadequate maps, and poorer intelligence. The airmen were forced to rely on outdated German intelligence materials and information from refugees, both grossly inadequate for the task at hand. One consistent concern throughout the Cold War was a fear of a Pearl Harbor–type attack, especially after the Soviets obtained nuclear weapons.

## Intelligence and War Plans

Intelligence was critical. The Chinese intervention in Korea heightened tensions and in December 1950 prompted President Harry Truman to authorize reconnaissance overflights of the Communist bloc nations. American allies, specifically the British and Nationalist Chinese, participated in this effort as well. The Americans used aircraft, as well as camera-equipped balloons in 1956 and 1958, to probe Communist air space. Western intelligence markedly improved with the introduction of the Lockheed U-2. The first U-2 overflight of the Soviet Union took place in early July 1956 and continued until a U-2 was shot down in May 1960. The spy plane made about two dozen penetrations of Soviet air space and flew fifty-one sorties worldwide. There were also photo and electronic reconnaissance missions along the periphery of the Communist bloc. On these missions only one American loss was clearly inside the Soviet Union, despite many attempted intercepts, some exchange of fire, and a few hits. However, through mid-1960 the Communists did shoot down approximately a dozen USAF and Navy aircraft near the coast.[6] Some of these aircraft may have inadvertently strayed inside Soviet territory while others were surely well outside the borders.

The manned overflights produced valuable intelligence for the airmen. Not only did they obtain accurate photographs for future bombing operations, they energized Soviet defenses, giving the Americans a wealth of data on these systems and their operation. These flights in the 1950s highlighted the weaknesses of Russian defenses, while the U-2 shootdown showed the capability of surface-to-air missiles against very high flying, slow, nonmaneuvering aircraft in 1960.

New technology aided the Americans. Within months of the U-2 incident and after a dozen failures, the United States recovered the first data from a Corona reconnaissance satellite, a system authorized in 1956. Satellites were not only invulnerable to Communist defenses but also did not risk pilots or international incidents, and they became increasingly more effective using a variety of sensors. Corona remained operational until 1972.

In contrast to the marked improvements in organization, delivery systems, and intelligence gathering, plans for employing the strategic bombers were little changed from World War II concepts. Initial U.S. air war plans called for an extended nuclear bombardment campaign as the atomic bombs became available, along with a much larger conventional one. The airmen intended to hit Soviet oil installations with night, radar-guided, solo bombing. The AAF's early plan envisaged a campaign that would last nine months and destroy three-quarters of the Soviet oil. An air plan in the summer of 1947 targeted forty-nine Soviet cities for destruction with one hundred bombs, an attack calculated to take out half of the country's industry. (The reader will recall that this exceeded the number of nuclear weapons then available.) The JCS agreed to a May 1948 plan that called for operations from Britain, Egypt, Okinawa, and possibly Iceland that focused on bombing twenty cities as well as oil and mining shipping lanes. LeMay quickly changed the SAC plan from a gradual delivery of nuclear weapons to one massive strike against the Soviets. In December 1949 the JCS approved a plan involving 292 nuclear weapons aimed at industrial, electric power, transportation, armament, and oil targets. SAC added more targets as more weapons and aircraft became available. Until the 1960s the United States planned to employ massive nuclear strikes.

## The Range Problem and Solution

Clearly one of the airmen's primary problems was range. While this had been of concern in World War II, it was a far greater problem with the Soviet Union as the most likely foe, particularly since the principal U.S. strategic bomber until 1952 was the B-29 with its range of three thousand nautical miles. The airmen considered a number of solutions to the problem. One idea was one-way missions, with the hope that after bombing the crews would reach friendly or neutral territory, a remote region, or ditch at sea near friendly vessels. In August 1948, Gen. Earle Partridge stated, "Expend the crew, expend the bomb, expend the airplane all at once. Kiss them goodbye and let them go. That is a pretty cold-blooded point of view, but I believe that it is economically best for the country."[7] With leaders considering the deaths of millions of civilians, it is not difficult to understand this attitude.[8] A technological solution was to develop the capacity to land and refuel on unprepared airfields, dry lakebeds, or on an Arctic icecap. Kenney suggested fitting bombers with tractor-type landing gear, and the Air Force tested such an apparatus. The airmen found more practical solutions.

The quickest solution was to obtain bases outside of the United States. The military realized, however, that bases closer to Russia were more vulnerable to attacks and those in non-American territory were subject to the dictates of the host nation. The Arctic region was one area of interest as cross polar routes were the shortest between the United States and Soviet Union; however, the USAF had inadequate facilities in Alaska and encountered extremely difficult conditions. The aircraft had to be winterized ("Arcticized" would be a more appropriate term) and men and equipment had to adapt to exceedingly cold temperatures and unique navigation conditions such as the featureless terrain, inoperative magnetic compasses, and periods of extended light or dark. Morale was a problem in the northern climes, both at the bases and because of the difficulty of survival in the event of going down in the inhospitable high latitudes. During World War II the AAF had established an emergency landing strip at Thule, Greenland, six hundred miles north of the Arctic Circle, which was extended to seventy-five hundred feet in 1951 and became operational two years later. After 1947, however, Air Force basing interest focused on Western Europe and North Africa. While American bombers could operate freely out of occupied Germany, getting there was difficult because Western European countries were hesitant to grant overflight permission. Britain was another matter; the AAF used British bases in 1946 and began rotating bomber units to these bases in 1948. Further, the United States obtained permission to build and use bases in North Africa, Iceland, Spain, and the Azores. SAC also built and upgraded stateside bases to accommodate the increasing numbers of heavier bombers.

In typical American fashion, the airmen developed a technical solution to the problem: air-to-air refueling. Americans had made some record-breaking endurance flights in the 1920s using this technique, the most famous of which was the January 1929 *Question Mark* flight, which stayed aloft for nearly 151 hours. But these were dramatic stunts. It was the British who pioneered a practical solution, which they demonstrated on sixteen transatlantic crossings in 1938. The AAF tested the British system in 1943 with B-24s refueling B-17s but went no further. While there was little need for bomber range extension in the European air war, certainly there was for bombers in the Pacific, although refueling hundreds of fighters and bombers was impractical at the time. Nevertheless, the British and Americans must be criticized for not developing this proven method to extend the range of their reconnaissance and antisubmarine patrol aircraft. After the war the British resumed the transatlantic refueled flights. Amazingly, despite the British flights and the clear need for range extension, it was not until October 1947 that SAC initiated a request for aerial refueling, which the USAF gave the highest priority the next January. The quickest way to satisfy the tardy requirement was to adopt the British system. Matters now moved rapidly as the USAF bought equipment sets (and rights) from the British and received its first converted B-29s from Boeing in May 1948. Flight tests began that month, with the 509th Bomb Group accomplishing a dozen hookups by the end of the year.

SAC quickly demonstrated its new capabilities. In December 1948 a B-50 flew from Fort Worth to Hawaii, dropped its bombs, and returned to its base, covering ninety-nine hundred miles aided by three aerial refuelings. Even more impressive was an operation only a few months later. The USAF planned an around-the-world mission, using five B-50s to be launched in sequence until one completed the circuit. The first that took off, in February 1949, was forced by engine problems to land in the Azores. The second, *Lucky Lady II*, completed the 23,500-mile flight in ninety-four hours with four in-air refuelings. U.S. bombers now could reach any point on the globe.

The USAF wanted a better system because the British cross-over method was complicated, limited to a maximum speed of 190 mph and maximum altitude of ten thousand feet. The Americans simplified the system and changed to a larger hose and pressurized system that tripled the pumping rate and allowed refueling at higher altitudes. Aerial refueling evolved further. Boeing developed a flying boom on the tanker that telescoped from twenty-seven to forty-five feet and was inserted into a receptacle on a bomber flying closely below and behind the tanker. This system could pump faster than the hose system and could operate at higher speeds and altitudes. Meanwhile the British went on to develop and demonstrate a probe and drogue scheme in April 1949. It involved a tanker extending a flexible hose with a basket (cone) at its end, into which the receiver inserted a short probe. The two systems had contrasting advantages and disadvantages. The boom system could pump faster (700 gpm versus 110 in the crossover system and 250 with the probe and drogue) and at a higher altitude. However, it required the two aircraft to fly a tight formation and could only refuel one aircraft at a time. The probe and drogue system was lighter, cheaper, and simpler and could use similar aircraft to refuel each other ("buddy system") or could simultaneously refuel as many as three aircraft using three baskets. In a SAC flyoff between the two systems in February 1952, bomber pilots preferred the boom system while fighter pilots preferred the probe and drogue. SAC wanted the boom system while the Air Force's research and development arm, Air Research and Development Command, recommended standardizing on the probe and drogue system, which was approved by the undersecretary of the Air Force. In the end SAC got the boom while USAF fighters went with the probe and drogue system, as did the Navy and the North Atlantic Treaty Organization (NATO). In 1959 the USAF fitted booms with an adapter to allow the refueling of probe-equipped aircraft (however, this can only be changed on the ground).

The USAF converted 92 B-29s into crossover tankers (KB-29M) and another 74 B-29s into receivers (B-29MR) for that system. SAC retired the tankers in 1954. The USAF also converted 116 B-29s into boom tankers (KB-29P), the first delivered to SAC in September 1950, which served with that command until 1957. Another Boeing aircraft became SAC's standard prop tanker, the KC-97, itself a development of the B-50. SAC got its first Stratotanker in July 1951, eventually receiving over eight hundred, which served in the regular Air Force until December 1965.

The disadvantages of aerial refueling are seldom noted but ever present. Negative aspects center on the reliability and vulnerability of the tankers and the difficulties of the operation. Every tanker employed meant another aircraft to maintain and operate, and possibly malfunction. Use of tankers presents an opponent with additional aerial and ground targets. It should be emphasized that aerial refueling is more complicated than gassing up the family car; it requires considerable skills as malfunctions and accidents can prevent refueling or damage or down the participating aircraft. In brief, while effective, aerial refueling is a complicated, expensive, and vulnerable way to extend range.

Aerial refueling saw action during the Korean War. The first combat aerial refueling was in July 1951, when a tanker refueled four RF-80s for reconnaissance missions over North Korea. This was followed a week later when an RB-45C was refueled on a similar operation, and in October tankers refueled F-84s on a combat mission.

## Flying Wing and Mixmaster

The airmen also developed longer range aircraft. One of the best known, well beyond its importance, is the Northrop Flying Wing. Jack Northrop was fascinated, if not obsessed, by the possibilities of an aircraft without the drag and weight of fuselage and tail. He built and flew a number of such designs, and in 1941 won a contract to build a bomber version. The XB-35 was designed to achieve a top speed of 380 mph, mount twenty .50-caliber guns, and reach a range of twenty-eight hundred miles with a maximum bomb load of fifty-one thousand pounds. Two piston engines turned four sets of counterrotating props in a pusher configuration. The futuristic aircraft, a giant with a 172-foot wing span, an empty weight of 90,000 pounds, and a maximum loaded weight of 209,000 pounds, made its first flight in June 1946.

The project was overcome by time and problems. The aircraft bomb bay was too small to carry the first atomic bombs internally, and a semiexternal carriage would decrease top speed by 7 percent and combat range, already considered inadequate, by 9 percent. Gear box and prop difficulties led the manufacturer to replace the counterrotating arrangement with single-rotation props, a version that first flew in February 1948, although with considerable vibration, less stability, and decreased performance.

(Right) Although Northrop won a contract to build the B-35 in 1941, the Flying Wing did not get airborne until June 1946. It had a strikingly different configuration than other aircraft, and although very large, it lacked space in its bomb bay to fit an atomic bomb. These constraints, along with propeller and stability problems, led to its demise. (National Museum of the USAF)

Another unique design was the Douglas XB-42, a small, quickly developed bomber. The Mixmaster was manned by a three-man crew and powered by twin engines that turned counterrotating propellers mounted on the tail. It did not meet performance expectations. (National Museum of the USAF)

These problems, along with the appearance of higher performing jets, finished the B-35. The Flying Wing would be revived within a few years powered by jets, as was the case with the Douglas XB-42.

The Douglas B-42 was another unusual design that sharply contrasted with the Northrop bomber. It was a small aircraft that was quickly designed, built, and flown; possessed unusual performance; and was conceived as a cheap substitute for the B-29. It had a wing span of seventy-one feet and weighed a mere twenty-one thousand pounds empty and had a takeoff gross weight of thirty-six thousand pounds. (The B-29's span was twice this, its empty weight over three times, and its maximum weight four times that of the Douglas bomber.) The Mixmaster was manned by a crew of three, the pilots seated side by side with "bug eye" canopies and powered by twin liquid-cooled engines in the fuselage that turned counter-rotating props mounted on the tail cone. (The bomber was defended by two rear facing, underwing turrets, each with twin guns, controlled by the copilot reversing his seat, and two fixed forward firing machine guns.) It first flew in May 1944 and was expected to exceed 450 mph and fly five thousand miles with a one-ton bomb load (1,800 miles with four tons). Flight tests disappointed the AAF as controls proved inadequate; speed, altitude, and range were less than desired; and there were vibration problems. The company built two, one of which crashed due to pilot error. The other was reincarnated as a jet.

## The B-29 (and B-50) Superfortress

Now to move from the fascinating to the important aircraft. For years after the war the Boeing B-29 was the backbone of the United States' strategic bombing. The airmen took Superforts out of storage and with little or no modification put them into service. Their numbers rose from 148 in SAC in December 1946, to a peak of 516 in December 1948, and numbered 435 in December 1952 (comprising 44 percent of the bombers and reconnaissance bombers in the USAF inventory), with the last bomber retired in November 1954. Two prop bombers supplanted the B-29.

The Boeing B-50 looked like, performed only somewhat better than, and was directly related to the B-29. It began life in 1944, when Boeing upgraded the B-29's engines with the more powerful R 4360 and redesignated the bomber XB-44, then B-29D, and finally in December 1945, B-50. The company claimed it was 75 percent different than the B-29, despite its looks. The main improvement was 60 percent greater power, an advantage somewhat mitigated by five tons of additional weight (a considerable amount due to the engine change) producing a small performance edge.[9]

The B-50 was a stop gap bomber, slightly superior to the workhorse B-29, a backup for the troubled B-36, and waiting to be replaced by the jet bombers then in the design and test stages. It was neither a new nor trailblazing aircraft, and had constant problems during its service, more than might be expected from a derivative design. The B-50 had difficulty carrying the new atomic bombs because the bomb developer, the Atomic Energy Commission (AEC), and the USAF were not in close coordination, the very tight secrecy surrounding the bomb, and because the bomb bay had been designed before the atomic bomb and proved too small for the later versions. Initially metal skin cracking required extensive modification, and throughout its service engine problems were chronic, turbosuperchargers failed, and valves and tanks leaked. It first flew in June 1947 and entered SAC service in February 1948; SAC had more than 200 on its books in December 1951, with the last leaving in October 1955. Boeing built 371 B-50s.

## The Last USAF Prop Bomber

More impressive was the massive B-36, which was the culmination of the effort to gain range by building larger and larger aircraft. The German conquest of Western Europe and the precarious position of Britain pushed the American airmen toward a transcontinental bomber. The Air Corps sent out invitations for design studies in April 1941 with requirements cut back in August to a ten-thousand-mile range with a five-ton bomb load. That November the airmen picked Consolidated to build the bomber.[10]

The aircraft's development experienced numerous delays during the war as the aircraft encountered problems with weight, the power plant, and faulty workmanship. Thus the XB-36 did not make its first flight until August 1946. The B-36 was a giant that indicated how far aviation had come in just one decade. The B-17 that first flew in October 1935 weighed 24,500 pounds empty, the B-29 that made its maiden flight in September 1942 weighed 70,000 pounds empty, and the empty B-36 weighed twice that. The bomber was distinguished by its size and its six engines turning pusher props. The B-36 had four bomb bays, which allowed the bomber to carry forty-one tons of bombs, more than the weight of a World War II B-24. This heavily defended bomber was capable of carrying a vast tonnage of bombs a long way, the dream of the bomber enthusiasts.

The aircraft had a crowd of initial problems, some of which persisted for years. The first concerned the landing gear, for the experimental bomber employed huge single-wheel landing gear that restricted operations to only three airfields in the United States that could handle that concentrated weight. The manufacturer changed this as well as substituting a dome canopy for the stepped (airline) configuration. Nevertheless the aircraft did not satisfy the user. SAC's first commander, Gen. George Kenney, considered the B-36 inferior to the Boeing B-50, and criticized the B-36's range, speed, and lack of protection of the fuel tanks. The top echelon disagreed, and the bomber's development continued. Throughout its service the B-36's chief advantage was its range and its major disadvantage its speed. Like the B-50, the B-36 would be an interim bomber until the jets appeared. The first combat capable B-36, the "B" model, joined SAC in November 1948.

The shift in leadership at SAC benefited the B-36. Unlike Kenney, LeMay supported the giant bomber. He was looking for aircraft with range, stating in January 1949 his basic belief that "the fundamental goal of the Air Force should be the creation of a strategic atomic striking force capable of attacking any target in Eurasia from bases in the United States and returning to the points of take-off."[11] The SAC commander had few illusions about prop propulsion in the future, favoring the trouble-plagued jet-powered B-47 over the B-50 and its planned development, the B-54, for the medium bomber role.

The B-36 carried a variety of defenses. It mounted the heaviest armament of any bomber to become operational, six retractable turrets each mounting twin 20-mm cannons as well as twin 20-mm guns in the nose and twin 20-mms in a tail turret. The armament encountered difficulties attributed to design, maintenance, and training, which just about covers everything, and these persisted, so as late as April 1952 armament was considered unsuitable.[12] The bomber's large volume allowed SAC to fit it with a number of ECM devices, which were increased during the aircraft's service. The USAF also glide tested an air-launched, rocket-powered decoy, Buck Duck, from a B-50 before canceling the project in January 1956 and tested air-to-surface missiles, specifically the Rascal aboard the B-36, which will be discussed below.

Another effort to defend and extend the bomber's effectiveness involved bringing along fighters, a concept pioneered in the early 1930s by U.S. Navy airships. The McDonnell XF-85 was a small fighter that could be carried internally and had to be flown by a pilot no taller than five foot nine

The last and largest U.S. piston-powered strategic bomber was the Convair B-36, a very big aircraft with an empty weight twice that of a B-29. The Peacemaker first flew in August 1946 as seen here. (Note the single wheel main landing gear.) Although the B-36 was very controversial, it provided SAC with a long-range bomber until superseded by the B-52. (National Museum of the USAF)

inches who had to straddle the jet engine.[13] Three of the egg-shaped fighters were to be carried, launched, and recovered by a B-36. Test flights of the diminutive fighter with a B-29 between August 1948 and April 1949 revealed that while the aircraft flew satisfactorily and could be launched, there were severe problems involving its recovery. The USAF cancelled the project in April 1949 before it could be mated to the B-36. A second system attached fighters to bomber wing tips. After several successful link ups, the Air Force cancelled Project Tip Tow after an April 1953 accident that destroyed the B-29 and one F-84 and killed all the aircrew. The B-36 Tom Tom project consisted of a claw-like clamp fitted on the fighter's wing tip that could be mated to a receptacle on the bomber's wing tip that allowed the fighter to hook up to the bomber and receive a tow along with fuel, pressurization, and heat. The USAF tested airborne linkups from November 1955 until September 1956, when an RF-84F was ripped from its wingtip position, dramatically demonstrating the hazards of this procedure. These systems were ingenious, difficult and dangerous, and proved to be impractical. Aside from that the idea that a stubby parasite fighter could overcome defending fighters and then recover aboard a battling bomber defies common sense. And the concept that fighter pilots attached to wing tips could successfully battle after long hours in their cockpit and recover aboard their carriers was just as farfetched.

A third parasite effort was more successful and went into service. Project Ficon (Fighter Conveyor) began in 1950 and involved an F-84 semienclosed in three of the B-36's four bomb bays. The bomber could launch and retrieve the fighter as well as takeoff mated. Ficon evolved from bombers carrying fighter escort, into a method to extend the range of high-speed

The U.S. Navy disputed the USAF on issues concerning strategic bombardment. In 1946 Cdr. Thomas Davies (seen here as RADM) commanded the Lockheed P2V *Truculent Turtle*, which flew nonstop and unrefueled over 11,200 miles from Australia to Ohio. Later, P2Vs were launched off of aircraft carriers. (U.S. Naval Institute Photo Archive)

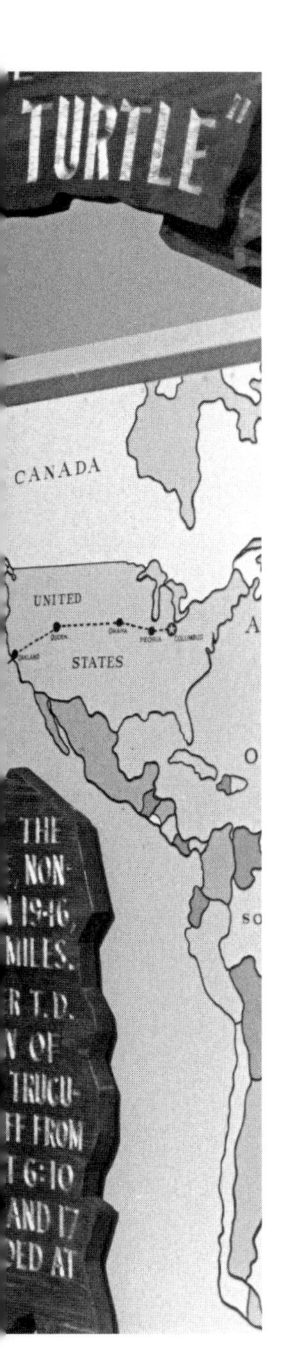


reconnaissance aircraft. The Air Force successfully tested the scheme in April 1952 with a straight-wing F-84E and went on to convert ten bombers and twenty-five swept-wing RF-84Ks and put the combination into operation for one year during 1955. The combat radius of the two aircraft was thirty-five hundred nautical miles.

Meanwhile the bomber demonstrated performance and encountered critics. In December 1948 a B-36 took off from its Texas base (Fort Worth), dropped a five-ton bomb load near Hawaii, and returned home, covering the eight thousand miles in 35.5 hours. In March 1949 a B-36B carried five tons of bombs half way on a ninety-six-hundred-mile mission. The USAF conducted these flights to show off the giant bomber and still its many critics for as late as 1948 the USAF considered canceling the bomber because of tight budgets. However, the heating up of the Cold War, especially the Berlin Blockade, which began in June 1948, changed many minds. Concurrently the B-36 only slowly overcame its problems, underwent considerable modification, and achieved true operational status in 1952.

The bomber faced its most serious challenge during the infamous "Revolt of the Admirals" incident, the most egregious example of interservice rivalry in U.S. history. The primary issue was funding, although the overall nuclear strategy, the condition of the naval air arm, and the capabilities of the B-36 were significant issues. The cancellation of the Navy's new supercarrier in April 1949 ignited the fracas. The two services squared off over the interpretation of the lessons from World War II strategic bombing experience and the nature of a future war. Many AAF and USAF airmen believed that strategic bombing could win future wars and that atomic bombers were adequate for defense. The Navy strongly disagreed, as summed up by Rear Adm. Ralph Ostie, who asserted, "We consider that strategic air warfare, as practiced in the past and as proposed for the future, is militarily unsound and of limited effect, is morally wrong, and is decidedly harmful to the stability of a postwar world."[14]

The two services were battling over turf. On its part the USAF believed and feared that the Navy was intruding on its strategic air mission, which was the basis of its existence. The Navy fanned the airmen's fears when in September 1946 a specially configured twin-engine Lockheed P2V Neptune naval patrol aircraft (the *Truculent Turtle*) flew 11,200 miles nonstop and unrefueled from Perth, Australia, to Columbus, Ohio.[15] The Navy modified a dozen P2Vs to carry a nuclear weapon and in April 1948 made a carrier takeoff. The sea service was also developing the AJ-1 bomber, which could carry a nuclear weapon and was modifying its three *Midway*-class carriers to operate heavier bombers that would allow nuclear operations. The Navy was thinking in terms of naval targets, targets on the periphery of the Soviet Union, and did not want to be shut off from nuclear weapons. Of course this looked different to the USAF. The Navy also feared it would lose naval aviation to an independent air force. Certainly the Air Force wanted control of all land-based aircraft, including naval patrol bombers and antisubmarine reconnaissance aircraft. These issues were overblown by zealots on both sides.

The supercarrier and B-36 became surrogates as the two services battled over money and missions, a dispute that spilled over into an unseemly, public interservice brawl. The Air Force took shots at the Navy's carriers, and the Navy reciprocated with assaults on the USAF's bomber. There were questions concerning the B-36's performance and the carrier's cost, vulnerability, and utility in a strategic war. One future chief of naval operations (1955–61), Arleigh Burke, called the B-36 a "billion dollar blunder."[16] In May 1949, shortly after the supercarrier cancellation, an anomymous document, later connected with a Navy official, was leaked to Congress claiming corruption in the procurement of the B-36, overstatement of the bomber's performance, and exaggeration of the importance of strategic bombardment. These allegations dragged the secretary of defense, secretary of the Air Force, the B-36, and the USAF into the critical eye of the press, public, and Congress. Charges flew, Congress investigated, and the document's author recanted. The hearings cleared the individuals and the Air Force, the chief of naval operations was fired, and the B-36 survived.

The giant bomber evolved into a more potent weapon as its various problems were solved. However, its chief deficiency of low speed remained. In 1948 the manufacturer proposed adding four jet engines mounted in two pods (essentially those from the B-47) outboard of the prop engines. These first appeared on the B-36D, and they increased takeoff and landing weights along with speed and ceiling.[17] The airmen converted most of the earlier B-36Bs to the "D" configuration.

The last major B-36 modification was the featherweight program, authorized in early 1954, which stripped the bomber of weight to improve performance. Later models ("F," "H," and "J") featured slightly more powerful engines and thus slightly higher speeds. SAC fielded a peak of 317 B-36s in December 1955, began junking the bomber in February 1956, and retired the last one in February 1959. Convair built 385 B-36s.

Another effort to extend range involved nuclear power. The idea first surfaced in 1944 and really started in 1946; however, it received only limited support. In 1953 the USAF cancelled a B-36 slated to test a propulsion system[18] but proceeded to test the effects of radiation on crew and airframe. A B-36 conversion, the NB-36H, was powered by its conventional engines and carried a nuclear power plant that was running on twenty-one of its forty-seven sorties, although it was neither intended nor used to power the aircraft. The aircraft had a new nose structure and carried a reactor weighing 17.5 tons and crew shielding weighing another 12 tons. First flying in September 1955, the NB-36H conducted successful tests until March 1957. In March 1961 the John F. Kennedy administration ended the project after an expenditure of over one billion dollars.

One last point regarding the B-36: It was the only U.S. prop bomber capable of carrying a hydrogen bomb. The Harry Truman administration authorized development of the super bomb in January 1950, and the first was tested in November 1952. It had a yield of 10 megatons, a tremendous increase in power over the first atomic

**(Right) The United States sought a long-range, great endurance aircraft and considered using nuclear propulsion. A nuclear reactor was fitted aboard a reconfigured B-36 to study the problem of protecting the crew from radiation. The NB-36 flew numerous flights but never was powered by the reactor. (National Museum of the USAF)**

bomb, which had a yield of 15 kilotons. However, the early H bombs were huge and heavy (measuring twenty feet in length and weighing twenty-five tons), demanding a large bomber. One major concern regarding the bomb's great power and the B-36's slow speed was crew survival from the bomb's explosion. Therefore a parachute was attached to the weapon to slow its descent and give the bomber more time to exit the scene.

The B-36 played a crucial role in providing U.S. nuclear deterrence in the 1950s, despite all of its problems. It earned the distinction of being the last prop bomber in the SAC inventory and the safest.[19] It is probably best remembered by the public, however, not for its importance in its assigned mission but

for the Air Force–Navy dust up and as the largest U.S. bomber that went into service.[20]

One Navy aircraft should be mentioned, if only in passing. The North American AJ Savage was a hybrid aircraft powered by two prop and one jet engine that first flew in July 1948. Designed as a heavy attack bomber, it carried a six-ton bomb load. The first carrier launch in June 1950 gave the Navy a nuclear delivery capability. The Navy equipped eight attack and two reconnaissance squadrons with the aircraft that served into the late 1950s. It was replaced by the jet-powered A-3, which first flew in 1952, with deliveries starting in March 1956.

A more serious challenge to the Air Force was the Navy's North American AJ-1 Savage, seen here aboard the aircraft carrier *Wasp*. The hybrid bomber, powered by two prop engines and one jet engine, first flew in June 1948 and gave the Navy a nuclear capability when it became operational two years later. (U.S. Naval Institute Photo Archive)

## The Korean War

The American focus on a conventional war in Europe and an all-out nuclear war was jarred by the Korean War, which erupted in June 1950. Nevertheless the United States and United Nations quickly responded, sending military forces into the battle. The fighting seesawed across the former border between the two Koreas, drew in the Chinese, settled into a two-year stalemate, created massive frustration from its cost and indecisiveness, and burned out after causing widespread destruction and massive casualties.

The fighting was confined to the Korean peninsula, although the Communists used Chinese soil for bases and supplies. As there were some fears that the conflict would permit or cover a Soviet advance in Europe or possibly trigger World War III, American decision makers limited the war, as did the Soviets. The United States used neither nuclear weapons, they were considered, nor their most modern bombers in the conflict. Strategic bombing was a mere footnote to the struggle on the ground, with the Air Force's primary roles clearly the gaining and maintaining of air superiority and supporting the ground troops. This was not the war the airmen anticipated.

In the conflict the USAF employed some of its long-range strategic bombers, mainly B-29s, against interdiction targets. The four-engine bombers had little trouble against the meager North Korean Air Force, and by mid-September they had taken out what few industrial targets there were in North Korea. The only key target set remaining was electric power, which was not struck because this largest installation was on the Yalu River close to the Chinese and Soviet borders and sheltered from American bombs by the rules of engagement.[21] Then the Chinese intervened in early November 1950 with a massive and effective ground force and with a large numbers of jet fighters. Everything changed on the ground and in the air.

There were numerous restrictions that constrained the bombers. The B-29s bombed only targets in Korea, few of which were strategic. Over the course of the war the numbers of bombers were also limited; the monthly average of B-29s in the theater inventory was 109. In addition, the USAF did not commit its newest bombers, the B-36, B-45, and B-47, although a few B-45s and B-50s saw action in reconnaissance roles. The Air Force did not want to risk either its small numbers of its best bombers or their secrets in this war. For the same reason the airmen restricted the use of electronic countermeasures.[22]

The war did send up warning signals to the bomber enthusiasts. Just as the Soviets had developed nuclear weapons more quickly than expected, they sprang another surprise by fielding large numbers of an excellent defensive fighter, the jet-powered MiG-15, which first saw action in November 1950. Although the Soviet fighter was defeated by American F-86s over MiG Alley, it proved successful in the role for which it was designed, killing American bombers. On April 12, 1951, the Air Force launched forty-six bombers, thirty-nine of which bombed the Yalu bridge at Sinuiju. The B-29s were hit by over seventy MiGs, which brushed off the F-86 and F-84 escorts, destroyed three bombers and damaged seven others.

The bombers claimed seven MiGs destroyed and escorting F-86s another four, while the Communists insist they downed ten B-29s without a loss. The Superfort's defensive armament, not tested by the Japanese as the Germans had tested the B-17 and B-24, was clearly inadequate against the MiGs.

As a result of high losses the USAF stopped the deep, daylight bomber attacks and transitioned to night bombing. Monthly night sorties exceeded day sorties for the first time in November 1951 and from then on essentially all B-29 bombing was done at night. Night bombing thwarted the Communists for about a year, but the defenders had the advantage between November 1952 and January 1953. Although the USAF had little night bombing experience, it responded with more ECM, glossy paint on the bombers' undersurfaces, flash suppressors on their .50s, tactics that varied altitudes and direction of the bomb run, and night fighter escorts.[23]

While strategic bombing was rare during the Korean War, two campaigns deserve note. Early in 1952 the USAF proposed to pressure the enemy to end the war by bomb ing the North Korean electric power system, a target set neglected in World War II, an error highlighted in the USSBS critique of the European bombing, and thus far protected by political restraints. (The Sui-Ho facility, the world's fourth largest generating facility, which supplied half of its electricity to China, was on the Yalu River, the border between Korea and China.) While Gen. Matthew Ridgway, the overall commander, would not approve this plan, when he rotated in May, his successor, Gen. Mark Clark, did. In June President Truman and the JCS approved the attack on Sui-Ho. In addition the joint Air Force, Navy, and Marine attack would hit three installations in central North Korea, to be followed that night by a B-29 attack on one of these.

On June 23, 1952, Navy AD dive bombers along with Air Force F-84 and F-80 fighter bombers bombed Sui-Ho under the watchful eyes of escorting F-86 and F9F pilots. The airmen feared the reaction of hundreds of MiGs based only a short distance from the target complex, but instead of engaging the Americans, the Communist fighters flew inland, apparently fearing an attack on their Chinese airfields. During the four-day campaign the three services flew almost eighteen hundred fighter-bomber and counter-air sorties and lost two Navy aircraft whose pilots were rescued. The attacks took out over 90 percent of North Korea's electricity potential, subjecting that nation to a blackout for two weeks and cutting Chinese industrial production. Far from the scene the bombing stirred reaction, as some members of the British Parliament cried provocation while some U.S. congressmen asked why these targets had not been hit earlier. The operation was remarkable for the successful bombing and low losses as well as its rare joint air effort.

The second campaign was waged in May and June 1953 against North Korean irrigation dams. The breaching of these dams would cause great destruction from flooding as well as threaten the North Korean rice supply, 70 percent of which required irrigation. The airmen began planning an attack on twenty dams in February 1953, but when Gen. Otto Weyland, commander of Far East Air Forces, was briefed, he reportedly was "skeptical of the feasibility and desirability" of the attack.[24] Although Weyland was unwilling to attack the Korean rice crop directly, he was willing to attack the dams because of the damage the consequential flooding would cause, a position that appears disingenuous.

On May 13 fifty-nine F-84s attacked a stone and earthen dam twenty miles north of Pyongyang; it broke that night. The flooding took out five bridges, six miles of one of two north-south rail lines serving the North Korean capital, two miles of highway, covered one airfield, and destroyed five square miles of rice. F-84 bombing on May 15 and 16 destroyed a second dam, the flooding from which cut the other north-south rail line. Although B-29 attacks scored hits on a third dam on May 21–22 and 29, the dam held as the Communists had quickly responded to the U.S. strikes by lowering the reservoir's water level. Similar action enabled other dams to withstand attacks by American aircraft in June. Nevertheless, the destruction was extensive (the rail lines did not reopen for two weeks) and the lack of water impacted negatively on the rice harvest. American timing deserves criticism, for had the attacks been compressed as originally planned, the result would have been far more effective. The two operations foreshadowed the future: the use of fighter-bombers and attack aircraft against strategic targets and the sensitivity of attacking dams and an enemy's food supply.[25]

During the war, B-29s flew almost twenty thousand effective sorties and dropped 168,000 tons of bombs, just over half at night. Only about 7 percent of this effort was directed at industrial targets. The crews of the B-29s suffered the greatest number of USAF casualties: seventy-two bombers lost, fifty seven on operations (of which twenty-four were attributed to enemy action).[26] In brief, this was about the same tonnage dropped by B-29s on Japan, the same number of sorties, with about one-sixth the losses.

The Korean War indicated that even escorted formations of B-29s could not survive against modern jet fighters, which the Soviets had in abundance. The change brought by jets was rapid for only five years after World War II, where the B-29 was clearly the outstanding bomber; jet fighters rendered it obsolete. The capabilities of the MiG-15 and the difficulties of the B-29s demonstrated the tough conditions American prop bombers would have faced had they been launched against the Soviet homeland. The future of bombers was tied to jet propulsion.

## The RAF's Last Prop Bombers

The British modified their best heavy bomber of World War II, the Avro Lancaster, during that war for operations against Japan, modifications that justified a new designation: Lincoln. It looked like its predecessor, but it was larger, heavier, better armed, and, most of all, had more powerful engines, giving it superior performance. It first flew in June 1944. The bomber went into service in 1945 and saw combat, albeit not strategic operations, against insurgents in Kenya, Malaya, and Yemen. Bomber Command received 529 Lincolns between 1945 and March 1951, and the bomber left British service in March 1963 after its last duties in an electronics unit. (It also flew in both

Australian and Argentine colors.) In the words of one author, the "Lincoln was effective, reliable and popular with its crews and quickly gained an enviable reputation."[27]

Special circumstances brought the Superfortress into British service. In the late 1940s Bomber Command was seeking an interim bomber to fill the gap between its prop-powered Lancasters and Lincolns and its oncoming jet-powered bombers. The idea of borrowing B-29s emerged when the United States considered extending large military aid to its allies in the West in the spring of 1948. Thus the RAF would not only get a better performing bomber than the Lincoln to tide it over until jets came into service, but one that was proven and would cost nothing for development or construction.

The first Washington arrived in Britain in March 1950, the eighty-seventh and last in June 1952, and most returned to the United States by the end of 1954. The Washington was more comfortable (it was pressurized unlike the Lancaster and Lincoln) and better performing than previous RAF four-engine bombers, but it proved expensive and difficult to maintain because of a shortage of spare parts. The Boeing bombers equipped eight RAF medium bomber squadrons and, with its range about twice that of the Lincoln, gave the British a delivery system that could reach into western Russia.

The Washingtons were not fitted with nuclear capabilities. This was not publicized at the time because it would have lessened the British deterrent, and it has confused writers ever since. It should be noted that the United States ungraciously ended nuclear cooperation with the British with the passage of the McMahon Act in August 1946, despite widespread wartime cooperation with the UK. The American move was prompted by suspicions of the socialist government, which initially was friendly toward the Soviets (and had sold them their most modern jet engine), revelations of Russian spies in Britain, and traditional American antipathy toward the British. This encouraged the British to build their own bomb despite the great expense and poor condition of the British economy. In October 1952 the British exploded their first atomic bomb. The cooling of relations and lack of cooperation was brief, world events pulled the two countries toward one another and by the end of the decade the United States amended its policy, allowing the two allies to cooperate in nuclear weapons development.

## The Soviet Threat

The Soviets presented a growing aerial threat to the U.S. heartland as the years rolled by. The Americans initially downgraded that threat, underestimating how long it would take their rival to develop nuclear weapons and nuclear delivery systems. Ironically, the first intelligence estimates were better than the later ones. In the fall of 1945 the JCS believed it would take the Soviets five years to develop an atomic bomb. Later this estimate was extended by consensus to 1952–53, although the nuclear project manager, Leslie Groves, believed it would take the Russians at least ten years. In August 1949 the Soviets exploded their first nuclear device.

The same was true with estimates of when the Soviets would field a long-range, nuclear delivery system. The Russians had requested long-range bombers from the United States during the war, which the Americans rejected. The Soviets were not to be denied, as they put a number of interned Forts and Libs into service and went even further with the Superfort. During the bombing of Japan three B-29s made forced landings in Russia. The Soviets returned the crews, kept the aircraft, and in a remarkably rapid fashion reverse engineered the bomber they designated Tu-4 (NATO code named Bull). Rumors of the Russian effort circulated as early as 1946, strengthened by Soviet attempts to buy B-29 tires, wheels, and brakes. The Tu-4 made its first flight in May 1947, and three flew in the August 1947 Tushino air show, which some Western observers discounted as the interned AAF B-29s. The concurrent appearance of a transport version of the B-29 undercut that interpretation.

The Tu-4 had a little less performance than its American sire partly due to the conversion from English measurements to the metric system, during which the dimensions were slightly increased which in turn increased weight. The only basic differences in the Russian version were to eventually change the armament from .50-caliber to 23-mm cannons and use their own slightly more powerful engines. With a three-thousand-mile range, the bomber could hit some U.S. cities on the West Coast, albeit on one-way missions. To reach further the Soviets would have to obtain forward bases or employ air-to-air refueling. Apparently there were trials with tankers in the early 1950s but no further progress at the time. This was some comfort, although the Tu-4s did threaten American overseas bases as well as American allies. The Soviets put the first Tu-4s into service in mid-1949 and by the end of that year had three hundred. In all they built between eight hundred and one thousand and even transferred twenty to twenty-five to the Chinese in the early 1950s. The Tu-4 remained in production until 1953 and in service into the early 1960s. The combat performance of the B-29 in the Korean War demonstrated to the Soviets that the era of the piston-powered bomber was over; instead they chose to build turboprop-powered bombers.

### The Bear

The Soviets mastered turboprop power plants to meet their requirement for an intercontinental-range bomber. They sought an aircraft with a ninety-three-hundred-mile range with bomb load, cruising at 466 to 509 mph, and a maximum speed of 571 to 590 mph. The response by the Tupolev design bureau was the Tu-95 Bear, which evolved from their earlier prop designs. (Tupolev's experience with the jet-powered Tu-16 Badger medium bomber convinced the designer that jet engines were inadequate for intercontinental bombers.) The project competed with and outlasted the jet-powered Myasishchyev M-4. The Bear was powered by four turboprop engines, each of which turned two giant, eighteen-foot-diameter counterrotating props and was defended by three turrets, each mounting twin 23-mm cannons. The Tu-95 first flew in November 1952 and encountered difficulties with both engines and propellers that were worked out. It could carry the largest Soviet bombs, and one dropped a 20-ton,

50-megaton hydrogen bomb in October 1961. The Russians introduced probe and drogue aerial refueling with the Tu-95 in 1964. The Bear entered operational service in April 1956, remains in the Russian long-range bomber force today, and has performed in a number of other roles during its long service.

The Soviets also used one aircraft in nuclear aircraft tests (similar to the NB-36) beginning in the mid-1950s and extending into the early 1960s. It flew thirty-four sorties before being terminated because of cost, weight, and the cancellation of the U.S. program. In the late 1950s a few (fifteen or so) improved variants appeared armed with air-to-surface missiles (ASM), the Kh-20 (AS-3 Kangaroo), a large missile requiring the entire weapons bay.[28] In the early 1980s the Soviets put the 6-ton Kh-22 into service. The Tu-95 could carry as many as three beneath its wings and fuselage. U.S. development of small cruise missiles triggered a Soviet response in 1977 that included not only similar missile development but also resumption of production of an updated

The Tu-95 Bear first flew in November 1952 and continues to fly missions today (2009). It has proven quite versatile as it can carry heavy loads over long distances. It is slightly smaller and lighter than the B-36 but has some performance advantages, especially a much higher speed. (National Museum of the USAF)

Tu-95 (designated Tu-95 MS). Initial versions could carry six 1.5-ton Kh-55 missiles in a rotary launcher, later versions could carry an additional ten externally, albeit with a decrease in range and speed.

The Soviet bomber outperformed the B-36, its U.S. counterpart, as might be expected because it was a decade younger. The Tupolev bomber was smaller in size and somewhat lighter than the B-36, with a comparable ceiling, slightly longer range, and much higher speed.[29] In 1961 there were two hundred Bears in bombing service, its peak inventory, which fell to sixty-five in 2001. The Bear is the longest serving and fastest prop-powered bomber.

In contrast the United States essentially skipped turboprop power for combat aircraft. There were plans for fitting the power plant to various bombers, the Air Force's B-47 and B-52, for example. Not that this power plant was not used, for the magnificent Lockheed C-130 Hercules remains in service as a transport and gunship almost half a century since its first flight. Instead the United States converted to jet power.

Chapter 8

# Between Korea and Vietnam

## THE TRANSITION TO JETS

Jet propulsion was a significant technological change that markedly transformed, if not revolutionized, aviation. The Germans introduced jets in World War II, and both the United Nations and Communist air forces employed jet fighters in the Korean War. Transitioning bombers to jet power lagged, so that jet-powered bombers did not enter service until after jet fighters.

### Early American Jet Bombers

The United States made a slow start with jet propulsion due to its focus on the immediate demands of World War II, while the slow transition to jet bombers was due mainly to technical problems. In September 1943 the AAF initiated studies to incorporate jet engines into an experimental bomber and encouraged a number of manufacturers to submit preliminary designs. The American airmen were pushing turboprops for heavy bombers, and in April 1945 they requested a design study from Boeing. The Seattle company declined, responding that the desired characteristics were "completely out of line with the state of the art."[1]

The first American jet-powered bomber was a modification of the Douglas XB-42 Mixmaster, a small pusher-prop bomber already discussed (see chapter 7). In January 1944 the AAF signed a contract with Douglas to substitute jet engines in two XB-42s to be redesignated XB-43. The AAF proposed in December 1944 to buy fifty of the aircraft to explore the capabilities of jet bombers, provide a test bed for the jet engine, establish an industrial production base, and educate the AAF in maintenance, flight, and combat operations, despite the fact that the performance estimates fell short of AAF bomber requirements. However, the rapid development of competing jet bombers, delays (caused mainly by the engine manufacturers), and the failure to meet military requirements limited procurement to two experimental bombers.

The two bombers differed in appearance. Douglas retained the prop engines and added two jets under the wing of one of the XB-42, an aircraft redesignated XB-42A. The jets caused problems yet could push the plane to a top speed of 488 mph. It would seem easier to add two externally mounted engines than replace the internally mounted piston engines with jets, but that was not the case. For the hybrid-powered aircraft did not make its first flight until May 1947, after the all-jet XB-43. The XB-43 was the first U.S. all-jet bomber to take flight, making its initial flight in May 1946. During these early tests the XB-43 achieved a top speed of 515 mph and indicated it would meet the performance requirements with the exception of takeoff and landing roll. Flight tests revealed airframe problems, however. Most troubling were cracking of the Plexiglas nose and bug eye canopies and problems with pressurization. Most important, the aircraft could not accommodate the 5-ton atomic bomb and was overtaken by other jet bombers.

The airmen developed a number of other jet bombers during this period, now mostly forgotten because only two went into service. Convair's B-46 was a beautiful aircraft that featured slender, very clean lines on its long fuselage and long, straight wing and a three-man crew with the pilots seated in tandem. It was powered by four engines mounted in two nacelles underneath the wing and made its maiden flight in April 1947. Convair's bomber had two major drawbacks, however. Its slender fuselage provided little room for equipment and bombs, and because it was heavier and had the same power as the B-45, it offered less performance. In addition, it encountered difficulties with engines and control surfaces during testing. The airmen cancelled the XB-46 in August 1947.

The first U.S. jet bomber was a conversion of the Douglas XB-42. After first adding two jet engines underwing (XB-42A), the USAF substituted two jet engines for the piston engines in the XB-43 version seen here. However, the bomber could not carry the five-ton atomic bomb and was outperformed by the rival B-45. (National Museum of the USAF)

The Martin B-48 also emerged from the AAF's August 1944 requirement. Compared with the B-46 it was chunkier in appearance, shorter in span and length, heavier, and powered by six engines mounted in two nacelles beneath its wing. The Martin bomber incorporated a number of unique features: a tandem main landing gear along with outrigger gear in the engine nacelles, a horizontal bomb bay that utilized clips for quicker and easier loading, and retractable bomb bay doors that lessened drag when opened. The bomber made its first flight in June 1947. It experienced control problems during flight testing, difficulties with government-furnished equipment (for example, requiring fourteen engine changes on its first forty-four flights), and could not meet the manufacturer's performance guarantees. More significant, a spring 1948 USAF study concluded that the B-47 had better performance and greater growth potential.

Of all the experimental bombers tested by the AAF/USAF, none has received more attention than the Northrop Flying Wing. Like the XB-43, it was a reincarnation of a prop design and was the most radical of the aircraft in the B-40 series. In June 1945 the airmen authorized Northrop to convert two of its prop-powered XB-35s to jet power with the designation YB-49. To compensate for the loss of stability when the propellers and propeller shaft housings were removed, Northrop added four trailing edge fins and four wing fences. The manufacturer deleted defensive armament, and while the aircraft could carry eight tons of bombs, the bomb bays could not accommodate the large nuclear weapons of the 1940s. The Northrop aircraft made its initial flight in October 1947. One of the two YB-49s crashed in June 1948, killing the crew of five, an accident probably caused by the aircraft's acknowledged stability and control problems.

The B-49 was Northrop's jet version of the B-35, in which the company substituted jets for the piston engines and added four trailing-edge fins and four wing fences. The resulting aircraft had control problems and lacked range and a bomb bay that could accommodate the atomic bomb. (National Museum of the USAF)

In April 1948 the Flying Wing competed for an Air Force's contract for a reconnaissance aircraft. It had two major advantages over its rivals: It had been extensively flight tested and the USAF desired to salvage some of its large investment in the aircraft. In August the Air Force awarded Northrop a contract for thirty RB-49As. However, in December 1948 the Air Force cancelled the RB-49 and instead purchased additional B-36s. The major factors in this decision were timing and performance. The Convair bomber was already coming off the production lines, while the Northrop aircraft would not be available until the appearance of the B-47, a bomber that was faster and able to carry the large nuclear weapons of the day. The USAF was especially concerned about the Flying Wing's stability during bomb runs. In October 1948 tests conducted along with a B-29 its pilots reported it was "extremely unstable" and that "it was impossible to hold a steady course or a constant airspeed and altitude."[2] The Flying Wing had twice the bombing error of the Superfort, which supported the airmen's general view that the Northrop aircraft was unsuitable for either bombing or reconnaissance work. Therefore the Air Force terminated the bomber program in March 1950, although testing continued. Too late for serious consideration, the last version, the YRB-49A, first flew in May 1950 powered by six engines, four housed internally and two underslung on separate pylons, marring its clean lines but providing more room for internal fuel.

For decades the Flying Wing has been celebrated by enthusiasts as a futuristic aircraft that lost out to an inferior aircraft (the B-36). There were charges at congressional hearings of corrupt and biased procurement, renewed by Jack Northrop in 1980 shortly before his death. Although the Flying Wing was faster than the B-36 (at least 50 mph top speed and 150 mph cruising speed), it had only about one-third the range, a restricted bomb-carrying capacity, and significant stability problems. Although these issues probably would have been solved, this would have taken both time and effort. The USAF wanted a long-range bomber as quickly as possible. As we shall see, the Northrop "design" would have a third opportunity, and three would prove to be a charm.

## The First American Service Jet Bombers

Only two of the jet aircraft in the "B-40" series went into production, the first of which was the not very successful North American B-45 Tornado. It illustrates the problems the airmen encountered converting to jet propulsion. The AAF pushed to get a jet bomber quickly into the field and ordered the immediate production of the Tornado in August 1946, months before the bomber's first flight the following March. The North American aircraft was powered by four jet engines mounted in two nacelles fitted into the straight wing and defended by a manned tail turret with twin .50-caliber machine guns. Aside from its bubble canopy and lack of propellers, it looked like a World War II bomber. The Air Force accepted the first B-45A in April and put the jet into service that November.[3] The company had to make "extensive post-completion modifications" involving strengthening the airframe, fixing the pressurization, and, as a result of a fatal crash in September 1948, increasing the span of the horizontal stabilizer and adding ejection seats for the pilots. The manufacturer did not solve the high-speed stability problems until May 1949. Other difficulties included component malfunctions, lack of spare parts and engines that created delays. One author has written that "unfortunately, the B-45 also earned an early reputation for being unreliable and a maintenance

The short-range North American B-45 was America's first operational jet bomber and, as a reconnaissance aircraft, the first U.S. jet bomber to see combat. The aircraft endured many problems, especially a very high accident rate. (National Museum of the USAF)

nightmare. Many of the aircraft's problems were directly related to new technology (the engines) or funding difficulties (radar and systems), but some were the simple result of the aircraft being inadequately engineered and rushed into production."[4]

North American modified fifty-five of the "A" models for delivery of tactical nuclear weapons in Europe. This project began in December 1950 and involved extensive modifications of the bomb bay (because of the secrecy of the atomic bomb project, the original bomb bay required rework) and adding extra fuel tanks and new electronic equipment. The first modified bombers arrived in the United Kingdom in May 1952.

The B-45 never dropped any bombs in combat, but the reconnaissance version served in the Korean War. In September 1950 three RB-45Cs arrived in Japan and in November went into action. One was lost in December to a MiG-15, the only B-45 combat loss. In January 1952 the unit was ordered to switch to night operations, for which it was ill suited. Throughout the rest of the war, Tornadoes flew over three hundred reconnaissance missions, including a dozen over China and Russia.

The Air Force had doubts about the B-45 early in its history, considering cancellation in June 1948 even before the first bomber went into service. In a fly-off competition for an USAF light bomber in February 1951, the Tornado lost out to the English Electric Canberra. The B-45 had a very high accident rate, over five times the major accident rate of the B-29, the least safe of the U.S. prop-driven strategic bombers of the era and almost ten times that of the jet-powered B-47. North American built 142 XB, RB, and B-45s. The USAF phased out the Tornado in 1959.

## The Boeing B-47 Stratojet

The Boeing B-47 Stratojet was also a troubled aircraft, but in the end it was a far more successful bomber. It began as a jet-powered B-29, but then Boeing, using captured German aerodynamics data, changed to a swept-back wing configuration in September 1945 that promised to increase speed from 400 to 555 mph. The company also moved the six engines from within the fuselage into four pods slung forward and below the wing, which eased maintenance and reduced the fire hazard. The tradeoff for the sleek aerodynamic wing was the absence of space for fuel, which limited range and required fuselage fuel storage that made fuel management crucial because of the large movement of the center of gravity as fuel was consumed. The slender wing also forced Boeing to adopt an innovative tandem (bicycle) main landing gear arrangement pioneered by Martin, which made for tricky touchdowns and the tendency to bounce and porpoise. The light and flexible wing also created control problems: decreasing aileron effectiveness at high speeds and, as speed increased, aileron reversal.[5] The bomber's high speed, clean airframe, and aileron issues meant that momentary inattention could result in an uncontrollable condition; if neglected, the B-47 quickly become very unforgiving. The XB-47 featured a bubble canopy over the pilots seated in tandem, resulting in a very streamlined ("clean") aircraft with superb looks and outstanding performance. The XB-47 made its first flight on December 17, 1947, the last of the forties series to fly, on the forty-forth anniversary of manned flight. Although the bomber initially fell short of expected performance due to its J 35 engines, it was seventy-five miles per hour faster than both the competing XB-46 and XB-48. The Air Force wanted

Aerial refueling was a major technical advance that extended the range of all types of aircraft. Three systems were put into service, the cross over, probe and drogue, and flying boom. Here a KC-97 with a flying boom refuels a B-47, giving the medium range bomber the ability to fly intercontinental distances. (National Museum of the USAF)

the bomber from the outset, although the production go-ahead was not given until September 1948.

The first B-47A reached SAC in October 1951 and proved a handful. The aircraft was underpowered and brakes were unsatisfactory, which meant that takeoff and landings required lots of runway and alert pilots. Landing approaches were also hazardous because the engines were slow to accelerate (spool up from idle to full power took seven to eight seconds), making go-arounds tricky. The "A" model was hardly an operational aircraft; it was unarmed, and only four of the ten built initially came equipped with the bombing-navigation system.

The "B" model addressed some of these issues. Boeing replaced the J 35 engines with higher power J 47s and mounted twin guns in an unmanned radar-directed tail turret. However, in a backward step pilot ejection seats were deleted, over strenuous SAC objections after an inflight canopy accident killed a pilot, in the belief that they were only marginally effective and to cut weight.[6] Boeing added vortex generators[7] that made handling more effective, two large external tanks, and air-to-air refueling capability.

Tests of the B-47B between August 1951 and April 1952 demonstrated that the bomber did not meet SAC requirements and revealed unsatisfactory handling during approach and landing, wing heaviness at high speeds to the point where full aileron use was required, and the necessity for close control of the center of gravity. The testers recommended a speed brake, improvement of the J 47 engine acceleration, and increased aileron effectiveness. Other problems emerged. The canopy was considered unsafe for high-altitude operations while the navigation-bombing system proved unreliable and difficult to maintain. An October 1954 inspector general report stated that the latter was the major limiting factor on the B-47's combat capability. In addition the Air Force cancelled the original tail gun system. In its place Boeing installed on the first 300 B-47Bs twin .50s fitted with radar warning but without radar gun laying and therefore capable of only harassing, not aimed, fire. Fuel boil off and purging presented problems along with leaking fuel tanks. Despite these problems production continued and the bomber was deployed to Great Britain in June 1953. Between that date and early 1958 the USAF kept at least one Stratojet wing based in England.

The four hundredth B-47 built featured ejection seats[8] and improved engines, a version designated B-47E. Later in the run the company added water-alcohol injection, which boosted engine power 25 to 33 percent, and strengthened the landing gear to allow greater operating weights (fourteen tons), which boosted combat radius to 2,050 nm, 300 more miles than the early "E" models. Defensive capabilities improved with the substitution of 20-mm guns for the .50s and the addition of radar tracking. More visible, the USAF replaced the internal rocket assist system for takeoff with an external horse collar that could be jettisoned. This increased power freed up internal space for twelve hundred more gallons of fuel and meant that the bomber did not have to carry twenty-two hundred pounds of rocket bottles for the entire flight. But as with so many other aspects of the bomber, the new device brought new problems. If

The Boeing B-47 was the United States' first operational strategic jet bomber. Although range limited and a handful to fly, it was fast for its day and reached a peak inventory of 1,560 in 1956, the most of any U.S. jet bomber. The B-47E was the ultimate model and, as shown here, Boeing substituted a jettisonable "horse collar" arrangement for the internally mounted rocket takeoff assist bottles. (National Museum of the USAF)

the bottles were dropped at speeds exceeding two hundred knots they hit the fuselage, and if all thirty-three bottles were fired simultaneously, their exhaust damaged the skin on the rear fuselage. Boeing also added a sixteen-foot-diameter approach chute and a thirty-two-foot-diameter braking chute. The first increased drag by one-quarter and allowed the pilot to carry more power on his final approach and thus ameliorated the problem of slow engine acceleration in the event of an abortive approach, while the second decreased landing roll 40 to 50 percent. The Air Force decided to upgrade most of the "B" fleet to this standard in April 1953, with the first renovated bomber delivered in March 1954.

The Air Force put additional demands on the bomber when the threat of surface-to-air missiles forced the USAF to adopt low-level bombing tactics. The command first publicly demonstrated the new tactics in 1957, the same year the Soviet SA-2 SAM became operational, and developed maneuvers for nuclear delivery. The more spectacular was the low-altitude bomb system (LABS), which caused both accidents and fatigue cracks. In response the Air Force initiated Operation Milk Bottle, May 1958 to June 1959, which inspected and repaired the bomber wing attachment to allow these maneuvers and devised a less stressful delivery tactic. Boeing delivered the last B-47E in 1957, the year the USAF began to phase out the Stratojet as the B-52 came on line.

In all the USAF accepted 2,041 B-47s. More than one thousand served in the operational inventory between 1954 and 1962, the last leaving SAC in December 1967. None bombed, but several penetrated Communist air space on reconnaissance missions and some were fired on.[9] One pilot summed up the B-47 by noting that although the bomber was "often admired, respected,

In the mid-1950s the U.S. Navy proposed an innovative concept of a sea-based strategic strike force featuring jet-powered seaplane bombers protected by jet-powered seaplane fighters and supported by seaplanes, ships, and submarines. Martin built two experimental bombers, the P6M-1 Sea Master, shown here, and later fourteen service and test aircraft. Cost, problems, and more promising alternatives led to the project's demise. (USN via Air University Press)

cursed, or even feared, [it] was almost never loved."[10] Others insisted the bomber was "a pleasure" and "a lovely aircraft to fly."[11] It did serve long and reasonably well in the transition between prop-powered U.S. bombers and the venerable B-52. Despite its problems, by 1954 the B-47 had the lowest accident rate of any Air Force jet, and over its lifetime its major accident rate was lower than any of the U.S. prop bombers.[12] The sleek jet was built in greater numbers than any other strategic jet bomber and served as the backbone of SAC for a number of years.

## A Navy Effort

The U.S. Navy sought to get into the nuclear delivery mission, as already noted with the P2V and AJ-1 prop-powered aircraft (see chapter 7). At the end of the decade the sea service took another tack. In March 1949 a Navy study advocated the use of long-range flying boats supported by ships and submarines in a variety of missions, including strategic bombing, a concept briefed to Congress late that year. This developed into a project known as Seaplane Striking Force (SSF), which would supplement aircraft carriers, using a striking force of large seaplanes supported by transport seaplanes, surface ships, and submarines and defended by fighter seaplanes. Its proponents claimed it would be cheaper and more flexible than land-based strategic bombers, as well as cheaper and more survivable than aircraft carriers.

The strike element was to be provided by a large, jet-powered seaplane. Martin won the contract in October 1952, and its Sea Master (XP6M-1) first flew in July 1955. The aircraft had four engines placed atop a high-mounted swept-back wing, tip tanks, and a horizontal stabilizer located atop the vertical stabilizer. Martin encountered numerous problems with engines, vibration, and controls and suffered the loss of two experimental aircraft. Nevertheless the company went on to build six service test models (YP6M-1) and eight production models (P6M-2). The latter had a gross weight of 160,000 pounds and a top speed of almost six hundred knots at sea level.[13] Cost escalation and technical problems undercut SSF, and at the same time the Navy made greater progress with carrier-based aviation and submarine-launched ballistic missiles. So while certainly innovative and initially promising, SSF could not compete with the alternatives, leading the Navy to cancel the project in August 1959 after spending more than a third of a billion dollars in this ten-year effort.

## The Enduring Bomber

One quick diversion is in order before discussing the B-52. In August 1950 Convair proposed an all-jet version of the B-36 and the following March was awarded a contract for the B-60. The metamorphous was achieved by sweeping back the wing and empennage, adding a retractable tail wheel to the tricycle arrangement, and substituting eight J 57 engines for the prop engines. The B-60's advantages included the potential to upgrade aircraft already in the inventory, interchangeable parts and manufacturing with the B-36 as the two aircraft were 72 percent common, as well as ease of maintenance. Convair converted two bombers, the first of which made its maiden flight in April 1952, three days after the first B-52 flight. As might be expected this conversion could not match the performance of the more recently designed B-52. In addition flight tests demonstrated a number of unfavorable aspects such as speed limits at low altitude (due to structural considerations), buffeting at high altitudes, unsatisfactory stability, and difficulties especially with the engines, auxiliary power, controls, and electric power. Air Force testers in 1953 summarized: "The YB-60 airplane did not prove to be a satisfactory airplane."[14] These flaws probably could have been worked out, but to no profit, as the B-60 was little improved over the B-36 (only a small air speed and no altitude advantage) and was markedly inferior to the B-52. In January 1953 the USAF cancelled the program.

When all is said and done, all of the jet bombers pale in comparison to Boeing's

Convair's B-60 was a B-36 fitted with swept-back wings and empennage and powered by eight jet engines. While boosting some economic advantages over the Boeing bomber, it could not match the B-52's flying performance and was little better than the B-36. The project was cancelled less than a year after its maiden flight in April 1952. (National Museum of the USAF)

B-52 Stratofortress, better known as the BUFF ("big ugly fat fellow" in the G-rated translation). It fought in many wars and is still in service over half a century after its first flight, a truly remarkable career. The B-52's concept dates back to AAF requirements in 1945 for an advanced bomber but not addressed by aircraft builders who believed it was "well beyond the state of the art."[15] The AAF realized design compromises would be necessary and that a proposal for an interim aircraft should emphasize high speed. Three companies submitted promising proposals in April 1946, and the AAF awarded Boeing a study contract in June 1946.

The Boeing design began as a straight-wing bomber powered by six turboprop engines, appearing much like a large B-29 except for a stepped up rather than glazed cockpit. The project continued despite concerns about size, weight, and range before Boeing arrived at a design that would take flight. This emerged from a storied meeting between Boeing and USAF personnel at Dayton in October 1948. According to some, the Boeing team went in with a turboprop design but was told that the Air Force was really interested in a pure jet-powered bomber and, over a weekend, submitted such a design. It was a notable achievement, but not quite as spectacular as the storytellers would have us believe. In fact the airmen considered changing to jets as early as the previous November and had asked the company in May 1948 to study a jet-powered B-52. Turboprops offered greater range while jets promised higher speed. What tipped the scale to turbojets were problems with the turboprop engines and propellers and the prospects that it would take four years to overcome these, by which time a jet-powered bomber would also be available. Headquarters USAF approved the jet-powered bomber in January 1949.

(Left) The Boeing B-52 could outperform the B-47, and its adaptability is indicated by its extraordinarily long service life beginning in 1955. It formed the bulk of the U.S. strategic bomber force from 1964 into the 1990s and remains in service today (2009). (National Museum of the USAF)

On a number of occasions the B-52 came close to cancellation. In early 1948, for example, the secretary of the Air Force stopped development of the B-52 as the Air Force looked carefully and approvingly on Northrop's jet-powered Flying Wing, and in January 1950 Headquarters USAF considered a number of alternatives to the B-52. LeMay championed the Stratofort, had a major influence on its design, first at Headquarters and later as commander of SAC, pushing its rapid development with 1954 as a target date because that was the projected "end of the useful operational life of the B-36" due to Soviet SAMs.[16] This led to production approval in January 1951.

The result was a bomber powered by eight engines in four pods, looking like an enlarged B-47. While the B-47 experience was of some help to the B-52 project, the two bombers were developed in parallel, not in sequence. The B-52 did resemble the B-47 with swept wings, underslung engines suspended on pylons, bicycle main landing gear, a braking parachute (but not the approach chute), and pilots positioned in tandem under a bubble canopy. The B-52's entire horizontal stabilizer moved for pitch control, a development just appearing on fighters, and a first for a large aircraft. The breakthrough was the emergence of the Pratt and Whitney J 57 engine, which provided adequate power and range. The engine suffered developmental problems, not unusual, and only passed its fifty-hour qualification five months before the initial B-52 flight tests. Engine surge, a condition analogous with backfire in a reciprocating engine sometimes resulting in engine flame-out, was another continuing problem that was not solved until November 1953. Nevertheless the bomber made its first flight in April 1952. Early testing uncovered no major problems, but it found some minor ones that led to a number of changes, which is the purpose of testing. Leaking fuel tanks were a principal and persistent problem and early on accounted for more than half the days lost to flight testing.[17] The bombing/navigational system proved unsuitable at the bomber's higher speeds and altitudes and there was a lack of lateral control in approach conditions. The bomber's control was enhanced by adding additional spoilers to the wing, which when actuated together acted as speed brakes and when employed individually aided roll. More serious was the finding that there was almost no stall warning in a landing configuration and a rather strong stall. These deficiencies were largely addressed in the YB-52.

The production models were altered based on this experience. The bicycle main landing gear gave way to a quadricycle gear (four wheels in tandem pairs forward and the same arrangement aft). These were installed with the ability to pivot 20 degrees either side of dead center, allowing the bomber to land in a crab configuration making cross wind landings easier and safer.[18] The tandem pilot seating, which made the B-47 difficult for the copilot to land as well as hindered cockpit coordination, was changed to side-by-side seating in mid-1951 at LeMay's direct insistence. Boeing added inflight refueling capability and two one-thousand-gallon auxiliary tanks, provision for ECM, and improved J 57 engines.

The airmen desired heavy defensive armament based on their World War II experience. The USAF considered numerous defensive measures for the B-52, including parasite fighters and manned wing-tip turrets. As late as 1952 Boeing was studying "all around" defenses, retractable 30 mm-cannon turrets, and a radar-directed (unmanned) tail turret. SAC considered these ideas to be an interim solution until air-to-air defensive missiles became available.[19] After seemingly constant delays and a number of problems, the decision makers settled on a manned tail turret armed with quadruple .50s that simplified sighting and was more reliable, although the installation added weight, complexity, and cost. These defenses should have been seriously questioned as guided air-to-air missiles for interceptors went into service in the late 1950s and rendered such defenses inadequate.

At the roll out of the initial production B-52A in March 1954, the chief of staff of the Air Force, Gen. Nathan Twining, remarked that just as "the long rifle was the great weapon of its day . . . today this B-52 is the long rifle of the air age."[20] Boeing built only three "A" models. SAC received its first B-52B in June 1955, a heavier aircraft propelled by more powerful engines with the first unit becoming operational in March 1956. The new bomber suffered a multitude of difficulties, most of all from a deficient bombing and fire-control systems. These were some of consequences of SAC pushing so hard for early deliveries of the new bomber over the objections of the developers, who wanted more time to work out the problems. Despite all of its warts, the "B" did have its moments of glory. In January 1957 three B-52Bs led by *Lucky Lady III* cut the time of the first nonstop around the world flight of the B-50A *Lucky Lady II* (February 1949) in half.[21] The "C" through "F" models were only modestly changed. Weights and range increased, and three-thousand-gallon external tanks replaced the existing ones. The "E" model also mounted two air-to-surface missiles (ASMs), Hound Dogs, on pylons between the fuselage and inboard engines.

In response to the increasingly effective Soviet air defenses, the B-52 changed to low-level tactics, not without problems, however, for the B-52 gave a rough ride near the ground and encountered new hazards. In addition to ground obstacles and bird strikes, turbulence could (and did) cause catastrophic failures as well as increased wear and tear on the aircraft.[22] The USAF assumed that low-level operations would increase structural deterioration by at least a factor of eight, a serious issue, for in contrast to the pervious Boeing bombers renown for their strength, the B-52 airframe was built light with a reduced permissible load limit factor to achieve higher speeds and longer ranges. The B-52's longevity exacerbated the problem. Whereas the prop-driven B-36 and B-50 had a fleet average of about twenty-five hundred flight hours before their phase out, Boeing designed the Stratofort for a five-thousand-flying-hour lifetime, a figure the B-52F fleet reached in 1967 and matched by the "G" and "H"models in 1970. Low-level flying (five hundred feet or less) led to bomber modifications, including changes to both the bombing-navigation system and the Doppler radar, strengthening the airframe, and adding terrain clearance radar and low-altitude altimeters. Most of these modifications were completed by October 1963. To further combat the growing defenses, SAC added additional ECM as well as standoff and decoy missiles.

The last two models in the B-52 series were somewhat different. The "G" had a wet wing (integral fuel tanks in place of bladder tanks) that cut almost three tons of weight and increased internal fuel, and although Boeing replaced the 3,000-gallon external tanks with 700-gallon tanks, fuel capacity increased from 41,600 gallons to 48,000 gallons and range increased. Spoilers replaced ailerons. Most visibly the vertical stabilizer was shortened eight feet from the (almost) triangular form to a cut down configuration and the tail gunner moved forward to join the rest of the crew.[23] The latter change saved one ton, and the other structural changes saved another six tons. Most of the "G" models were fitted to carry two Hound Dogs as well as two Quail decoy missiles. In the early 1980s the "G" was also modified to employ the air-launched cruise missiles. (The Hound Dog and Quail are discussed in chapter 9).

The Air Force modified the B-52 fleet with new electronics, for both offensive and defensive purposes. These included an electro-optical viewing system (EVS) for low-level flying (combining infrared and low-light TV sensors to allow flying as low as 250 feet above the ground), which was very useful and effective, and various ECM devices. In the 1980s Boeing installed the offensive avionics system (OAS) on the "G" and "H" bombers, an expensive program ($1.7 billion) that updated and integrated the aircraft's avionics. Both the "G" and "H" were modified to employ both short-range attack missiles (SRAMs) and air-launched cruise missiles (ALCMs) (see chapter 10).

The last in the B-52 line was the "H," which featured turbofan engines and a new defensive fire-control system. The engines increased power 30 percent yet reduced fuel consumption and increased range 10 to 15 percent, thus making the "H" the longest range B-52. In January 1962 a B-52H broke the unrefueled distance record with a 12,500-mile flight from Okinawa to Spain, breaking the 1946 record of the Navy's P2V *Truculent Turtle*. In March 1980 two B-52Hs completed an around-the-world flight of 19,500 miles in forty-two and a half hours, a feat duplicated by two other "H" models in August 1994 (20,000 miles in forty-seven-plus hours). Defensive firepower was bolstered by a six-barrel (Gatling) 20-mm gun in the tail.

The USAF received a total of 744 B-52s. During the two decades after 1966, and before the arrival of the B-1Bs, the B-52s dominated SAC, accounting for more than 80 percent of its bomber inventory during this period. In 1994 the "G" models were retired, reducing the B-52 inventory to ninety-five "H" models. Presently the USAF plans to retain the B-52 into midcentury. The Boeing bomber served not only as a major deterrent in the Cold War but also in combat, notably the Vietnam War and the Gulf War.

Along with the acquisition of the B-52, SAC received another valuable addition, the long-lived Boeing KC-135 Stratotanker. It took an unlikely track toward success. In a 1954 design competition it had ranked third; however, the advanced stage of the Boeing prototype promised an earlier delivery date compared with the paper planes of the

Convair's delta-wing B-58 was the first American supersonic bomber. One unique feature was its jettisonable, underslung pod, which carried both the weapon and part of the fuel. While a technological marvel, its decade of SAC service proved troublesome, expensive, and accident prone. (National Museum of the USAF)

winner and the runner up and thus earned a procurement order as an interim tanker. It was developed in parallel with the Boeing commercial jet transport, yet despite its appearance, the KC-135 shared less than 22 percent commonality of tooling and manufacturing with the 707 commercial transports. It made its first flight in August 1956 and in June 1957 became operational with SAC. The Stratotanker had higher performance than the prop-powered KC-97 and closer to that of SAC's jet-powered bombers. The KC-135 was especially important because it had four times the offload capacity of the KC-97s and the B-52B required twice the fuel of the B-47 for a full load. Boeing built 732 tanker versions along with others that served in transport, reconnaissance, and command and control roles. The number of SAC tankers peaked in 1961 at almost eleven hundred (three-fifths KC-97s and the remainder KC-135s).

## Supersonic Strategic Bomber

The Convair B-58 Hustler's career is strikingly different from that of the B-52. This was a remarkable aircraft, a radical design in comparison to the B-36 or the B-52. Despite its high performance it was not an effective military weapon because it brought little but expense and woes to the operators. The original concept was especially daring: a two-seat delta-wing bomber powered by three engines, one of which was to be jettisoned along with a fuel/weapon's pod at the target. Another unique feature was that the aircraft was to be carried, launched, and retrieved by a B-36. The parasite arrangement would shed considerable weight from the bomber, allow supersonic speed, and give it long legs. Thus it was half the weight of the competing, more conventional, and losing Boeing design. Although LeMay wanted a large bomber with great range, Headquarters USAF wanted a small bomber with supersonic speed.

By December 1951 Convair deleted the parasite concept and the jettisonable engine. In addition the company added a third crewmember (in tandem seating) with which it won a contract in November 1952. By March of the next year the design evolved in response to higher than expected drag at supersonic speed and the discovery by NACA of a solution, the area rule. Using the company's experience with the F-102, Convair pinched in the waist to give the fuselage its "wasp-waist" or "coke bottle" shape. Other elements included a 60-degree delta wing with a conical camber leading edge and four separate, underwing engine nacelles on pylons. The original ninety-foot-long pod was shortened to thirty feet and transformed from a conformal configuration to a pylon mount.[24]

Air Force opinion was divided on the Hustler. Although both the company and the Air Force developers were pleased with the progress of the design, the B-58 had numerous detractors and as late as the end of 1954 some of the top brass considered the design marginal. More serious, the intended user did not want the bomber. In January 1955 LeMay wrote the Air Force chief of staff, General Twining, that the B-58 lacked both the range and defensive capability required of an acceptable bomber and was "not desired in the SAC inventory."[25] In 1955 the USAF acknowledged that the B-58 would not meet its range and altitude requirements. With the program's high cost, SAC's objections, and the aircraft's performance difficulties it appeared it would be cancelled. Three factors saved the program. First, the airmen had already spent a considerable sum on the aircraft, three-quarters of a billion dollars by mid-1958, making cancellation difficult to justify to Congress. Second, Senate majority leader Lyndon Johnson was a strong advocate because the B-58 was to be built in Texas. Third, the USAF wanted a bomber to bridge the gap between the B-52 and its dream bomber, the B-70. Thus in August 1955 Twining ordered one wing to be equipped with the delta-winged bomber.

The bomber featured a number of innovations. The aircraft's supersonic speeds required new materials to deal with the heat generated, resulting in a metal and fiberglass sandwich that made up 80 percent of the airframe and 90 percent of the wing. The high speed also demanded new egress provisions for the crew. The answer was not easy because of its late addition to the bomber's already tight cockpit. The cockpit would enclose the crewmember in the event of cabin depressurization or egress and detach from the airframe to permit safe ejection at supersonic speed, absorb landing shock, and provide a flotation device in the event of a water landing. In late 1962 the USAF began to retrofit the Hustler fleet with the new system. In service, 90 percent of ejections were successful. The manufacturer also fitted the Hustler with advanced avionics built around a twelve-hundred-pound analog computer, and the navigation aids included not only search radar and a radar altimeter, but also an astrotracker and inertial system. The aircraft's defensive armament consisted of a tail-mounted 20-mm Gatling gun. The Convair bomber made its initial flight in November 1956, went supersonic the next month, and in June 1957 reached Mach 2. During its career it broke nineteen world records, setting six during one mission in January 1961. Although its top speed was listed at 1,000 knots, it set one record at 1,133 knots.

SAC was not enthusiastic about the speedy bomber, LeMay in June 1957 pointing out that it was not substantially superior to the B-52. Analysis concluded that the penetration ability and thus attrition of the two bombers would be about the same, but the B-52 could carry a greater bomb load farther. The range factor weighed heavy because air-to-air refueling was considered unreliable and costly. On the other hand the B-58 would replace the obsolete B-47 and complicate Soviet air defenses. The USAF planned to buy 290 B-58s but in the end got 116. President Dwight Eisenhower noted in his 1959 state of the union address that the bomber literally cost its weight in gold, clearly much more costly to buy and operate than SAC's previous bombers. The Hustler reached the command in August 1960 but did not become combat ready for two more years.

The USAF got a fast and troublesome bomber that had some dangerous and deadly characteristics. At high angles of attack it could pitch up and enter an uncontrollable spin. Landing was complicated by high landing speeds and a landing attitude from which the pilot could not see the runway when the aircraft was less than two hundred feet above it. Throughout its service it experienced tire and brake problems. Perhaps its most serious failing was the "occasional unreliability" of the flight controls, linked with a series of accidents that persisted for at least six years. A SAC general in May 1959 opined that the command was getting an aircraft that "will have limited operational capabilities, questionable reliability, and dubious maintainability."[26] He was correct. Throughout the bomber's service there was a long list of malfunctioning components and serious fatigue problems. SAC's new, low-level tactics cut range, introduced low-altitude control problems, increased structural fatigue, and along with the lack of terrain following/avoidance radar limited the B-58 to subsonic speeds. Designed for a different operational environment, the Hustler ended up range-limited with its primary virtue, supersonic speed, nullified. Convair proposed a number of improvements to enhance the bomber's value; however, the Air Force rejected them due to cost and because they posed a threat to the airmen's preferred bomber, the B-70. Not surprisingly, the B-58 had the highest accident rate of all of SAC's jet bombers except the B-45.

The B-58 did not see service in the Vietnam War. It lacked a conventional bomb capability, the USAF had other aircraft to do the job, its wing tanks would be vulnerable at low altitudes, and the Convair speedster had problems at low level, where it would be forced to operate. One source claims the aircraft flew one reconnaissance mission over Cuba during the October 1962 Missile Crisis.

It can be argued that the B-70 has garnered more attention than any other experimental U.S. jet and many operational aircraft. SAC wanted this supersonic, high-altitude, long-range bomber, but before it was ready, surface-to-air missiles made it highly vulnerable. The Mach 3 B-70 was an expensive technological wonder that had no military utility. Note the bomber's canards and downturned wing tips. (National Museum of the USAF)

In December 1965 Secretary of Defense Robert McNamara ordered the Hustler out of service by July 1970. The reasons given were the aircraft's problems, the prospect of the oncoming FB-111, modernization of the B-52, and the introduction of solid-fuel ballistic missiles. Ironically, after its longtime opposition to the Hustler, SAC wanted it retained. That was not to be. In November 1969 the first bomber went to the bone yard, where all were located by the end of January 1971, having served fewer than ten years. The B-58 was an innovative aircraft that added little positive to America's strategic force, was dangerous to operate, and expensive to buy and keep in service.

## Supersonic Failure

If the B-58 was a troubled bomber that did not satisfy SAC, the B-70 was exactly what the command wanted. Nevertheless it also was a troubled bomber and had the bad fortune to arrive at the wrong time. The B-70 turned out to be a technical wonder that has mesmerized the public with its dazzling performance—which obscures the fact that it was a failure.

In March 1953 SAC wrote Headquarters Air Force that the new bomber could embody "the longest range, highest altitude, and greatest speed (in that order of priority), capable of attainment in the time period under consideration."[27] In July 1955 the Air Force proposed six contractors for consideration, of which only Boeing and North American submitted proposals and both were granted contracts in November. To achieve the desired range, both manufacturers came up with a "floating wing tip," that is, disposable wings outboard of the fixed wing. LeMay wryly commented, "This is not an airplane. This is a three plane formation."[28] The Air Force rejected the submission in March 1956. The requirements at the time of the final competition in 1957 were a speed of Mach 3 to 3.2, an altitude of seventy to seventy-five thousand feet, and a range of 6,000 to 10,500 miles. Or simply put: faster, higher, farther. In December 1957 the Air Force announced North American as the contract winner.

The North American design was to use a high-energy fuel (boron) in its afterburner to reach the required range. The fuel had two major drawbacks, extreme toxicity (ten times that of cyanide) that required special handling and storage and great expense. In August 1959 the Air Force cancelled the high-energy program. As the efficiency of the standard jet fuel (JP-4) increased this was not a fatal blow, although it did force North American to use a different version of the engine. The cancellation of the nuclear bomber (essentially by December 1957) and F-108 interceptor (September 1959), however, were much more serious because the cost of common systems (bombing and navigation, air-to-surface missiles, bomber defenses, decoys, and engines) would now fall entirely on the B-70 program. The money crunch was such that Headquarters redirected the program in December 1959 from an operational wing to a single test aircraft.

The aircraft was not only large but also unique.[29] It mounted a canard control surface forward of the cockpit, a delta wing with a very sharp sweep back, and twin vertical stabilizers at the wing's trailing edge. Six side-by-side engines underneath the rear fuselage and wing powered the B-70. At high speeds the wing tips could be deflected downward to 60 or so degrees to increase stability as well as add 5 percent to lift. The bomber was to be manned by a crew of four and carry only ECM for defense.

New technology moved the aircraft toward its ambitious goals. More powerful engines and new materials helped, as did the discovery by the National Advisory Committee for Aeronautics (NACA), in March 1956, of the principle of supersonic wedge or compression lift. (At supersonic speeds the shock waves created as much as one-third more lift.) As the B-70 gained weight, it added a number of beneficial changes, such as a movable windshield and ramp that retracted in supersonic flight to decrease drag, and extended during subsonic flight to increase visibility, variable geometry inlets, and a trim able canard.

Despite the promised performance and technical success, a number of factors undercut the giant aircraft. Costs spiraled at the same time the bomber's strategic importance declined with the advent of intercontinental and sea-launched ballistic missiles. Soviet defensive improvements, especially the appearance of SAMs, made high-speed and high-altitude tactics untenable. This was known by the Air Force when North American won the contract in late 1957. There are some who would claim the success of the Mach 3 SR-70 reconnaissance aircraft indicates that the B-70 would have been a viable bomber. But this overlooks the fact that the B-70 was a much larger aircraft, and certainly not designed with stealth in mind as was the SR-70. Then the bomber got entangled in politics. Eisenhower was for economy and not especially for bombers, likening them in the missile age to bows and arrows in the age of gunpowder. The B-70 became an issue in the 1960 election. Presidential candidate John Kennedy repeatedly advocated bombers in general and the B-70 in particular. He criticized the Republicans for their neglect of bombers and stated, "I wholeheartedly endorse the B-70 manned aircraft."[30] Nevertheless, shortly after taking office, Kennedy cut the program to research and to a buy of three aircraft.

North American built and tested two aircraft; the XB-70A made its maiden flight in September 1964, and one was lost in a midair collision on a photo shoot in June 1966. The aircraft gathered valuable aerodynamic data despite difficulties with engine inlets, fuel and hydraulic leaks, landing gear and brakes, and corrosion of panels. In the late 1960s the National Air and Space Administration (NASA) used the sole B-70 in support of the abortive supersonic transport program. The B-70s made a total of 129 flights and reached a top speed of Mach 3.08 and an altitude of seventy-four thousand feet.

In the end the B-70 turned out to be a magnificent technology in search of a purpose. Times had changed as the aircraft developed, there was now no need for a "manned missile." More significant, the cancellation of the B-70 and the Skybolt air-to-surface missile (see chapter 9) in the early 1960s demonstrated the shift in strategic bombardment away from bombers toward ballistic missiles.

## RAF Jet Bombers

Other air forces also transitioned from piston to jet-propelled strategic bombers. The British needed a delivery system to make their deterrent creditable and to retain major power status but faced severe economic problems. Financial constraints ended the limited British ballistic missile research in 1948, which essentially consisted of firing only a few captured V-2s. The withdrawal of U.S. cooperation in developing nuclear weapons only made matters more difficult. In response to this need, the RAF fielded three jet bombers that formed the V-Force.

In late 1946 the Royal Air Force circulated a draft specification for a medium bomber capable of operating as high as fifty thousand feet at five hundred knots with a radius of fifteen hundred nautical miles. The same day authorization for nuclear development was approved in January 1947 the Air Ministry issued a specification that resulted in contracts for the Avro Vulcan and Handley Page Victor. Doubts about these two aircraft led to backup efforts. Because the authorities believed it would require eight years to meet these requirements and field such an aircraft, they added an insurance bomber, the Short Sperrin. The realization that the Short design would not meet specifications led the government to build yet a fourth bomber, the Vickers Valiant, with a performance between the insurance bomber and the two more advanced types.

The Short S.A.4 Sperrin was designed in response to an October 1948 specification for a four-engine bomber that could carry a 5-ton bomb to a target fifteen hundred nautical miles distant at an altitude as high as forty-five thousand feet and a cruising speed of 435 knots. Two were built with the first flying in August 1951. It was a boxy, straight-wing, unattractive aircraft with its two long and large engine nacelles enclosing engines stacked vertically and mounted midwing. Production plans were nixed before its first flight because of the much more promising performance offered by the Vickers Valiant.

The first of Britain's trio of V-bombers was the Vickers Valiant. While it did not distinguish itself in the ill-fated Anglo-French Suez campaign of 1956, it was the RAF's first nuclear bomber. The graceful bomber entered service in February 1955 and was prematurely retired ten years later due to severe metal fatigue, the high cost of repairs, and the advent of other V-bombers. (Air University Press)

Vickers also had competed for the high performance role and met all the requirements except range. It was a less daring and less risky design with lower performance than either the Vulcan or Victor designs but could be put into service sooner. In early 1948 Vickers got the go-ahead. The Valiant was distinguished by its swept wing positioned high on the fuselage, large oval (later "spectacle") intakes positioned on the leading edge near the wing roots, bulging dome cockpit, and four engines buried in the wing. The three V-bombers shared two common elements: a lack of defensive armament and a questionable crew escape system. As with the other V-bombers, the Valiant designers originally planned a jettisonable cockpit for emergency purposes, but this proved impractical for structural reasons. Instead all three V-bombers provided the pilots with ejection seats, while the other crewmembers were to exit through the entry hatch. This was controversial throughout the service of these aircraft as there was a clear difference in survival rates between those who used and did not use ejection seats in egress. The Valiant made its maiden flight in May 1951. The Valiant program was hindered by the crash of the first article, aileron flutter, and acoustic cracks in the rear fuselage; nevertheless the RAF began receiving the new bomber in February 1955.

Two dozen Valiants flew 131 sorties and delivered 942 tons of bombs during the week long October 1956 Suez operation. The aircraft's bombing and navigation equipment did not perform well in the Valiant's only combat operation and it is an understatement to note that "Suez did little to enhance the prestige of the British bomber force."[31] However, the aircraft did successfully fulfill its primary mission, strategic deterrence. By the end of January 1956 Bomber Command had a force consisting of eight Valiants and a number of nuclear weapons. In October of that year a Valiant dropped a nuclear bomb in Australian tests, the first dropped by an RAF bomber, and in May 1957 it dropped Britain's first hydrogen bomb, demonstrating both the weapon and the delivery system. The Valiants served in the nuclear deterrent role between 1956 and 1960, until displaced by the Vulcans and Victors. The RAF then shifted the Valiants to a tactical nuclear bombing role under NATO, replacing Canberras. Later the Valiants served in tactical, ECM, photo reconnaissance, and refueling roles.

The RAF discovered cracks in the Valiants following two crashes in May 1964. These were caused by wing spar metal fatigue exacerbated by the low-level flying, although the aircraft had flown only half of its designed fatigue life. Inspection revealed a severe problem that led the RAF to retire the entire Valiant fleet ahead of schedule in January 1965 because of the high costs of a fix and the aircraft's limited remaining life. Vickers delivered 107 Valiants to the RAF, a bomber considered by one historian to be an "exceedingly elegant aircraft, and one of the most graceful and refined of all British bombers."[32]

The June 1948 specification for which the Valiant unsuccessfully competed was for a medium bomber capable of carrying a

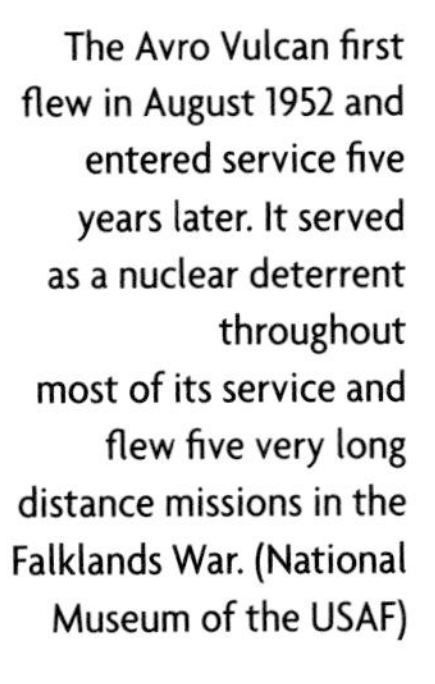

The Avro Vulcan first flew in August 1952 and entered service five years later. It served as a nuclear deterrent throughout most of its service and flew five very long distance missions in the Falklands War. (National Museum of the USAF)

5-ton bomb load half of a required distance of 3,350 nm. The Avro delta-wing design was the most radical of those proposed and won the contract for two prototypes in July 1947. The bomber prototype made its initial flight in August 1952, and the first production model reached a combat unit in July 1957. The close cooperation between the two English speaking nations was made clear the next year when SAC included the RAF bomber in its war plans. In 1959 the bomber conducted successful aerial refueling experiments with Valiant tankers.

The production bomber had a compound (curved) swept leading edge to cure problems of buffet near its maximum speed, which increased its redline speed from Mach .95 to Mach .98. The bomber also had a bubble canopy, mounted its four engines in the wing root, and was manned by a crew of five. The British quickly built an improved version, Mark 2, armed with a stand-off weapon, the Blue Steel supersonic air-to-ground, inertially guided missile, which became operational in February 1963 and stayed in service until the end of the decade. This bomber version also featured an enlarged and modified wing, more powerful engines, strengthened landing gear, and a tail cone housing ECM. It first flew in August 1957 and entered service in July 1960 with improved flying and combat performance. Like the other British and American strategic bombers of the day, it was adapted to low-level tactics and proved better able than the Victor to withstand low-level operations. However, the resulting reduction in range put Moscow outside its reach.

The Vulcans were slated to leave RAF service by the end of June 1982, but hostilities in the far off Falklands Islands intruded. For that conflict the RAF modified five Vulcans with a good maintenance history and hardpoints (originally for the cancelled Skybolt missile), which allowed carriage of additional ECM and Shrike antiradiation missiles. The RAF added an inertial navigation system and modified the bomb bay to handle a bomb load of twenty-one 1,000-pound bombs. Restoring the aerial refueling system proved a problem because of its long disuse and therefore required refurbishing the equipment and retraining the crews.[33] The retraining also included dropping conventional bombs, something not done in years.

The Vulcans flew very long distance missions to deny the Argentines the use of the Falkland's one airfield suitable for high-performance aircraft. The round-trip distance between the Falklands and Ascension Island was seventy-five hundred miles, the longest combat mission flown to that date. This required eleven Victor tankers inbound and another seven on the return to refuel one Vulcan six or so times, as well as tanker-to-tanker refueling, all to deliver a stick of twenty-one bombs. The RAF flew three missions against the Port Stanley airfield; on the first raid one bomb cut the runway, the second registered no hits, and a third used an airburst over the airfield. This was a very expensive undertaking but well worth it because the attacks deterred the Argentines from deploying their best aircraft to the islands. The Argentines saw the long-range attacks as a threat not only to the Port

The crescent-wing Handley Page Victor's chief advantage over its counterparts was its large bomb bay and ability to carry a heavy bomb load. After a relatively brief service as a bomber, the RAF converted the aircraft into an aerial tanker. (Air University Press)

Stanley airfield but also to the mainland and redeployed their best jet fighters to defend strategic targets in the homeland. (The British considered such attacks but decided instead to keep the war limited to the Falklands.) The Vulcans also flew two missions trolling for Argentine radars, using Shrike antiradiation missiles that caused some minor damage. The second of these resulted in one bomber landing and being interned in Rio de Janeiro, Brazil, when a damaged refueling probe prevented aerial refueling.

The British retained a force on Ascension Island after the war that stretched the RAF tanker force. Therefore six Vulcans were converted to tanker duties between 1982 and 1984. The RAF began to retire the Vulcans (Mk. 1) in 1966 and the Mk. 2s in 1981. Avro built forty-five Mk 1s and eighty-nine Mk 2s.

The Handley Page Victor was the most sophisticated of the three V-bombers. It was the last ordered and featured a large wing with a curved leading edge, four engines buried in the wing root, and a horizontal stabilizer mounted atop the vertical tail. The "crescent" wing was a novel approach, combining the advantages of a swept-back and delta-wing form without their disadvantages. The bomber made its first flight in December 1952 and entered combat service in April 1958. In March 1962 the RAF modified the first Victor as a receiver for aerial refueling and in 1964 converted some of the initial versions (Victor B.1) to tanker duties, first with a two-point and later with a three-point system.

The builder mounted an extended wing and bigger engines on the B.2 version, which went to the first operational unit in February 1962. Both versions could employ the Blue Steel ASM. One advantage the Victor had over its two older siblings was

its larger bomb bay and capacity to carry a greater bomb load. (It also had an altitude advantage over the Vulcan.) The Victor began to leave the bomber role in 1968 to take up tanker duties. The Mark 2 Victor had a ceiling of fifty-six thousand feet and a top speed of Mach .95; however, it did (accidentally) exceed the speed of sound in a dive. The Victor soldiered on as a tanker until the end of 1993. Handley Page built fifty Mark 1s and thirty-four Mark 2s.

The RAF intended to replace these bombers with the TSR.2 (Tactical Strike Reconnaissance Mach 2). It, like the American B-35, B-49, B-70, and F-111, was very controversial, and like the first three, has garnered more press than more successful, more numerous operational aircraft. One aviation historian writes that it "is likely to be remembered for all time as a textbook lesson in how not to arm an air force. The very mention of it arouses passions, some thinking that this aircraft should never have been cancelled and others that is should never have been started. Like so much of Britain's aerospace history, it was an amalgam of indecisiveness by the customer, massive and chaotic bureaucracy and disastrous involvement by the politicians."[34]

It is a classic understatement to note that the timing of this aircraft was most unfortunate as it got caught up in a major change

(Left) The British planned to replace the V-bombers with the supersonic TSR.2. Instead, a major controversy erupted and the combination of surface-to-air missiles, the lure of cheaper American weapons, and British politics shot it down. (U.S. Naval Institute Photo Archive)

in aircraft requirements, a monumental shift in British industry, and a turnover in politics. By the late 1950s the quest for higher and faster encountered the uncomfortable reality that surface-to-air missiles rendered such tactics problematic, mandating a transition to low and fast. The overall environment was laid out by the April 1957 British white paper that held that manned aircraft were obsolete and the future belonged to ballistic missiles for offensive and surface-to-air missiles for defensive needs.

Nevertheless, in the late 1950s a requirement emerged for a long-range aircraft capable of supersonic speed at altitude and for target dash, as well as high subsonic speeds at low altitude. The specific requirement called for a replacement of the Canberra tactical bomber with an aircraft with a combat radius of one thousand nautical miles, the ability to deliver a nuclear weapon at low altitude, and the capability to carry sophisticated reconnaissance equipment. One obstacle to the new aircraft was the Royal Navy's NA.39 Buccaneer attack bomber, which first flew in April 1958. It was a fine aircraft, but the RAF objected to its lack of supersonic speed, short range, limited avionics, and need for a long runway.[35]

In January 1959 Vickers-Armstrong and English Electric won a contract for the TSR.2, with the latter taking responsibility for the wings, rear fuselage, and tail and the former handling the rest. Multiple contractors were part of a push by the government to rationalize the British aviation airframe and engine industry and adopt new managerial processes. These moves led to poor management from both industry and government. If this arrangement was not enough of a challenge, the government overruled the project team and forced them to use an engine they did not want. One consequence of this arrangement was that initially the bomber's engine did not fit into the airframe! To complicate matters further, the out-of-power Labor Party took on the aircraft as a political issue, and as the only RAF combat aircraft under development, the TSR.2 became the focus of media attention.

The TSR.2 had a long fuselage and relatively small, delta-shaped wing with down-turned wing tips, along with a conventional empennage. It was powered by twin afterburning engines and crewed by two men seated in tandem. The TSR.2 made its first flight in September 1964 and went supersonic the next February. The bomber encountered problems with its landing gear, air conditioning, and especially with its engines. Although the British intended to buy one hundred of the advanced bombers, in April 1965 the project was cancelled. The most visible factors in this decision were high costs and the opportunity to buy the General Dynamic F-111 at half the price. Other factors in the aircraft's demise were engine problems, delays, the competition with the Royal Navy's Buccaneer and Polaris missile, and the election of the opposition Labor Party in October 1964, which favored buying American aircraft rather than developing British ones. In February 1966 the RAF ordered fifty F-111Ks, but two years later it cancelled this contract. The TSR.2 was the last all-British bomber and marked the end of the independent British military aviation industry, a sore point to some Britons to this day.

The Soviet equivalent to the B-47 was the Tu-16 Badger. It first flew five years after its American counterpart but served much longer, into the 1990s. The Russians built more than fifteen hundred, some of which were sold to other countries. Shown here is a Badger C carrying a Kingfish AS-6 air-to-surface missile. (U.S. Naval Institute Photo Archive)

The V-bombers gave the British a nuclear deterrent until Polaris submarines came on line and took over that responsibility in June 1969. These aircraft were fine machines that could outperform the B-47, their older, American medium bomber contemporary. However, during the early 1960s the V-bombers numbered only about 140 (October 1962) and were dwarfed by SAC, which at this point had 1,500 bombers on full alert. The V-bomber story demonstrated the prowess of the British aviation industry, the limits of the British economy, as well as the constraints of finances and politics on aircraft development and deployment.

## Soviet Jet Bombers

If the mainstay of the Russian long-range bomber force was the prop-powered Bear, the Tupolev Tu-16 Badger was its best known and most numerous jet bomber. This aircraft emerged a little later than the B-47 but served in a similar role. In 1948 Stalin ordered the Tupolev bureau to build a bomber with the range and bomb load of the Tu-4 powered by jet engines and therefore capable of higher performance. The engines were the key, for the arrival of the AM-3, the world's most powerful jet engine in 1951, allowed the designers to substitute two engines for four pylon-mounted engines. What became known as the Tu-16 (NATO code named Badger) first flew in April 1952, clocked 550 knots, but fell short of the range requirement. It competed with the Ilyushin Il-46, which had approximately the same performance, but as the Tupolev had greater range it was ordered into production in December.

Tupolev's production aircraft was four tons lighter than the prototype and thus had better performance and exceeded the range specification. It had a long fuselage and swept-back wings, with two engines in the wing roots and main landing gear fitted in two wing nacelles (pods). Manned by a crew of seven, the bomber was defended by three turrets, dorsal, ventral, and tail, each with twin 23-mm cannons, along with a fixed single nose gun. It entered service in early 1954. The Soviets began testing air-to-air refueling of the Tu-16 in 1955 and installed this equipment into the fleet in 1957, modifying 570 as receivers and 110 as tankers. Other Tu-16s were fitted with missiles, some for antiship duty and others for strategic service. (Initially the Badgers were deployed almost equally between strategic and naval forces.) The Soviets built over 1,500 of these bombers. As 1982 began, strategic forces had 490 Badgers, and it was the standard Soviet bomber into the mid-1980s, when it began to be replaced by the Tu-22M. In 1993 the last Tu-16s left the combat units.

The Badger served and fought in several different air forces. The Soviets employed them in the Afghan War and sold a number to Egypt, India, Indonesia, and Iraqi, some of which also saw action. The Egyptians procured twenty Badgers that were destroyed in the 1967 war, then got another thirty or so, using some to launch cruise missiles in the 1973 Middle East War and employing them in the 1977 four-day war with Libya. The Indonesians obtained twenty to twenty-five Tu-16s, which served in their 1960s war with Malaysia. The Indonesians broke off relations with the Soviets in 1965, neglected the bombers, and soon scrapped them.

The Myasishchev M-4 Bison was much less successful than either the Badger or its U.S. rival, the B-52. The bomber first flew less than a year after the Boeing bomber, but fewer than 100 were built, compared with 750 B-52s. Note the F-4 escort. (U.S. Naval Institute Photo Archive)

The Soviets sold eight Tu-16s to the Iraqis, who employed them against the Kurds and in their war with Iran. The last were destroyed in the 1991 Gulf War. The Soviets licensed the Chinese to build the bomber in September 1957, and in December 1968 a Chinese-built Badger flew for the first time. In one decade the Chinese built over 120, exporting some to Egypt and Iraq.

Just as the Tu-16 was a contemporary and rival of the B-47, The Myasishchev M-4 Bison was the same for the B-52, although much less successful. Its development began in the early 1950s with its initial flight in January 1953. The swept-wing Bison was powered by four jet engines buried in the wing roots, and like the B-52 it had a tandem landing gear and outrigger wheels (in this case mounted on the wing tips). It exceeded its speed requirement but fell short of the range requirement. Nevertheless, production began in 1955. In 1956 some were converted to tanker duties to refuel Badgers and other Bisons. Attempts to add air-to-surface missiles to the bomber failed because of the landing gear arrangement. A modification with more powerful engines helped satisfy the basic range requirement, and along with aerial refueling gave it intercontinental range. The Soviets made a few further modifications that employed different engines and permitted both bombing and tanker duties. Bison production stopped in 1963 after the Russians had built 93. The bombers remained in service into the late 1980s when they were removed in accordance with arms-control agreements. It made its last flight in the mid-1990s.

In a similar fashion the Myasishchev M-50/M-52 Bounder was the contemporary of the B-58. It had a relatively long fuselage compared to the span of its delta wing, a conventional empennage, and contrary to the normal Soviet designs which buried engines in the wing roots, mounted four engines on the wing, two on underslung pylons and two on the wing tips. The bomber made its maiden flight in October 1959. But the aircraft encountered aerodynamic problems as well as competition from Tupolev's Tu-22 and was cancelled in December 1960. Only a few were built, and these made but twenty-three flights.

The Tupolev Tu-22 Blinder was similar to the B-58 in terms of performance, roles, rewards (record-breaking flights and great publicity), and risks (high accident rates). However, although both developed at about the same time they were much different in appearance. Tupolev's entry into the supersonic bomber field began with an August 1954 study for a supersonic bomber. The final version had a sharply swept wing, needle nose, and twin engines mounted on either side of its vertical tail. The engine arrangement was a departure from the traditional, and while it had some advantages, it added weight (15 percent) and complicated maintenance by its location. Three men positioned in tandem crewed the bomber, which was defended by a remotely controlled tail turret mounting first two 23-mm cannons and later one 23 mm.[36]

Although the prototype first flew in June 1958, a few months earlier the Soviets allowed a basic redesign to incorporate the area rule. This model made its first flight in September 1959 and testing encouraged the Soviets to authorize production before the year was out. Pilots found the bomber

**(Right) The Tu-22 Blinder was similar to the B-58, as both set many records and had high accident rates. The Blinder was distinguished by its unusual engine configuration, two engines mounted on the vertical tail. However, the air and ground crews disliked the bomber because of its poorly designed cockpit, dangerous flying characteristics, and unreliability. (U.S. Naval Institute Photo Archive)**

difficult to handle in some flight regimes as it possessed radically different handling characteristics, experienced aileron reversal at high speeds, and problems with high landing speeds. They also severely criticized the cockpit layout for its lack of attention to ergonomics and limited visibility. Overall, the ground crews found the aircraft unreliable and the aircrews found it unsafe. Its downward ejection seats and frequent accidents fed its notorious reputation, leading some to call the aircraft "unflyable," and apparently some crews refused to fly the bomber.[37] Nevertheless, the Blinder went into service in 1962. Tupolev reportedly regarded his bold design as one of his "less fortunate creations."

The aircraft served in bombing, reconnaissance, ECM, and missile carrying (with the Kh-22 cruise missile semienclosed) roles, with the latter becoming its primary function. In 1965 the Soviets added aileron fences to solve the aileron reversal problem, changed to slightly more powerful engines, and fitted aerial refueling equipment to a number of the aircraft. Despite the accidents, short range, and limited payload capacity, the Soviets built 311 Blinders, and prior to the Soviet breakup in 1991, they had 155 to 180 Blinders in service. In the end, however, the bomber was a failure. It was intended as a replacement for the Badger, but it had little performance advantage, could carry only one missile while the Badger could carry three, and proved unreliable and accident prone.

Although about two dozen Blinders provided ECM support for Backfire bombers in Afghanistan, most of the bombers saw combat in non-Russian colors. The Soviets sold Tu-22s to the Libyans, who employed them in March 1979 against a town in Tanzania, in an off-and-on seven-year campaign over Chad during which SAMs downed two, and in strikes against targets in Sudan in 1984 and 1985. The Iraqis used Tu-22s in their war against Iran in the 1980s, hitting a variety of targets, including Iranian cities. The Iranians claimed to have downed at least three Blinders, and there are indications that as many as seven were destroyed during the conflict. The remainder were destroyed in the 1991 Iraq War.

Its successor, the Tu-22M Backfire, was perhaps the most controversial Soviet bomber, not for its aerodynamics but over arms-control issues. At the same time it proved to be much more important and a considerable improvement over its predecessor. The Backfire was designed to have a twelve-hundred-mile combat radius, a maximum speed of Mach 3, a cruising speed of Mach 2.8, and the ability to carry two Kh-45 ASMs for operations against U.S. aircraft carriers as well as targets in Europe and Pacific regions. Because Nikita Krushchev

The Soviets employed a variable-sweep (swing) wing on the Tu-22M (the Backfire C shown here), an aircraft that became the focus of contentious arms-control efforts. The supersonic bomber went into service in the mid-1970s and saw action in Afghanistan and Chechnya, but unlike other Russian bombers, it was not exported. (U.S. Naval Institute Photo Archive)

was unhappy with Tupolev, the Russians sent the requirement to other design bureaus. However, the dictator's ouster brought the aircraft designer back into favor, who then offered his design, along with the existing and tried Kh-22 ASM, as an alternative to the struggling and more expensive Sukhoi effort. In November 1967 the Soviets opted for the Tupolev aircraft. Despite its similar designation, the Tu-22M "Backfire" was entirely different from the Tu-22.

The Tu-22M Backfire first flew in August 1969. It reverted to the traditional Soviet engine arrangement, two engines buried in the wing roots, but did offer a novel feature, variable geometry (swing) wings. (Unlike its American counterpart, the F-111, the Soviet bomber's wing could only be locked in four positions, between 20 and 60 degrees.) The pilots were seated side by side, with the other two crewmembers positioned abreast behind them. Tupolev mounted a four-barreled single gun in the tail. While the Backfire was better regarded than the Blinder, it also had its woes. Engines had a short life, the ECM system caused problems, and the aircraft had difficulties with vibrations, leaky hydraulics, and flying low-level missions. The biggest issue, however, was political, regarding the aircraft's range. Although American intelligence sources were divided, some thought the bomber was capable of intercontinental missions, a significant issue as the Americans and Russians attempted to hammer out an arms limitation treaty. Consequently the Soviets agreed to remove the refueling probe from the Backfire to preclude it from serving as an intercontinental bomber. (To allay fears that the bomber could be quickly restored to aerial refueling status, the Soviets agreed not to enlarge its tanker fleet.)

The bomber went into service in the mid-1970s. The Russians fielded an improved version of the Backfire (M2) with different armament and avionics in 1978. A year earlier the M3 model first flew and introduced significant changes: more powerful and more fuel efficient engines, new air intakes, a single two-barreled tail gun, and the ability to sweep the wings to a fifth setting, back another 5 degrees. These improvements increased speed from about 830 knots to over 1,220 knots, extended range by one-third and doubled combat load. The Backfire C version went into operations in 1983. When the Soviet Union dissolved the air force had 210 Backfires in service and the navy 160, and by the end of the century, at least 125 and 45, respectively. The Backfire saw combat in both Afghanistan and Chechnya. The Tu-22M has not been sold abroad despite interest expressed by a number of countries.

Khrushchev and the Soviet missile establishment were instrumental in cutting Soviet long-range bombers and canceling a number of promising aircraft. However, the development of the American B-1 bomber pushed the Soviets to build a corresponding aircraft. (However, the Russian bomber remained in development even after the United States cancelled the B-1 in 1977, lending credence to the American lament that when the United States built arms, the Soviets built arms, and when the United States stopped building arms, the Soviets built arms.) The insistence on long range and the need to be able to use existing airfields forced the Russians to use a variable geometry design despite the weight penalty. The authorities did not believe that Myasishchev could produce his design, so it was given to Tupolev to develop, which by 1973 became the basis for the Tu-160.

The bomber featured a variable geometry wing, a blended wing and fuselage, and would be the first large Soviet aircraft to employ fly-by-wire controls. Tupolev considered a number of different configurations for the four engines, before adopting pairs mounted under the wing. The Soviets put the bomber into production prior to its first flight in December 1981. The sleek aircraft first went supersonic in February 1982, deliveries began in April 1987, and it entered service the next month. The Soviets authorized building one hundred in 1985, but production stopped at thirty-five when the Communist state imploded.

The Tu-160 is a huge aircraft, the heaviest combat aircraft of them all, yet it is said to be easier to fly than the Tu-22M. The Russian bomber has a crew of four and was to be armed with (Kh-45) ballistic missiles, two carried internally and the ability to carry two more externally. However, the American development of cruise missiles (AGM-86B) pushed the Russian designers to reconfigure the bomber to carry six to twelve Kh-55 cruise missiles in one or two rotary launchers or, as an alternative, twelve to twenty-four Kh-15 short-range ballistic missiles. Defensive armament initially was to consist of a tail turret mounting a six-barreled 30-mm cannon, but this was displaced by a radar warning device, ECM, and a Chaff and flare dispenser. The bomber encountered problems with the avionics although there were also crew comfort and maintenance issues. The designers decreased its radar signature with the use of radar-absorbing materials on the intakes, special paint on the forward fuselage, screens on the engines, and wire mesh glazing on the canopy. It was fitted with aerial refueling capability, although SALT restrictions rendered it unused. The Blackjack set forty-four world records for speed, payload, and distance during 1989–90.

The Tu-160 was not only spurred by the B-1 bomber but also ended up looking very much like it. The American bomber was the first into the air (1974 versus 1981), was cancelled in 1977, but then was revived as the B-1B, which flew for the first time in 1983. There are, of course, major differences between the two. Whereas the United States built one hundred, the Russians built only thirty-five. (The B-1 is discussed below.) The Russian bomber has variable geometry intakes and more powerful engines, and is heavier, larger, and faster than its counterpart.[38] There is some dispute over which has the smallest radar signature.

The breakup of the Soviet Union found the bulk of the Tu-160s (nineteen) in two squadrons stationed in the Ukraine. As the Ukrainians had the Blackjack but neither needed nor could afford it, they began talks with the Russians to make a deal. However, no agreement was reached and scrapping began. There matters stood until NATO took action in Yugoslavia in 1999, prompting resumption of talks and an agreement by which the Ukrainians traded eight Tu-160s (and three Tu-95s), as well as six hundred air-launched missiles, for Russian natural gas. These bombers formed part of the Russian strategic bombing force consisting of one

The Soviet effort to match, or supersede, the American B-1 produced the Tu-160. It featured a variable-sweep wing, blended wing and fuselage, and the first Soviet use of fly-by-wire controls. The Blackjack is larger, heavier, and faster than the B-1B but built in smaller numbers. Here it is seen in 1994 in Ukrainian service. (U.S. Naval Institute Photo Archive) unit of Blackjacks with fifteen bombers, one of Bears, and three with Backfires. An indication of the state of this force is that in 2000 the average annual flying time of its pilots was a mere ten hours. More recently (2008) the decline in the Russian military appears to have stabilized and perhaps reversed and that the Russians are upgrading their sixteen operational Tu-160s. The press reports that the Russians would like to resume production of the bomber and field a force of thirty by 2025 or 2030.

The two superpowers created different delivery strategies in the late 1950s. Whereas the United States adopted a triad consisting of manned, penetrating bombers, land-based intercontinental ballistic missiles (ICBMs), and submarine-launched ballistic missiles (SLBMs), the Soviets essentially phased out bombers to rely on a dyad consisting of the other two. This decision was fostered by both economic and military factors, the desire to cut costs along with the increasing lethality of defenses (longer range radars and surface-to-air missiles) and the long time to target as compared to the less vulnerable and much faster ballistic missiles. The USAF, along with the aviation industry, were strong lobbyists for manned bombers with much greater influence in the American process than their counterparts in Russia. And while the Soviets had an advantage with ballistic missiles, certainly initially, they trailed the United States in bomber technology. The Soviets were more comfortable with ballistic missiles, consistent with their strong artillery tradition, than with strategic bombing, with which they had essentially no experience. In addition, employment of ballistic missiles allowed greater centralized control than the use of manned bombers, of greater importance to them than to the Americans. As a result the Soviet leader pushed for missiles at the expense of bombers.

Chapter 9

# The 1960s and 1970s

## VIETNAM, NEW AIRCRAFT, MUNITIONS, TACTICS, AND TRAINING

The 1960s are remembered for a number of different reasons; however, in the context of United States and world history, two events stand out. One was a serious crisis, like no other, that frightened the world and fortunately ended peacefully and left the world a safer place. The other was a war that scarred America deeply and devastated a country. Strategic bombardment played a role in both of these epic events, the Cuban Missile Crisis and the Vietnam War.

### The Cuban Missile Crisis

In the fall of 1962 the world experienced a dangerous standoff between nuclear powers and came as close to a nuclear exchange as it ever has in its history. The crisis was brought about by the convergence of two elements of the Cold War, a Communist takeover of Cuba in the late 1950s and the arms race between the United States and Soviet Union. Fidel Castro's inflammatory rhetoric, America's fear of communism, and Cuba's proximity to the United States made him a particular irritant and seemingly a potential threat. At the same time the competition for military superiority and the near hysteria following the launches of the *Sputnik* satellites led to American fears of bomber and missile gaps. The Soviets saw an opportunity to offset the balance of power by capitalizing on Cuba's location, the inexperience of the new American president, John F. Kennedy, and the lack of will evidenced by the reverses in Cuba and inaction to the building of the Berlin Wall by emplacing ballistic missiles in Cuba that would menace a major portion of the United States. In mid-October U.S. intelligence found the telltale signs of this activity. Some American decision makers called for air strikes or invasion, but cooler heads prevailed and instead the United States clamped a blockade (termed "quarantine" to satisfy international law) on Cuba and demanded the missiles' removal. As Soviet merchant ships and submarines approached the quarantine line, there were fears of escalation.

The United States prepared for war, deploying aircraft, troops, and ships for a possible invasion of the Caribbean island. SAC cancelled leaves, recalled personnel, dispersed bombers to civilian airfields, increased the number of aircraft on ground alert, expanded the B-52 airborne alert, and put some two hundred ICBMs on alert. These measures improved the war-fighting posture of the strategic forces and sent a visible signal to the Soviets.

Within a week the Soviets backed down in the face of overwhelming American strategic and local tactical superiority and agreed to withdraw their missiles and medium bombers from Cuba.[1] In turn the United States pledged not to invade Cuba and shortly withdrew similar missiles from Europe. We now know matters were even worse than believed at the time as the Soviets had already deployed nuclear warheads to the island and had given local commanders the authority to use them. This was the most visible impact of strategic bombardment on world events in the Cold War. The Cuban Missile Crisis scared both powers, led to the "hot line" and the above-ground nuclear test ban treaty, and drew a line that prohibited similar provocative situations in the future.

## Vietnam

The second event took place a half a world away. The U.S. military built up its forces during the 1950s, reequipped with more modern weapons, yet found itself ill prepared for the next challenge. The military's assumption that if you could fight a big war you could also successfully fight a little one proved false. For the war machine designed to deter, fight, and win a nuclear war against a nuclear armed superpower, failed to achieve victory in a (mostly) nonconventional war against a small Asian country. It was another frustrating war like Korea, only more costly and less unsuccessful.

After the French defeat in and withdrawal from Southeast Asia in the early 1950s, American involvement increased steadily, yet the situation grew worse. The United States awaited an opportunity to increase the pressure and found it in the controversial August 1964 Gulf of Tonkin incident. This prompted (or better yet, provided an excuse for) a carrier strike on North Vietnam and the passage of the Gulf of Tonkin Resolution, which became the authorization for American intervention in Vietnam, as there never was a declaration of war. Following the American presidential election of November 1964 (electing Lyndon Johnson) and an attack on an American camp at Pleiku in February 1965, the United States responded with air strikes of the North that became a full-scale bombing offensive code named Rolling Thunder. In March the first U.S. ground units landed in South Vietnam, ostensibly to defend American air bases in the South, but in short order involved in offensive operations. America was now directly engaged in the war.

The goals of American intervention were unclear. Certainly they did not include a military victory over the North Vietnamese or an overthrow of that regime; this was not

to be a total war akin to the "last good war," World War II. Apparently one objective was to show the Communist that military actions could not win, that wars of national liberation (the Communist catch phrase of the day) and guerrilla war tactics could be defeated. At the same time American leaders wanted to avoid a larger confrontation. In brief, then, the goals were negative: not to lose and not to get into a larger conflict. Throughout there was a difference between the objectives sought by the military leaders, who wanted to undermine Communist capacity for war, and the civilian leaders, who wanted to influence North Vietnamese leadership to end their aid to the insurgents in the south. The objectives of the bombing campaign were set out in a March 1965 memo from Assistant Secretary of Defense John McNaughton to Secretary of Defense Robert McNamara that stated that the purpose of Rolling Thunder was to (1) reduce North Vietnamese will, (2) improve the morale of the South Vietnamese, (3) provide a bargaining chip, (4) reduce North Vietnamese infiltration, and (5) show the world U.S. determination to help an ally.

Seen here in November 1961 are (*left to right*) USAF Chief of Staff Curtis LeMay, Secretary of Defense Robert McNamara, and Chairman of the Joint Chiefs of Staff Lyman Leminitzer. McNamara served in this position under President Kennedy and President Johnson and was perhaps the country's most active, strongest, and most controversial secretary of defense. In war and peace, LeMay was the epitome of strategic bombing. (U.S. Naval Institute Photo Archive)

## Rolling Thunder

The air war in Vietnam was long and enormous, during which the United States dropped almost two and half times the bomb tonnage in Southeast Asia as it did in World War II and Korea, and about the same tonnage on North Vietnam as on Germany and Japan. Paradoxically the United States' major strategic weapon, the B-52, was used in South Vietnam against tactical targets, while fighter-bombers, F-105s and F-4s, were used in the North. It also should be noted that SAC was against using the B-52s in this war, arguing that this would expose its hardware and tactics to the Communists and diminish its effectiveness in its primary role, nuclear deterrence. American airmen applied air power with advanced technology, considerable skill, and much courage, yet this effort was thwarted by a third-rate military power.

This air campaign remains controversial. There are some (especially in the Air Force) who firmly believe that the bombing could have been decisive. "With the benefit of hindsight," a 1996 book boldly states, "it is clear that 'Rolling Thunder' could have been successful only if the civilian leaders in Washington had made a policy decision to use military power effectively."[2] The military wanted to administer a quick, sharp, overwhelming blow to the North Vietnamese. In August 1964 the JCS approved a plan to attack ninety-four targets in the North in sixteen days, which was later revised into an eleven-week plan to destroy most of the ninety-four targets. Surely this would have been easier in 1965 than later, because at that time the North Vietnamese had essentially no air defense system. But two contrary points stand out. First, the decision makers feared such actions might provoke a severe Chinese or Soviet response that might turn the conflict into another Korean War—or worse. Second, the Vietnamese Communists had already demonstrated great stamina in enduring the losses and punishment over the course of the ten-year war against France and would demonstrate the same resolve against the Americans over the next decade. While America was capable of bombing Vietnam back into the stone age, as Gen. Curtis LeMay's suggested in 1965, there is every indication the Vietnamese would have continued to fight.

In any event the air war was executed far differently than the military wished. Instead of a sharp massive blow, civilian leaders gradually increased sorties and expanded the target list as Rolling Thunder moved slowly northward, not attacking targets in the key Hanoi-Haiphong area until June 1966. The decision makers established a thirty-mile buffer zone south of the Vietnam-Chinese border and restricted zones centering on both Hanoi and Haiphong that could not be attacked without specific permission from the White House. Initially North Vietnamese airfields and SAM sites were left untouched, and throughout the war dikes and the population were off limits. Targets were authorized for two-week periods along with a specific number of sorties, and if the targets or sorties could not be serviced, the military had to go back to the White House for reauthorization, which was not always forthcoming ("use it or lose it"). There were also bombing halts ("pauses") for diplomatic purposes that permitted the North Vietnamese to rebuild, resupply, and regroup.

The graduated and sporadic application of force allowed the North Vietnamese to adjust to the air assault. When Rolling Thunder began the North Vietnamese had few fighters and experienced pilots and no SAMs. But over a period of years they were able to integrate small numbers of dated technology into an effective air defense system, and while their "guerrilla tactics of the air" were unable to stop American air power, they made it costly. The air war over the North provided the Vietnamese with American prisoners who became hostages and pawns in the propaganda war. Most important, the bombing allowed the North Vietnamese to rally their people to support the war and the regime as gradual American bombing and effective North Vietnamese defenses bolstered civilian morale, not the contrary. American restraint lessened the impact of the air campaign while still inflicting enough intentional and unintentional harm to anger the population. The Communists were also able to muster world support by presenting Rolling Thunder as a David-like struggle of a peaceful people against a cruel, aggressive Goliath. Further, it is now clear that until the 1968 Tet Offensive the war was an unconventional conflict fought by South Vietnamese (Vietcong) guerrillas requiring minimal help from the North. Fervent Vietnamese nationalism, fierce pride, and a strict authoritarian state bolstered the Vietnamese will to fight despite heavy losses, disruption, and destruction.

The American airmen also must share responsibility for the failure of the bombing campaign in terms of aircraft, equipment, tactics, and training. Aircraft designed for a short nuclear war proved unsatisfactory in a long conventional one. The B-52 was the only Air Force aircraft at the beginning of the conflict with true day-night, all-weather capability, but it would not be used over North Vietnam in force until 1972. Unguided bombs lacked sufficient accuracy and precision-guided munitions did not go into action until late in the war. Nuclear delivery tactics proved unsatisfactory in the air war over North Vietnam. ECM for fighters was inadequate and marginal for the B-52s.

A final factor that undermined the bombing campaign was the lack of strategic targets in North Vietnam. Even if the bombing of Germany's and Japan's economy was decisive in World War II, a dubious and certainly debatable assertion, the Vietnam conflict was decidedly not a total war between equivalent industrial countries. The Korean and Vietnam Wars indicated how air power could be thwarted by a Third World country, with a simple agrarian economy, a high tolerance for casualties, abundant manpower, tight authoritarian control, and outside supply. Because of a fear of Chinese and Soviet intervention, political repercussions, and morality, some key targets were not hit: Vietnamese ports were not blockaded until 1972 and dikes and people were not deliberately attacked.

When the bombing stopped in mid-1968, America had been defeated. While the indigenous insurgency in the South had been beaten, increased numbers of North Vietnamese took up the slack while the American will for the war had been shattered. The considerable American bombing of North Vietnam (during Rolling Thunder the United States dropped three-quarters of the tonnage it unloaded on Germany and Japan during World War II) had not achieved McNaughton's five objectives. North Vietnamese morale did not decline, on the contrary it was enhanced by the bombing. South Vietnamese morale may have been helped, but not enough to positively affect the war effort. The bombing proved to be an ineffective bargaining chip; in fact, it was turned against the United States in both domestic and international politics. Infiltration was reduced from what it might have been, but the Communist supply requirements were meager and they demonstrated their ability to mass resources in the country-wide Tet Offensive in early 1968. Finally, Rolling Thunder may have shown U.S. determination, but to many it showed a wrong-headed policy. Along with the draft, American casualties, critical news coverage, and an apparent lack of progress, the bombing made the war increasingly unpopular. If anything, North Vietnam appeared stronger in body and mind during and after the bombing. A CIA report noted that it cost the United States eight dollars (in operating costs and aircraft lost) for every dollar of damage it inflicted in the North. More important, another intelligence report concluded that for every dollar of damage the North Vietnamese suffered, the Chinese and Soviets shipped in six dollars of assistance. A December 1968 study concluded that while Rolling Thunder had caused severe strain in the North, overall it was a net benefit to the government.

## Linebacker I

The Vietnam War was just a bad memory for most Americans by the early 1970s. The U.S. military personnel in South Vietnam had peaked above half a million in April 1969 and by March 1972 had fallen below one hundred thousand, of which only six thousand were combat troops. The draft was history. The USAF had fewer than four hundred combat aircraft in theater, and the South Vietnamese flew more than three-quarters of the four thousand Allied air strikes flown in South Vietnam in February 1972.

Allied reconnaissance and intelligence detected signs of an upcoming North Vietnamese offensive in 1971, but as with the 1968 Tet Offensive, the timing and extent were misjudged. The North Vietnamese blow fell early on March 30, 1972. What is important to emphasize is that this was a massed conventional invasion that eventually involved almost the entire North Vietnamese army, uniformed regular forces, not guerrillas. These Communist units employed a full range of modern equipment with the conspicuous exception of aircraft.

How would the United States react? The situation had drastically changed from what it had been in the 1960s. There was a different administration in Washington; perhaps more important, the Chinese and Soviets had less interest in the Vietnam War and were seeking to work out an arrangement with the Americans, which the administration was able to exploit. When the invasion exploded, this diplomatic effort had progressed to the point where the president had met with the Chinese in Beijing in February and was scheduled to meet with the Soviets in Moscow in May, thus the fear of Chinese or Soviet intervention was largely removed. On the other hand the American people had no stomach for war in 1972; clearly the United States would not commit ground forces. Therefore the American reaction would be air power or nothing. President Richard Nixon exclaimed, "The bastards have never been bombed like they're going to be bombed."[3] Initially he suggested a three-day bombing campaign against Hanoi and Haiphong. After the JCS disappointed him with a plan he called "too timid," the president's assistant, Gen. Alexander Haig, came up with a wider air campaign that included the long-sought-after mining of North Vietnamese ports. The air effort was code named Linebacker, later Linebacker I, and would last from May 10 to October 22, 1972.

In May the Navy planted mines that blockaded North Vietnamese ports. Meanwhile the United States rapidly reinforced its air forces in Southeast Asia, including sending over half its B-52s to Guam and Thailand and boosting B-52s from 42 aircraft and 57 crews to 200 and 275, respectively. During the first six months of 1972 half of the B-52 flying hours were flown in Southeast Asia, a buildup that strained SAC to the extent that some of the nuclear war targets could not be covered.

Initially most of the B-52 effort was in South Vietnam and only gradually did it cross the border. While SAC pushed for bombing North Vietnam, the theater commander strenuously insisted on keeping the Stratoforts bombing in South Vietnam. In April B-52s attacked targets in the North for the first time. By late May the Communists' assault began to falter, and the South Vietnamese began small counter attacks. There was some movement in July toward hitting airfields and later rail yards in the North, but SAC presented a wider plan in September. As the Communist drive stalled, diplomatic activity increased and the wider bombing was put on hold.

Linebacker I differed from Rolling Thunder in two ways. For one, it had clear goals: It was aimed at North Vietnam's ability to make war, its army in the field, its supplies, and its transportation infrastructure. Another difference was that the various prohibited zones were reduced. Nevertheless the bulk of the targets were interdiction. There were few strategic targets, and because of fears of collateral damage and the resulting bad press, a number of prime targets were not attacked. As the Communist ground offensive stalled, interest on both sides shifted to the diplomatic arena and in late October the United States stopped bombing north of the twentieth parallel.

During the period April through October the airmen dropped 156,000 tons of bombs on North Vietnam, and because of the use of guided bombs, the bombing was more effective and cheaper than it had previously been. This was significant as the Communist invaders required large amounts of supplies to conduct their conventional attack, a far different situation from the minimal logistical requirements of the previous guerrilla war. Linebacker I was an overwhelming success and indicated what modern air power could do in a conventional conflict. Coupled with the resistance of the South Vietnamese ground forces, air power inflicted heavy casualties on the invading Communists and interdicted their much needed supplies and halted the offensive.

## Linebacker II

The diplomatic negotiations did not yield peace as first the South Vietnamese, then the North Vietnamese balked. A complicating factor was the November 1972 American election, which overwhelmingly reelected Nixon yet brought in a Democratic Congress intent on pulling out of Vietnam. On December 12 the North Vietnamese negotiators left the peace talks, the Communists apparently counting on bad weather and the incoming American Congress to reap victory for them. They seriously misjudged Nixon and underestimated the capabilities of the American airmen. On December 14 the president demanded positive action from the North in seventy-two hours—or else. The "or else" was a concentrated and powerful air assault on North Vietnam.

For some time the Air Force had pushed for B-52 strikes on Hanoi, a move resisted by both the commanders in the theater and the JCS, who feared the compromise of U.S. equipment, the addition of American captives, a domestic and international outcry, and heavy civilian casualties. The secretary of defense and secretary of state were opposed, the chairman of the JCS, Adm. Thomas Moorer, was ambivalent, and even top security advisor Henry Kissinger was hesitant. Nevertheless the president made up his mind and ordered the bombing. Unlike Linebacker I, which was aimed at the North's war making capability, this bombing intended to break the will of the North Vietnamese by hitting targets in the Hanoi/Haiphong area. Admiral Moorer told the SAC commander, "I want the people of Hanoi to hear the bombs." He added, however, that he intended to "minimize damage to the civilian populace."[4]

Although the Air Force may have wanted to bomb the North with maximum effort for some time, SAC proved ill prepared for the campaign, lulled by its seven-year uncontested bombing experience. It had not run exercises against captured Soviet radars or SAMs or evaluated the viability of its tactics long used successfully in the benign skies over South Vietnam. The proposed strikes with one hundred bombers would be far different than those that employed no more than thirty bombers at a time and required coordination with escorting fighters, defense suppression aircraft, and ECM. The problem was complicated by the fact that the operation was run from Omaha, far from the field in distance, time, and perspective.

The Air Force sent all of its conventionally configured B-52Ds to the theater as well as one hundred B-52Gs. The B-52G had a range advantage over the "D" model, allowing it to complete a mission from Guam without aerial refueling, while the "D" required at least one refueling. This advantage was counterbalanced by the fact that the "D" had been modified for conventional warfare, which allowed it to carry many more bombs both internally ("Big Belly" modification) as well as with racks for external carriage: thirty tons versus "only" nine tons on the "G." There was also a question concerning the B-52G's vulnerability compared to its older teammate because of its lighter structure and "wet" wing, and as only half the "G" models had the most advanced electronic jammers.

Another self-inflicted wound concerned target selection. The JCS sent SAC a target list that showed an ignorance of the B-52's strengths and weaknesses. (An alternative explanation is that the top military leaders put political considerations, hitting certain targets, above military considerations.) The Stratofort's advantages were its large bomb load and all-weather capability. However, its accuracy was measured in hundreds of feet, for the bombers were not configured to employ guided munitions, which achieved accuracies measured in tens of feet. The JCS listed airfields as a B-52 target although other attackers and fighter escort could deal with Vietnamese fighters. At the same time no effort was made to hit SAM sites or storage facilities. The costly consequences of these decisions would be clearly demonstrated during the campaign.

The air offensive, code named Linebacker II, began on the night of December 18. At long last the United States had the opportunity to use its air power rather freely in a major strategic effort. Operations on each of the first three nights were essentially the same, F-111s hit North Vietnamese airfields before about one hundred B-52s flying at high altitude, all on the same heading, in three separate waves, bombed targets in Hanoi and Haiphong at night. In some respects these bomber streams ("elephant walks," as they sometimes were called) resembled the World War II bombing of Germany and Japan, with one critical difference; in 1972 the airmen used identical tactics, in the same area, three nights in a row. The Communists launched more than two hundred SAMs and downed three Boeing bombers on the first night. MiGs attacked the bombers, one of which was claimed destroyed by a B-52 tail gunner. On the second night the North Vietnamese fired over 180 SAMs, but there were no losses. The third night was much different. The USAF launched ninety-nine sorties, of which ninety were effective, and lost six bombers, all to SAMs. When SAC learned of the loss of three bombers in the first wave, it considered canceling the two in-bound waves. Such a move would have major consequences; therefore, the chief of staff of the Air Force and the JCS weighed in on the issue. In the end the decision makers considered the loss of SAC's perceived ability to penetrate enemy defenses more important than possible B-52 losses. (A matter of pride was also involved as no AAF/USAF mission had ever been turned back by fear of losses.)

The stereotyped tactics help explain the losses. The separation between the three waves split up the support aircraft and allowed the defenders to reload their missiles. The same course permitted the Vietnamese to anticipate the routing of the successive bombers. And the routing, which reversed direction shortly after bomb release, turned the bombers into a head wind that slowed the bombers, and that steep turn blanked out some of the ECM, rendering the bombers more vulnerable to SAMs. The B-52 crews strongly objected to the order banning evasive maneuvering on the bomb run, although that had been allowed previously. (One commander threatened courts-martial for crews who took evasive action on the bomb run.) A complicating factor

**During the Vietnam War B-52s delivered huge tonnages of bombs in South Vietnam, facing no enemy opposition but achieving questionable results. In 1972 B-52s helped stop the Communist offensive (Linebacker I) and then bombed and brought the North back to the peace table (Linebacker II). Here a B-52D unloads thirty tons of bombs. (National Museum of the USAF)**

was SAC's micromanagement from Omaha and the difficulty of coordinating a large number of bombers and support aircraft from afar. SAC headquarters was slow in getting plans to the units, and more seriously, disregarded the requests, recommendations, and protests from the field for changes. The irony is that the USAF had criticized micromanagement from Washington throughout the war, yet when given its head, substituted SAC micromanagement from Omaha.[5]

The stiff losses forced the airmen to alter both their targets and tactics. One shift was to target the SAM sites and storage facilities, beginning on the fifth day with Navy and Air Force tactical aircraft and with B-52s on the sixth mission on December 23.[6] The airmen also dramatically revised tactics, for the next four nights limiting bombing to one wave of thirty B-52Ds attacking targets away from Hanoi. The airmen also changed from attempting to lay a Chaff corridor, which the winds dispersed or moved, to laying a Chaff blanket. (This was successful for while the former had covered about 15 percent of the aircraft, the latter gave protection to 85 percent.) After a thirty-six-hour stand-down over Christmas, the B-52 bombing resumed with larger numbers and a return to the Hanoi area with tactics that featured compressed attacks from multiple headings, attacks of SAM sites, and more effective Chaff tactics. On December 27 the North Vietnamese sent a message to

Washington that they were ready to restart the talks. One reason for the North's change of heart was that they were running out of SAMs; on the night of the twenty-sixth they had only half the number they had had available for action on the first night of the campaign. They were essentially defenseless. On the other side, American airmen were tired and had run out of significant targets. (All that remained would be direct attacks on the population or dike system.) On December 29 the airmen flew the last of the eleven Linebacker missions.

The bombing achieved its objectives of getting the North Vietnamese back to the peace table and an agreement was signed at the end of January. During Linebacker II American airmen dropped twenty thousand tons of bombs (three-quarters by the B-52s) on 708 B-52, 657 Air Force tactical, and 505 naval sorties.[7] Civilian casualties were remarkably light considering the intensity of the operation, a result of planning, luck, and Vietnamese evacuations. The North Vietnamese claim that nineteen hundred civilians were killed, a very small number since World War II experience would predict a loss of ten thousand to forty thousand lives from this bomb tonnage. The cost, according to U.S. sources, was twenty-seven American aircraft, of which fifteen were B-52s downed by SAMs. (SAC had predicted a higher toll, the loss of 3 percent of the attackers.)[8] The North Vietnamese claim to have downed eighty-one American aircraft, of which thirty-four were B-52s (all but four of these attributed to SAMs).

The North Vietnamese were defeated, not by superior strategy, tactics, or technology, but by numbers and courage. Both bombing accuracy and bombing damage were poor: a CEP of twenty-seven hundred feet (not the predicted eight hundred feet) and damage was less than a quarter of that expected. The Stratoforts had suffered significant losses against a dense, although dated and limited air defense system.

Nevertheless Linebacker II has been held up as air power at its finest, fairly unrestricted bombing achieving success. Some have gone as far to proclaim it as a war winning strategy that could have brought victory if it had only been used at the beginning of the conflict. But to be clear, while Linebacker I was a clear success, air power playing a major role in defeating a ground assault, Linebacker II gained the same terms as what had been agreed to in October. It did demonstrate American resolve (certainly Nixon's) and did reassure the South Vietnamese. However, what is overlooked is that the diplomatic situation had radically changed from the beginning of the war. In 1965 there were genuine fears that American actions might trigger a response from the Chinese or Soviets that would spill over the borders and enlarge the problem. Freed of these worries, the 1972 actions were less restrained. Most often overlooked is how close Linebacker II came to failure.

During the entire war the USAF lost seventeen Boeing bombers in combat, wrote off two due to battle damage, and lost another dozen to operational causes. SAC flew almost 125,500 sorties, of which only 6 percent were over North Vietnam. An important consequence of the Vietnam War was its impact on the USAF. SAC had dominated the Air Force and the U.S. military during the 1950s, but now its influence receded. The dominance of bomber generals diminished in favor of the fighter generals. It now would be a different Air Force.

## The F-111

It was during the Vietnam War that the Air Force's next bomber first saw action, albeit a tactical version in a tactical role. The F-111 was probably the most controversial USAF aircraft of all time from the contentious selection of General Dynamics as its builder, through its problems, and especially its troubled early combat service. These problems and high cost generated a flood of criticism, prompted considerable press attention, and congressional hearings. Despite all of this, two decades after its less than glorious combat debut in Vietnam it flew in the Gulf War with advantages over other USAF aircraft operating in that conflict.

The F-111 story begins in March 1958, when the USAF put together specifications for a Mach 2-plus, all-weather fighter capable of vertical or at least short takeoffs and landings. Tactical Air Command (TAC) wanted an aircraft capable of getting to Europe without refueling, operating out of small and rough fields, flying at supersonic speeds at altitude, and engaging in low-level, high-speed attack. Despite these daunting requirements, in 1959 NASA convinced TAC that this could be done using a new technology: a variable geometry (moveable or swing) wing. This resulted in a specific operational requirement released the next summer. The most important requirement that would shape (some would saw doom) the project was to achieve Mach 1.2 at low level for four hundred nautical miles. Meanwhile the Navy was looking for a replacement for its F-4 as a fleet air defense fighter that would have extended loiter and the ability to carry a number of heavy air-to-air missiles. The Dwight Eisenhower administration believed that one aircraft would be able to do both jobs and in so doing save money and therefore directed the services to work together. However, it was the next administration's secretary of defense, Robert McNamara, who pushed the concept of a common aircraft. In October 1961 the request for proposal was issued and the defense secretary announced that he was thinking in terms of an aircraft that was 85 percent common with a combined Air Force–Navy production buy of over three thousand.

The key to combining the ability to fly far and long with high speed was the variable geometry wing. The range goals required a large wing, while the speed goals mandated a small wing; a variable wing could do both. The United States got two such aircraft into the air, the Bell X-5 and the Grumman XF10F. The wings of both pivoted within the fuselage of the aircraft, and as the wing moved it caused a shift of both the center of lift and center of gravity, which induced aerodynamic changes. NASA modified the concept, pivoting only the outer portion of the wing and thus finessing the aerodynamic problems. This represented a giant breakthrough, as engineers calculated that to achieve the range and speed requirements with a fixed wing would require an aircraft of over 100,000 pounds, while a variable sweep- wing aircraft could weigh about 60,000 pounds. Other new technologies would also be useful. Turbofan engines with afterburners promised high thrust with increased fuel efficiency. New materials such as titanium were lighter and stronger, albeit, more expensive than existing materials. Advances in avionics enabled improved navigation and bombing as well as increased range of detection of hostile threats. Clearly this aircraft would be on the cutting edge of technology.

In December 1961 six companies responded, with the proposals by Boeing and General Dynamics judged far above the rest. The military, with the Air Force as the lead service because it was getting the bulk of the buy, concluded that the Boeing design had a number of performance advantages and a lower cost, while the GD design was less daring and thus less risky and offered higher speed. In the final analysis, the most important advantage of the GD design was that its Air Force and Navy versions were 84 percent common by weight contrasted with Boeing's 34 percent. The military picked the Boeing design four times, and each time was rebuffed by their civilian superiors. It came down to the fact that the Air Force had more confidence in Boeing (having built the very successful B-17, B-47, and B-52) and stressed the superior performance promised by the company. The Navy really didn't want either design. The civilian decision makers had a different view, believing that the GD design was less risky, more realistic in cost estimates, and, most of all, more common. As the military agreed that either design could meet the basic requirements (and that the performance gap between the two designs had narrowed during the course of the four competitions), that was enough for the decision makers. In retrospect it appears that Boeing designed an aircraft for the Air Force, while GD designed one for the secretary of defense. Therefore it should come as no surprise that in November 1963 McNamara announced that GD had won the Tactical Fighter Experimental (TFX) contract.

McNamara's four reversals of the military bothered some. More serious were allegations that the decision was politically inspired. GD needed the work and its Fort Worth facility was located in Texas, home to the vice president and a politically important state. There were also charges of illegal campaign contributions to the Kennedy forces, which prompted Senate hearings that found no decisive evidence of wrongdoing.

The hearings ended but not the aircraft's controversies and problems. The very premise of the TFX, a common airframe for two services, each with a different mission, soon fell apart. The Navy never wanted such a machine, and fought the concept all the way. Certainly there were problems of operating such a large aircraft from a carrier, but the Navy certainly didn't try as hard with the F-111 as it later did with its own design, the F-14. Within six months of Secretary McNamara's resignation in February 1968, the Navy withdrew from the program. This left the Air Force with an aircraft that had been compromised in its design, some say ruined, by Navy requirements that shortened the aircraft's length to fit carrier elevators; mandated side-by-side seating for the two-man crew (and also permitted a larger radar antenna), which increased drag; and beefed up the airframe for carrier operations and to carry the heavy air-to-air (Eagle later Phoenix) missile, which increased weight.

The F-111A first flew in December 1964. It was not a pretty aircraft on the ground and was nicknamed "Aardvark" because of its long nose. In flight it was a much different story in both appearance and fact, for it had outstanding performance and for all of its problems and critics, the crews loved it. Most visible, and most important, was its variable swept wing. The wing was moved full forward for takeoff and landings, rearward to obtain best cruise performance, and full rearward for maximum speed. The F-111 was the first military aircraft to use turbofan engines, two boosted by afterburners. But there were other technological features as well. GD fitted it with an escape capsule in place of ejection seats. This system could operate from zero air speed up to at least 800 knots, one hundred knots higher than the B-58's similar system. Another new technology was the terrain following radar system that permitted the bomber to fly at high speeds as low as two hundred feet, although more typically at five hundred feet above ground level, regardless of the hour or weather. This would give the crew a tremendous advantage in penetrating enemy defenses.

The new aircraft with its glittering technology came at a high cost and a variety of problems. One of the most severe involved mating the wing, fuselage, and engine inlets. The latter proved to be a persistent and serious issue causing engine stalls and seriously restricting air speed. GD installed new inlets in early 1968 that solved the stall problems but increased drag and reduced range. The aircraft also experienced unexpected high drag from the engine exhaust nozzle. There were manufacturing problems never encountered by the aviation industry: flawless holes had to be drilled, but fatigue tests in 1968 found cracks at less than half their expected life. Even the windshield proved inadequate. Drag and weight estimates were excessively optimistic, with weight rising from the specified sixty-nine thousand pounds to almost eighty-six thousand pounds. The net result was that the aircraft did not achieve its specified performance. Maximum speed declined from Mach 2.5 to Mach 2.2, takeoff distance increased 28 percent, landing roll increased 3 percent, and ferry range fell 34 percent. A number of crashes, eight by May 1968, did not help the aircraft's reputation.

The F-111 entered combat in March 1968 in an operation code named Combat Lancer. The Air Force had high hopes for the new bomber as it promised to deliver bombs more accurately than any but laser equipped aircraft. In addition it was expected to be less vulnerable because of its high speed and low-altitude tactics and not require the large number of support aircraft as did conventional aircraft. Results, however, were disappointing. The unit lost three bombers within a month[9] and bombing accuracy was not good, an average error of 1,050 feet.[10] On the positive side the crews reported only four cases of ground fire, none when flying at five hundred feet or less.

The F-111 returned to combat four years later during Linebacker I, during which four were lost. SAMs were observed on eight missions, although only one of the sixteen fired caused any damage, and it was minor. Bombing accuracy was much better on this

tour, although there was very little bomb damage assessment. The low-altitude radar altimeter (LARA) was the major cause of aircraft out of commission and believed to have contributed to two losses.

During Linebacker II the GD bombers flew a total of 154 sorties, the bulk against airfields. At least six SAMs fired at the bombers missed, while ground fire inflicted minor damage on four aircraft. Two F-111s were lost, one believed to have flown into the sea when its navigational equipment failed and one to small-arms fire that ruptured the hydraulic system, a case of a "golden BB." So while the aircraft's combat performance improved from the 1968 debut, it still was not up to what might be expected from such a sophisticated machine. But there would be other days and other wars.

GD built a bomber version of the F-111 for SAC. The slow development of the advanced manned strategic aircraft (AMSA, later B-1), insurance against a major B-52 "catastrophe," along with the prospect of early B-52 retirement, pushed the USAF as early as 1963 to consider acquiring an interim bomber, a decision the secretary of defense announced in December 1965. SAC was hostile to the F-111 because it wanted the B-70. Two colonels sent by LeMay to brief the SAC commander on the GD aircraft were admonished and kicked off the base. That did not change events. Although SAC sought a bomber with minimum

The General Dynamics FB-111 was an enlarged version of the controversial F-111. Initially SAC did not want this bomber, although it had some flying and tactical advantages over its serving bombers. Here the bomber has its wings in the full forward position with four SRAM missiles underwing. (National Museum of the USAF)

changes from the fighter version in order to get it into service as quickly as possible, the secretary of defense wanted better than that. Thus compared with the F-111 the FB-111 had longer wings and fuselage, larger fuel tanks, more powerful engines, and a strengthened landing gear to allow greater weights.[11] While it had capabilities superior to the B-47 and B-58, it lacked the bomb load, range, and crew the Air Force desired.

The FB-111 made its maiden flight in July 1967. The bomber showed its promise when in its first appearance it took top honors in bombing at SAC's 1970 combat competition and went on to win a number of these annual competitions. However, as with the other F-111 versions, it met considerable maintenance difficulties with engines, landing gear, and electronics. It was not until October 1971 that SAC got its first FB-111 bomb wing combat ready. The aircraft also experienced flying safety issues. For the bulk of the period prior to 1979 the F-111 and FB-111 accident rate exceeded that of the USAF, although over its lifetime, the FB-111 record was only bettered by the B-52 and B-2 and slightly by the B-1. The GD bomber could carry six munitions (SRAMs or free fall bombs), four underwing and two in an internal bomb bay. A June 1965 Air Force plan called for the acquisition of 263 FB-111s to provide the same capability as the B-52Cs through the "F" models they would replace. But in March 1969 this was cut to 76 FB-111s of the 563 F-111s built. The Australians bought two dozen F-111Cs and the RAF ordered, then cancelled, 50 F-111Ks.

## Penetration Aids

The airmen pushed for devices to aid bomber penetration. Some of this was seen in World War II, most notably ECM, but also glide, guided, and powered bombs. One new avenue was to turn radar against the defenders by using decoys to mimic the radar signature of the bomber and proliferate radar returns and thus degrade enemy defenses. Convair developed the straight-wing Buck Duck to protect its B-36 bomber; however, due to slippages in its schedule and the approaching phase-out of the giant bomber, the USAF cancelled the project in January 1956.

The USAF began work on the intercontinental-range, ground-launched Bull Goose (XSM-73) decoy project in December 1952. The concept was rather grand, calling for the fielding of 2,300 missiles with the first unit becoming operational in the last half of 1960. The delta-winged missile, which weighed seventy-seven hundred pounds including its five-hundred-pound payload, was to be launched with the aid of a rocket booster and powered over its four-thousand-mile range by a jet engine at a speed of Mach .85. The project was delayed by funding problems along with difficulties with the fiberglass-resin bonded wing, booster rockets, and sustainer engine. The Air Force began flight tests in June 1957 and construction of launch sites in August 1958. But that December the USAF cancelled the program because of finances, but most of all, because the missile was unable to produce a radar simulation of a B-52.

The USAF did deploy one decoy missile, the McDonnell Quail (GAM-72, later ADM-20). The Air Force conducted flight and glide tests in 1957, the first powered test in August 1958, and awarded a production contract in December. The Quail was a squat, tailless delta-wing aircraft with four vertical fins. Its relatively small size allowed eight to be carried on a B-52, four on a B-47 although the normal loading was half that. The Quail was designed to operate at thirty-five thousand to fifty thousand feet between Mach .75 and .9, and, depending on launch altitude, reach between 360 and 450 nm. It could be programmed to make at least two changes in direction and one in speed during its forty-five- to fifty-five-minute flight. Its configuration, along with a one-hundred-pound ECM package, that later included Chaff and heat emitters, produced the image of the bomber. The first arrived at SAC in September 1960 and built to a peak in 1963 of almost five hundred. In that year the Air Force modified the decoy for low-level operations. As time passed the decoy became less effective because of reliability problems and improvements in Soviet defenses. In 1971 the SAC commander wrote that the Quail was only slightly better than nothing. During a test the next year pitting American radar against B-52s and Quails, the radar operators correctly identified the B-52 twenty-one out of twenty-three times. Nevertheless, the Quail served in SAC until 1978.

Bomber effectiveness was also aided by the addition of standoff weapons. Throughout most of its history the weapon of destruction of strategic bombing has been unguided, free fall bombs delivered by aircraft. The airmen sought a better way to destroy their targets. During World War II the Germans used ballistic and cruise missiles and both the Americans and Germans experimented with and, to a limited degree, employed guided and unguided, powered and unpowered, air-to-surface missiles. These devices were designed to permit standoff attack and thus lessen vulnerability, but most of all to improve accuracy.

After the war Radioplane adapted its Crossbow target drone to aid bomber penetration. The straight-wing, jet-propelled craft carried a half-ton warhead close to the speed of sound to destroy enemy radars from beyond their range. A B-47 could carry four on underslung pylons. Tests began in December 1955 with the first powered flight in March 1956 and the first guided flight in May 1957. But the weapon could not meet speed specifications and Russian radar soon outranged it. These problems, along with the financial pressures, led to its cancellation in June 1957, just three weeks before it achieved its greatest success by flying through its target's radar antenna.

The Rascal program began in April 1946 intended for use on the B-36, B-47, and B-52. The 9-ton Bell rocket-powered missile was to have a range of one hundred nautical miles, have a speed of Mach 1.5 to 2.5, and carry a 3,000- to 5,000-pound warhead. It would use inertial guidance to get near the target, then in the final phase transmit radar images back to the mother ship for terminal guidance, and was expected to have a CEP of about five hundred feet. From the start SAC had doubts about the program and recommended its cancellation in February 1957. LeMay wanted a better missile but was rebuffed by Headquarters USAF, which held that this was the best currently available. SAC conducted its first air launch in February 1958, and although it was a success, later attempts were not. Meanwhile the Air Force trained crews and converted B-47s into missile carriers (DB-47). SAC was never enthusiastic about the missile because of reliability and guidance issues, the degradation in aircraft performance due to its external carriage, space required by its electronics in the bomb bay, and the improving Soviet defenses. Cost was also a factor in the November 1958 decision to cancel Rascal in favor of the Quail and Hound Dog.

The Air Force had a somewhat better experience with its next ASM project, which began in March 1956. North American won the contract in August 1957, and the next February the program was accelerated in the wake of the *Sputnik* launches. The weapon was named Hound Dog after the Elvis Presley song of the day and designated GAM-77, then AGM-28. The weapon was large, heavy (five tons), and had an unusual delta-wing and canard configuration. Powered by a jet engine mounted beneath the rear fuselage, it could reach a speed of Mach 2.1 and 650 nm from a high-altitude launch, and 340 nm from a low altitude. The missile was guided by an inertial system that gave it an accuracy of no better than 1 nm, which was not that significant since the missile carried a 4-megaton warhead. The B-52 could carry two of these weapons mounted on pylons between the fuselage and inboard engines, and use the missile engines to shorten takeoff. SAC launched its first Hound Dog in February 1960 and by the end of the year encountered major problems, centering on erratic performance, maintenance, and guidance reliability. Nevertheless the missile went on alert in July 1962, armed B-52G and H bombers, growing to a peak inventory of just under six hundred missiles in 1963. One major complaint was that Hound Dog's external carriage detracted from the bomber's flying performance. The North American ASM served between 1961 and 1976 when it was displaced by the SRAM. But that was not the plan.

Ironically the best known of all of the USAF bomber munitions was never fielded. In the late 1950s the Air Force sought a long-range, air-launched ballistic missile, and in May 1959 Douglas won the contract. The GAM-87A Skybolt was a 5.5-ton, two-stage ballistic missile that featured a star tracker navigation system and combined supersonic speed, with long range (1,150 miles) and a 1.2-megaton warhead. The Air Force planned to buy twelve hundred of these missiles and mount two under each wing of a B-52H. The USAF made the production decision in February 1960 and slatted it to become operational in 1964. The British ordered one hundred Skybolts for their Vulcan bombers in June 1960. The missile failed its first five launch attempts and proved costly and troublesome. Progress with ballistic missile along with its high cost and growing questions about bomber utility led President Kennedy to cancel the project in December 1962. (The administration was embarrassed when the missile had its first successful flight on the day the termination was announced.) The cancellation caused a considerable diplomatic row with the British who understandably were disappointed, felt misled, and had to scramble to keep their nuclear deterrent viable until the Polaris submarines and missile systems came into service.

The short-range attack missile (SRAM AGM-69) emerged after Skybolt's cancellation. Boeing won the contract in October 1966 and the missile made its first powered flight in July 1969. The SRAM was a 2,200-pound, inertially guided, two-stage ballistic missile armed with a 200-kiloton warhead and a top speed of Mach 3.5. It could reach more than thirty nautical miles when launched from low level and perhaps one hundred nautical miles from a high-altitude launch with a CEP of twelve hundred feet. Initially the missile was intended to attack Soviet air defense systems; later its targets were expanded. B-52G and H models could carry twenty SRAMs (eight in an internal rotary launcher and twelve on two pylons mounted between the fuselage and inboard engines) and the FB-111A six (two internally and four on wing pylons).[12] The former became operational with the missile in January 1973 and the latter in August 1975. AGM-69A numbers peaked in the late 1970s at around fourteen hundred, and the missile served into 1990, when it was retired because of reliability (safety) issues with both motor and warhead. The Air Force cancelled an improved version for the B-1, as well as SRAM II in 1991.

The United States was not the only country to develop standoff weapons. The British were forced to revise their plans when Skybolt was cancelled. The best they had in development was an inferior weapon, but a British one that could be made operational. The Blue Steel was an 8-ton rocket-propelled missile with a range of one hundred or so miles. The missile was inertially guided and carried a 1.1-megaton warhead at a speed of Mach 3. It flew its first flight in early 1962 and went into service that same year. But it was difficult to handle because of its highly volatile fuel, and it proved difficult to integrate with the RAF bombers. In mid-1964 Blue Steel was adapted for low-level (below one thousand feet) launch, which considerably cut its range.

In contrast, the Soviets put a number of standoff weapons into service. Most of the Russian winged missiles came in at least two versions, an antishipping model armed with a conventional warhead and a land attack version fitted with a nuclear warhead. The AS-1 (U.S. code name) Kennel (NATO code name) was a scaled-down MiG-15 (without a canopy) that used command guidance, had a range of about sixty miles, and went into service on Badgers as early as 1956. It was followed in 1961 by the AS-2 Kipper that was also turbojet powered but with supersonic speed. It was fielded with either a conventional or nuclear warhead, and had a range of between 30 and 100 nm. In 1960 the AS-3 Kangaroo entered service, armed with a nuclear warhead. It was a 12-ton supersonic missile powered by turbojet with a range of 100 to 350 nm that was carried by the Bear. The AS-4 Kitchen became operational in 1964, was powered by rockets, and could reach Mach 3.5 over a 240-mile distance. It was fielded in three versions (nuclear, antishipping, and antiradiation) and was carried by both Blinders and Backfires. The rocket-powered, supersonic (Mach 1.2) AS-5 Kelt could carry either a conventional or nuclear warhead and was fielded in an antiradiation version. It went into service in 1965 with a range of 125 nm on the Badger, which could carry two. The AS-6 Kingfish was faster (Mach 3.5) and longer ranged (300 nm) than the AS-5 and was built in numerous versions, including both an inertially guided nuclear and antiradiation model. It entered service in the early 1970s, two being carried by a Badger. The AS-15 Kent was a subsonic missile that looked and performed very much like the U.S. cruise missiles, that is, it carried a 200-kiloton nuclear warhead some eighteen hundred miles guided by an inertial-Doppler navigation and a terrain comparison system. It went into service on the Bear in the mid-1980s and later on the Blackjack. The AS-16 Kickback family had a version that was similar to the SRAM. The Tu-160 carried twenty-four in four launchers. The 2,600-pound rocket-propelled missile was inertially guided, carried a 350-kiloton warhead, had a top speed of Mach 5, and had a range of sixty miles; it was deployed in the late 1980s.

## New Tactics

SAC had to deal with not only the growing effectiveness of Soviet air defenses but also the threat of offensive strikes on its bases. One SAC response to the reduced warning time was to change its overall attack concept. Into the mid-1950s the airmen planned to base their bombers in the United States and, when ordered to war, stage them through overseas bases. This was reversed in a 1954 exercise which simulated launches from stateside bases directly to the targets and then recovery at overseas bases. Concern with the survival of U.S. bases led SAC to explore the poststrike recovery of bombers on dry lake beds. SAC found that two hundred locations met its minimum standards, and the command ran tests with the scheme in March and April 1962 that demonstrated it was viable for B-47, B-52, and KC-135 operations. However, SAC adopted other methods.

SAC devised two tactics to lessen its vulnerability on the ground in response to Soviet ICBMs that took less than an hour between launch and detonation. In August 1950 LeMay stated that he wanted no more than fifteen aircraft on each SAC base, thus increasing the enemy's targeting problem and making more runways available and speeding the takeoff of the bomber force. On the debit side operating more bases lessened maintenance efficiency and increased costs. In any case it was not until 1954 that SAC submitted a dispersal plan for B-52s and their tankers to be completed by mid-1965. SAC began the dispersion in March 1958 but it was not completed until 1963 because base construction lagged behind aircraft production. In the event of increased international tension or strategic warning, the bombers would be further dispersed to civilian airfields and other military bases, a tactic first exercised in June 1960.

A second tactic was to have the bombers and tankers prepared to launch on very short notice in order to get airborne before a Soviet attack. SAC proposed the ground-alert system in October 1955, began tests in November 1956, and began implementation in October 1957. In July1957 SAC began sending B-47s to overseas bases (Reflex Action), to stand two weeks' alert.[13] The goal was by July 1960 to have one-third of SAC's force able to get airborne within fifteen minutes. One method to get the aircraft off in the mandated fifteen minutes was to decrease the interval between the takeoffs. Starting in 1960 B-47s were allowed to make rolling takeoffs on three-hundred-foot-wide runways in staggered (alternate sides) takeoffs with ten-second intervals between bombers and on two-hundred-foot-wide runways with twenty-second intervals. By the end of 1959 B-52s could be launched with fifteen-second intervals as long as the "F" and "G" models were not mixed with the earlier models. By July 1959 SAC had 16 percent of its force on ground alert and intended to raise this to one-quarter of the force. In 1961, however, President Kennedy ordered the ground-alert level increased to 50 percent, which SAC achieved in July of that year. Concurrently the Air Force began putting its ICBMs on alert. In April 1964 the number of ICBMs on alert equaled the number of bombers on ground alert and the growing proportion of ICBMs continued.

The fifteen-minute launch goal was difficult to achieve, leading to airborne alert, which SAC commander Gen. Thomas Power recommended in October 1958 (with tests the next year). The B-52s flew for twenty-four hours with the aid of two aerial refuelings. SAC proposed putting a sixteenth of the force on airborne alert, Headquarters USAF wanted one-quarter of the force on airborne alert beginning in July 1960, and the secretary of defense approved an eighth.

Airborne alert was not only a costly tactic it also was a dangerous one as the bombers carried nuclear weapons. Inevitably there were accidents although fortunately none resulted in a nuclear detonation. In January 1966 a refueling B-52 collided with a KC-135, both aircraft crashed, seven of the eleven crew were killed, and four nuclear weapons fell near the town of Palomares, Spain, and into the ocean. Two bombs underwent a nonnuclear explosion, contaminating the soil, and two went deep into the ocean. All were recovered after a difficult eighty-day operation. Two years later a nuclear-armed B-52 crashed and burned in an attempted emergency landing at Thule, spreading radioactive material. Airborne alert was terminated shortly after the second incident. Other factors in the decision included its high cost ($.8 billion a year) along with the increased number of U.S. land- and sea-based ballistic missiles.[14]

To further protect against a sudden air strike, in July 1960 SAC began the ground alert of a KC-135 equipped and manned as an airborne command post able to control SAC forces in the event that the land-based headquarters were unable to function. In February 1961 these aircraft began continuous airborne operations. The number of such aircraft increased in April 1962, when three auxiliary commands were put into service. SAC took further steps to ensure communications in the event of a nuclear war, putting three Blue Scout rockets on alert in July 1963 that carried a prerecorded force execution message as a payload. In late 1967 they were replaced by Minuteman II missiles fitted with a communications package that could transmit control messages to SAC forces. A further innovation was demonstrated in April 1967, when an airborne command plane successfully launched a Minuteman missile.

Another effort to protect the retaliatory force was to provide radar warning of hostile threats. SAC believed that an extra ten minutes of warning would increase the percentage of bombers it could get airborne from 14 to 66 percent. Radars deployed within and off the coast of the United States to protect targets in the continental United States would not give adequate warning time for American bombers to reach relative safety. To push the radar line farther from SAC bases, the Americans concluded an agreement with the Canadians in 1951 to build a chain of thirty radar stations across southern Canada known as the Pinetree Line, which was completed by the end of 1955. An agreement in 1954 pushed the radar warning farther north with the Mid-Canada Line, built along the fifty-fifth parallel, and the Distant Early Warning (DEW) Line, north of the Arctic Circle along the sixty-ninth parallel and extending from Alaska across Greenland. The three-thousand-mile DEW Line, which became operational in August 1957, consisted of sixty-three stations. The advent of the ballistic missiles, both ICBMs and SLBMs, changed the equation, for while flight times of bombers from detection to target could be measured in hours, flight time for ballistic missiles was measured in minutes. More powerful radars were fielded with the first very long radar, ballistic missile early warning system (BMEWS), sited in Thule, becoming operational in February 1961. Along with two other later installations built in Britain and Alaska, BMEWS gave over-the-horizon warning of ballistic missiles launched from the Soviet Union. This system was supplemented by the missile defense alarm system (Midas), a satellite that used infrared sensors to detect missile launches and began tests in the early 1960s.

SAC made other tactical changes as well. It adopted low-level tactics to counter Soviet SAMs and reduce the range of radar detection.[15] Early warning radar could detect high-altitude bombers at 150 to 200 nm and low-flying aircraft at no more than 30 nm. Low-flying aircraft rendered airborne interception ineffective, degraded AAA, and largely nullified air-to-air missiles. By mid-1960 three-quarters of SAC strike sorties were planned at low level. These tactics put additional demands on the bombers and crews. Hazards such as bird and insect strikes had to be overcome, along with the challenges of high-speed, low-altitude navigation (50 to 60 percent of which was visual), crew fatigue and airsickness (as many as 15 percent of the crews suffering the latter), and increased structural fatigue.

The Air Force developed two maneuvers for nuclear delivery. Beginning in January 1956 it tested the spectacular low-altitude bomb system maneuver that had been developed for fighter bombers delivering nuclear weapons. This involved a low-level run at around 425 knots and five hundred feet, a 2.5-g pull-up with bomb release at the 45- to 60-degree point, and then an Immelman reverse maneuver. These maneuvers may have been militarily effective but they also led to flying accidents and fatigue cracks in the wings. In just over thirty days in March–April 1958 there were six B-47 accidents. In response the Air Force suspended the tactic in April and initiated

Operation Milk Bottle, May 1958 through June 1959, the goal of which was to inspect and repair the wing attachment to allow these high-stress maneuvers. Meanwhile SAC incorporated a less stressful tactic in October 1957. In the pop-up maneuver (code named Long Look) the bomber approached the target at normal altitude and then, outside of radar range, descended to five hundred feet until fifty miles from the target, where it pulled up, released its bomb at eighteen thousand feet, and exited the target with a steep dive to five hundred feet. SAC introduced a variation in September 1958, code named Short Look, in which the bomber bombed from five thousand feet. Combined with a drogue chute to slow the nuclear weapon and allow the bomber to exit the target, these were effective delivery tactics. With the first method the command achieved a CEP of 1,370 feet, and with the second 2,150 feet, compared with SAC's high-level accuracy in that same year (1960) of 1,300 feet. By March 1960 all B-47 and most B-52 crews were qualified for low-level operations.

## SAC Training

Training in the SAC was never ending, demanding, as realistic as possible, and strictly scrutinized. The objective was to ensure getting the weapons accurately on the target, despite weather, mechanical difficulties, and enemy defenses. The command attempted to instill realism into its training and evaluations with no-notice inspections and simulated exercises.[16] Concerns with safety and inconveniencing the American public (by using ECM or shutting down civilian air lanes) inhibited these efforts. To monitor accuracy SAC initially employed fixed radar bomb scoring (RBS) sites that could accurately plot where bombs would impact upon simulated release based on the bomber's location, course, air speed, and altitude.

To diversify targets, in 1961 SAC deployed scoring radars on railroad trains that were moved from place to place and by the end of the year had three RBS trains in service. SAC also used annual bombing competition to encourage its crews and indicate what the very best could do.

SAC also conducted a number of exercises. Some of these were to perfect means by which American air defenses could distinguish between SAC bombers entering and exiting the United States and enemy bombers as well as give SAC a measure of its tactics. Sky Shield I, an exercise flown in September 1960, involved 306 bomber sorties and indicated that even though low-level bombers flew above their planned wartime altitude (for safety purposes), 90 percent were undetected. Sky Shield III, in September 1962, involved over 530 bomber and tanker sorties. One portion of the exercise that SAC thought was relatively realistic consisted of ten B-58 and three B-52H bombers operating at high altitude and forty B-52s at low level protected by two dozen EB-47 ECM aircraft and sixteen simulated Quails (flown by B-47s). The defenses only intercepted seven of the bombers before they reached their targets, indicating SAC's ability to penetrate presumably lesser Soviet defenses.

Chapter 10

# Missiles

## WINGED AND BALLISTIC

The story of strategic bombardment is dominated by aircraft. Aircraft were the first delivery system and have delivered essentially all of the warheads aimed at cities, people, and factories over the history of this mode of warfare. More recently two other systems (ballistic and cruise missiles) have emerged and pushed bombers into a subordinate position, at least for strategic targets, especially for the delivery of nuclear weapons. However, while missiles have increasingly armed powers large and small, and become increasingly important in a deterrent role, they have seen little combat.

### Nuclear Strategy

Before proceeding further some context is required. Immediately following World War II there were few nuclear weapons in the U.S. arsenal and U.S. war policy was little changed from that of World War II. War plans called for a combination of nuclear strikes along with a conventional strategic bombing campaign based on the World War II experience. Nuclear weapons were seen as just bigger bombs. But this changed when the Soviets got nuclear weapons and as the Americans both acquired growing numbers of weapons and developed enhanced atomic bombs (1951) and hydrogen bombs (1952) that dramatically increased power from the Hiroshima bomb's 15 kilotons to many megatons. The strategy became one of mass nuclear strikes, mostly of cities, a policy called "massive retaliation." These changes coincided with President Eisenhower's desire to balance the budget by cutting government spending. He focused on the military, slashed conventional forces (Army and Navy), and emphasized the USAF in his "New Look" strategy. Nuclear warfare requires fewer men, machines, and money than conventional warfare. Thus the defense share of federal spending dropped from 64 percent to 47 percent, and personnel fell from 3.5 million to 2.5 million. This was the period of Air Force, SAC, and bomber dominance. Another aspect of the 1950s was the rise of civilian theorists of nuclear policy and the corresponding decline of military influence on the theoretical level.[1] The peak of this strategy is perhaps indicated by the February 1961 Single Integrated Operational Plan (SIOP, the U.S. nuclear targeting plan), which called for the launch of all of America's nuclear weapons (2,164 megatons) with a projected death toll of 175 million Russians and Chinese.

The incoming Kennedy administration pushed in a different direction to better counter nonconventional (guerrilla) and lower level conventional warfare, a policy other than an "all-or-nothing" nuclear response that became known as "flexible response." This led to an emphasis on counterinsurgency warfare, a buildup of conventional forces (increasing by a quarter of a million men), and a shift in U.S. nuclear policy. In June 1962 Secretary of Defense Robert McNamara gave a speech in Ann Arbor, Michigan, that set out a policy of not attacking cities but instead attacking enemy military forces; it came to be known as the "counterforce strategy." But pushed by financial factors as well as strategic ones, in short order (December 1963) this developed into a policy that emphasized deterrence over counterforce. By the mid-1960s U.S. war policy called for both inflicting "unacceptable" damage on Russian industry and population ("assured destruction") in addition to hitting Russian military forces ("damage limitation"). As the Soviets neared nuclear parity this grew into a policy of mutual assured destruction, with both adversaries having the capability to destroy the other, with the fitting but unfortunate acronym, MAD. This involved deploying a force that could weather a first strike from the Soviets with the ability to counter and destroy (or mortally wound) the Soviets. (The goal has been variously stated to ensure the destruction of half to two-thirds of Soviet industry and one-fifth to one-third of their population.) Today U.S. policy is based on nuclear deterrence and the ability to absorb a first strike and retaliate with an overwhelming counterblow.

**(Right) The United States cautiously added intercontinental missiles to its inventory and initially was more comfortable with cruise rather than ballistic missiles, as they shared many of the characteristics of manned aircraft. With considerable difficulty the USAF developed the tailless, subsonic, ground-launched Northrop Snark. Although reliability and guidance were serious problems, the missile was operational for a year. (USAF)**

## Cruise Missiles

The first attempts with strategic missiles were winged missiles, or pilotless aircraft, what have become known as cruise missiles. The Americans had a long acquaintance with these devices. During World War I the Army and Navy had separate programs that produced a "flying bomb," a piston-powered, unmanned aircraft. Both services continued minor efforts with the concept between the wars. The Army Air Forces tested another device (General Motors A-1) into mid-1943, again with little success. There were other experiments with radio-controlled aircraft during the war and some combat trials. The Navy tested drones equipped with TV seekers and radio control in the Pacific theater. A more ambitious project, Aphrodite, was conducted by both the AAF and Navy in Europe. This used worn-out heavy bombers, stripped of nonessential

flight equipment, crammed with nine tons of explosive, and equipped with radio control gear and in some cases television sensors. A crew of two got the Aphrodite aircraft airborne, adjusted the equipment, set the fuses, and bailed out; then a mother ship guided the bomber into its target. The missions, between August 1944 and January 1945, were rendered ineffective by mechanical problems, weather, vulnerability, and inaccuracy. The top airmen also noted that Aphrodite was a terror weapon, which ran counter to AAF doctrine.[2] Nevertheless, the airmen experimented after the war with a modified B-29, but problems led to its cancellation in April 1949.

The Germans made a direct contribution to American programs. The United States quickly reverse engineered the V-1, began flight tests in October 1944, and planned to launch one thousand a day by January 1946. Practical matters intervened. The program would have taken up huge resources, cutting artillery production by 25 percent and bomb production by 17 percent as well as absorbing vast amounts of shipping. When production terminated in September 1945, fourteen hundred had been built. The Americans did improve the missile's accuracy, achieving an average error of 5 miles at a 150-mile range in a postwar test. In addition the airmen equipped their version

(JB-2) with a radar-controlled system that did much better, one-quarter mile accuracy at a one-hundred-mile range. The Navy also ran tests with the V-1 they named "Loon," accomplishing the first launch in January 1946. The sea service considered a variety of launch platforms for the missile, making the most efforts with surfaced submarines.

The AAF began a more ambitious project in August 1945 with a requirement for a six-hundred-mile-per-hour, five-thousand-mile missile. Early the next year it awarded Northrop a study contract for two fifteen-hundred- to five-thousand-mile missiles, one subsonic and the other supersonic. The latter did not make it off the drawing board; the subsonic version first flew in April 1951. Snark was a sleek-looking craft with sharply swept wings, notable for its lack of cockpit, crew, and horizontal tail, that was launched by two solid-propellant rocket boosters, powered by a turbojet engine, and was recoverable. The USAF conducted twenty-one flights over the next year that proved the missile's airworthiness. Northrop installed an inertial navigation system monitored by stellar navigation, which it claimed could achieve a 1.4-nm CEP to meet the major challenge, accurate guidance over intercontinental distances.

Meanwhile the Air Force increased the requirements, which considerably complicated the project. The USAF now wanted a missile able to reach a supersonic speed at the end of a 5,500-nm mission and achieve a .25-nm CEP.[3] This required Northrop to redesign the missile, and although it looked like the original, takeoff weight rose from twenty-eight thousand to forty-nine thousand pounds. There were numerous troubles and failures during testing, and it was not until June 1954 that the new version had a successful flight. Criticism grew along with the aerodynamic, cost, and scheduling problems. SAC, the intended user, had concerns over the missile's ground and air vulnerability and in 1959 recommended the missile's cancellation but was overruled by Headquarters Air Force. The key issues were guidance and reliability. On seven tests in the late 1950s the best CEP achieved was just over four nautical miles, but that was the only missile to reach the target area and one of only two to exceed forty-four hundred nautical miles. On the last ten launches of the program the guidance system demonstrated only a 50 percent chance of performing to specifications and only one of these missiles flew the planned distance.

Nevertheless the USAF put the missile on alert in March 1960. Snark's service was brief, for in March 1961, shortly after taking office, the new president ordered its immediate phase-out and in June the USAF deactivated the Snark unit. Although the task of adapting aircraft technology (airframe and engine) would seem much easier than developing the ballistic missile, it was not. In the end the rapid development of ballistic missiles coupled with the slow and troubled development of the cruise missile spelled the end of the Snark.

Meanwhile the Air Force was developing a more ambitious cruise missile. In July 1947 the airmen added to their missile program a fifteen-hundred-mile, supersonic ramjet missile, which was to follow the subsonic Snark into service. A scaled-down test version (X-10), powered by two turbojets, first flew in October 1953. It was radio

(Right) The shorter range Martin Matador was more successful than the Snark. Also powered by a turbojet, it could reach 620 miles at a top speed of 650 mph. The Matador, shown here dropping its booster rocket after launch, and its derivative, the Mace, served in the European and Pacific theaters between 1959 and 1969. (USAF)

controlled and recoverable, and it featured a canard and delta wing. On twenty-seven flights it reached a top speed of Mach 2.05. The follow-on full-scale XSM-64, named Navaho, substituted two ramjets for the turbojets and added a seventy-six-foot booster beneath the missile, allowing a vertical launch. It did not do as well as its predecessor. A multitude of problems, affecting about everything except the airframe, caused considerable delays. After eleven failed attempts the missile was able to fly just over four and a half minutes in March 1957. Little wonder the Air Force cancelled the project that July. Testing continued after the cancellation which saw the missile exceed Mach 3 and remain aloft forty-two minutes.

Why this dismal story? As with the Snark, the program could not produce the promised performance, reliability was awful, and cost overruns and schedule slippages proved fatal. Immature technology along with poor management were responsible for these failures. The delays put the cruise missiles into direct competition with ballistic missiles, a contest in which they could not compete. In the end these long-range cruise missiles proved inferior to both the bomber and ballistic missile. In fairness the country did benefit from the $800 million Navaho project, gaining, among other benefits, new materials and manufacturing techniques demanded by the aerodynamic heating at high speeds, the booster engines used in a number of ballistic missiles, and the inertial auto navigation system employed in the Hound Dog, *Nautilus* submarine, and Navy A3J.

The shorter range Martin Matador was more successful. It was developed in response to an August 1945 AAF requirement for a surface-to-surface missile with a range of 175 to 500 miles at 600 mph. Martin won a study contract in March 1946. The project survived several cutbacks that eliminated other programs and received top priority in September 1950. The missile was about the same size and appearance as contemporary fighters (albeit without a cockpit), sporting swept wings and a horizontal tail mounted atop the vertical stabilizer. The turbojet-powered craft employed a boosted launch after which it could carry a 1.5-ton warhead over 650 mph to a range of 620 miles. It initially was controlled by ground-based radar, a mismatch because the radar was limited to a 250-mile range, well short of the missile's maximum range. Therefore in late 1954 the Air Force installed the Shanicle guidance system, which transmitted signals to the missile, allowing it to navigate along grid lines. Tests of the TM-61C indicated an overall reliability of 71 percent and a CEP of twenty-seven hundred feet (instructors achieved a CEP of sixteen hundred feet). Because Shanicle was limited in range and could be jammed, the Air Force fitted a third guidance system, automatic terrain recognition and navigation (ATRAN), a radar map reading system, into the TM-61B. Named Mace, the missile's fuselage, wing, engine, booster, and weight differed from the "A" and "C" models and produced better performance. The USAF deployed both Matador and Mace to Europe and the Far East between 1959 and 1969. Throughout their service the two missiles demonstrated reliability and accuracy problems, as well as control difficulties over long distances.

## Ballistic Missiles

In contrast to the disappointing and difficult experience with these postwar cruise missiles, ballistic missiles were much more successful. U.S. forces brought scores of V-2s stateside and between 1946 and 1952 fired sixty-seven of them. But the real prize were the Germans associated with the program, most notably Wernher von Braun and 127 other missile experts who came to the United States in early 1946 and formed the initial core of the Army missile (and later space) program. The AAF was especially interested in the device pushed by its chief, the futuristic minded Henry "Hap" Arnold. After much infighting the Air Force was granted control over air launched, winged, and intercontinental-range missiles; the Army was given control of shorter range missiles to support its forces; and the Navy was given control over ship- and submarine-launched missiles. In late 1945 the AAF had twenty-six surface-to-surface missile projects with ranges between twenty and five thousand miles.

In early 1946 Convair received a contract for a missile able to carry a five-thousand-pound payload up to five thousand miles; it developed into the first American intercontinental ballistic missile. The Atlas introduced three innovations that improved its performance over that of the V-2. The first was a detachable warhead that made

The first American intercontinental ballistic missile was the Convair Atlas. The liquid-fueled missile had early difficulties, steadily improved, and was operational between 1958 and 1965. (USAF)

reentry much less a problem. To save weight the missile used integral tanks and had a thin outer skin. The third improvement was swiveling (gimbaled) rocket engines. It was a good start, but the cutback in military spending led to many project cancellations, including the Atlas in July 1947. Problems cited included the heavy payload, inadequate fuels, and high reentry temperatures that indicated the missile would require eight to ten years development. A June 1947 Air Staff missile memo put the long-range missile fourth in priority. (Between fiscal years 1951 and 1954 the Atlas received $26 million, the Snark and Navaho $450 million.)

Some attribute this imbalance to the airmen's bias toward aircraft. This has an element of truth, as aircraft represented a known technology and demonstrated weapon of war, but can be easily overemphasized. Curtis LeMay, the epitome of a bomber general, held a cautious yet realistic view. "I consider an ICBM with capability of instantaneous launch and with acceptable reliability, accuracy and yield," he wrote in November 1955, "to be the ultimate weapon in the strategic inventory."[4] In the interim, however, he saw missiles as aids to bombers. LeMay considered the missile's greatest military value was political and psychological, as ICBMs alone were incapable of destroying an enemy.

Clearly the ballistic missile had to overcome a variety of severe technical challenges to demonstrate its ability to reliably and accurately deliver a thermonuclear warhead over intercontinental distances. This was not a simple task. With limited funds the ballistic missile faced a "Catch 22" situation of not having enough money to develop and demonstrate its capabilities, and not being able to get money because it had not demonstrated its capabilities.

The Korean War opened the money spigot, and in January 1951 the USAF restarted the program that Convair had been sparsely supporting on its own. The delay was beneficial as great progress had been made with thermonuclear warheads that had many times the power of the Hiroshima bomb, with reduced size and weight.[5] The greater yield allowed accuracy requirements to be reduced from .25 nm at 5,000 nm to 2 to 3 nm. More significant was the reduction of the payload requirement, from 8,000 to 1,500 pounds, which in turn permitted a reduction in the missile's gross weight from 440,000 pounds to 240,000 pounds. Air Force opinion was split on the program, with the development people favoring it while the Air Staff showing indifference and skepticism. But in March 1954 the Air Force accelerated the Atlas and in May elevated it to its highest research and development priority. (Interservice rivalry may have played prompted this action, as the Air Force did not want to lose ICBMs to the other services.) In September 1955 President Eisenhower gave the program the highest national priority.

The Atlas was a one-and-a-half-stage missile, that is, its two boosters and one sustainer engine were burning at liftoff, then after the boosters burned out, they were jettisoned. In June 1957 the first attempted Atlas launch failed, as did others, with only three of the first eight attempts going reasonably well. The first successful flight was in December 1957, just months after the first shocking *Sputnik* launch. The U.S. military

Martin's Titan missile, another liquid-fueled missile, followed the Atlas and was a considerable improvement. Here a Titan I is seen in its launch sequence. (USAF)

was concerned not so much by the Soviet satellite capability as by their demonstrated ability with large boosters, which could also hurl nuclear warheads a long way. The influential Gaither report in December 1957 acknowledged a Soviet lead in ICBMs, which accentuated the fear of a "missile gap." *Sputnik* ignited a panic. President Eisenhower responded by increasing the Atlas deployment plan from four to nine squadrons, building two intermediate-range ballistic missiles (IRBMs), accelerating the Navy's ballistic missile program, and beginning work on a third Air Force ICBM (Minuteman).

The first Atlas B flew full range in November 1958, over a year after the Soviets accomplished that feat. The USAF got its first Atlas ("D") operational in late 1959, although it lacked reliability and was vulnerable because it was stored above ground and had to be erected to a vertical position then fueled (requiring about fifteen minutes) for an above-ground launch. Later the "D" and "E" models were stored horizontally in a coffin-like structure and erected to the vertical position for fueling and launch, while the last Atlas version ("F") was stored in a silo, fueled, and brought above ground for launch.[6] The "E" and "F" models were somewhat improved over the "D" with upgraded boosters, a more powerful warhead, and all inertial guidance system that made possible a CEP of two nautical miles over a distance of sixty-three hundred nautical miles. The Atlas ended its alert duties in June 1965 and went on to serve as a booster for space operations. For all of its faults the Atlas did establish the United States in the strategic missile business.

The Titan quickly followed. After calls in 1954 for an alternate ICBM, in April 1955 the secretary of the Air Force authorized development of a second and parallel ICBM program, granting Martin the contract that September. Although Titan was similar to the Atlas with its liquid fuel, silo emplacement,[7] and above-ground fueling and launch, it was more sophisticated. It was a two-stage missile that was larger, capable of more reliably carrying twice the payload a longer distance, and required less maintenance than the Atlas. However, the missile did not have an easy path to deployment because critics saw the Titan as a costly duplication of the Atlas and not as impressive as the emerging Minuteman. The *Sputnik* crisis changed all that. Titan made its first flight in February 1959 and had mixed success in testing and harsh public criticism, although two-thirds of the missiles achieved full success through June 1961. The missile became operational in May 1962 and served until June 1965, when the transition to the Titan II was completed.

Development of the Titan II was approved by the secretary of the Air Force in September 1959. It was considerably different than its predecessor: 10 percent longer, 50 percent heavier, and with 40 percent more power in the first stage and 25 percent more in the second. This gave the Titan II a range of nine thousand miles compared with the sixty-three hundred miles of the Titan I, permitting the Air Force to base it farther south than the Titan I. It also featured all inertial guidance that allowed the USAF to disperse the missile silos at least seven miles from each other, increasing an attacker's targeting problem. The Martin missile achieved a CEP of .5 to .8 nm. However, the most important change was the use of a storable fuel. The early liquid-fueled missiles were difficult to handle, had to be fueled immediately before flight, and were dangerous.[8] The storable fuel was easier and safer to handle and allowed the missile to remain fueled, which cut reaction time. Titan II could be fired directly out of its silo in one minute compared with fifteen minutes required by Titan I. It carried a 9-megaton warhead (compared with 4 megatons on the Titan I), the most powerful deployed on an American missile. The Air Force launched the first Titan II in March 1962 and, unlike so many other missiles, got it right the first time with a flawless five-thousand-mile flight. It achieved operational status in March 1963, and fifty-four served for a quarter of a century until the last was deactivated in the summer of 1987. They were later employed as space boosters.

## Intermediate-Range Ballistic Missiles

At about the same time the Air Force began developing the Titan, the United States was also developing shorter range ballistic missiles that would serve in a strategic role. The genesis of the intermediate-range ballistic missile program was the February 1955 Killian committee report, which recommended ICBMs and pushed the development of both land-based IRBMs sited in Europe and sea-based IRBMs stationed off the Soviet coasts as a response to similar developments in Russia. In November 1955 the Department of Defense authorized development of the Thor, a land-based missile by the Air Force, and the Jupiter, a land- and sea-based system by the Army and Navy. (The Navy, concerned about the danger of liquid-fueled missiles and desiring autonomy, withdrew from the program in 1956, and went on to develop its own solid-propellant ballistic missiles as described below.) In November 1956 the secretary of defense granted the Air Force responsibility for all surface-launched missiles with a range over two hundred miles, which should have ended the Army program. However, the Army successfully pleaded its case, undoubtedly aided by the woes of the Thor program. The following January the IRBM programs were elevated to the same development priority as the ICBM, generating USAF fears that its ICBM effort would be diluted. Some assert that the twin IRBMs were a creature of interservice rivalry, others that the effort was favored more by the politicians and scientists than by the military and builders. The major factor behind the IRBM was the desire to get a nuclear-armed ballistic missile into service as rapidly as possible to reassure both the American public and American allies.

Both programs built on the progress of other missiles and were similar, both using

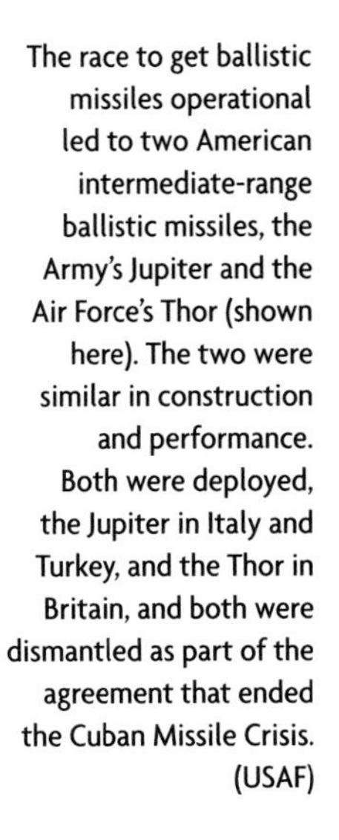

The race to get ballistic missiles operational led to two American intermediate-range ballistic missiles, the Army's Jupiter and the Air Force's Thor (shown here). The two were similar in construction and performance. Both were deployed, the Jupiter in Italy and Turkey, and the Thor in Britain, and both were dismantled as part of the agreement that ended the Cuban Missile Crisis. (USAF)

(Right) The U.S. Navy also developed cruise missiles and deployed the Chance Vought Regulus I. It looked and performed like the USAF's Matador, and in fact used the same turbojet engine. Regulus became operational in 1955 and served for almost ten years. (National Museum of the USAF)

the same liquid-fueled Navaho engine. In 1951 the USAF, concerned over the vulnerability of its Matador missile, ordered the development of a tactical-range (initially six hundred to one thousand miles, then in 1955 increased to twelve hundred to fifteen hundred miles) ballistic missile. The Air Force selected Douglas to build the missile in December 1955. To speed its progress the program used the Atlas' booster, guidance system, and reentry vehicle. This should have eased the road for the Thor, but the Air Force adopted the risky procedure of concurrency, as it had with both the Atlas and Titan, in the race to field the missile before its two rivals, the Russians and the Army.

The Army built the Jupiter on its Redstone program, which had its start in 1951 and was well along and provided a test vehicle in addition to test facilities. The Jupiter drew ahead of its Air Force rival and completed a twelve-hundred-mile flight in May 1957. The Thor had testing problems with successive failures: its first attempted flight exploded on the launch pad in January 1957, and its was not until its fifth attempt, only weeks before the *Sputnik* flight, that it achieved a successful flight of thirteen hundred miles. It appeared that one of the two IRBM programs would be axed, but the urgency following the *Sputnik* launch changed everything, encouraging the president to authorize production of both missiles.

The two missiles had comparable performance. Both had a range of fifteen hundred miles, carried the same 1.4-megaton warhead, and were erected vertically, fueled, and then fired, a procedure that took about fifteen minutes. There were differences, however. The Jupiter was a mobile missile, and secondary sources list its CEP as .9 miles compared with the Thor's 2 miles. The Army turned over the first operational Jupiter to the USAF in August 1958.

The two IRBMs were quickly deployed overseas. The British-based Thor became operational in June 1959, while the Jupiter became operational in Italy in July 1960 and in Turkey in April 1962.[9] The missiles worked under a dual key arrangement, with the missile manned by crews of the host nation and the warheads under U.S. control. The two missiles were only briefly in service, for as part of the resolution of the Cuban Missile Crisis and with the introduction of ICBMs into the strategic inventory, the two IRBMs were deactivated by the end of 1963. The Thor later served as a launch vehicle in the U.S. space program.

## The Navy Programs

The Navy brought unique attributes to the strategic weapons arsenal. It offered mobile platforms that were difficult to locate and track. Sea-based platforms also avoided the diplomatic problems of base rights associated with aircraft and shorter range missiles. Submarines added stealth to the mix. Quiet, deep cruising, mobile submarines are considerably more difficult to detect than surface ships. The submarine's advantage greatly increased with the advent of nuclear energy, which gave essentially unlimited range and underwater endurance to the submerged launch platforms. The Navy's problem was to surmount problems involved in mating two cutting-edge technologies, nuclear-powered submarines

and nuclear-armed missiles, into a reliable strategic weapons system.

Similar to the AAF/USAF, immediately after World War II the Navy experimented with winged subsonic missiles and like the Air Force, it favored winged missiles over ballistic missiles. These included the modified German V-1 (code named Loon), and its own Gorgon and Pollux, missiles powered by a variety of power plants—rocket, turbojet, and pulsejet. These led nowhere. However, in response to the AAF contract to Martin in May 1947 for a winged missile (later named Matador), the Navy quickly responded with a counterpart that could be launched from submarines. The sea service was desperately seeking a nuclear delivery device because it was being pushed aside in the development of strategic weapons by the Air Force with its monopoly of the nuclear delivery system of choice, the long-range bomber. Although the Navy's attempts with carrier-based aircraft (the P2V and AJ) and seaplanes (P6M-2) proved unsuccessful, its new attempt was another story.

The Chance Vought Regulus looked very much like the Martin Matador and was powered by the same J 33 turbojet engine, and thus it is not surprising that their performance and costs were similar. Although the Matador was about a year ahead of the Navy missile and there was pressure from the Department of Defense (DoD) to adapt the USAF missile for sea service, the Navy successfully fought off such an effort, confirmed by a June 1950 decision.[10] The Navy missile initially flew in March 1951, with its first submarine launch in July 1953. The Regulus was surface launched from a submarine and then guided to impact by two other submarines. The missile could also be launched from surface ships and guided by aircraft and became operational aboard submarines,

The breakthrough that pushed the U.S. Navy into a primary strategic bombardment role was the mating of ballistic missiles with nuclear-powered submarines. Here the USS *Stonewall Jackson* is loaded with Polaris missiles. It was armed with sixteen Polaris ballistic missiles that could be launched when the submarine was submerged. (U.S. Naval Institute Photo Archive)

carriers, and cruisers in 1955. The last version of the device carried a 3.8-megaton warhead, almost six hundred miles at Mach .87. The Navy bought 514 of the missiles, which it kept in service until August 1964.

A follow on missile, the Rigel, was planned but encountered "insoluble problems" and was cancelled in August 1953. Thus the Regulus II, another Chance Vought product, was slated to succeed its namesake. This project began in June 1953 and first flew three years later. With the same six-hundred-mile range as the Regulus I, its primary advantage over the older missile was its Mach 2 speed, although it could reach over eleven hundred miles at reduced speeds. The Navy cancelled the canard configured missile in December 1958. The Triton, a twelve-thousand-mile-range, Mach 3.5 missile, got as far as full-scale development in 1955 but not into production. Cruise missiles gave way to another technology, ballistic missiles.

The Navy left the Jupiter program in September 1956, correctly evaluating the danger of liquid-fueled missiles aboard naval vessels, to pursue a new and safer missile technology, solid fuels, which became a viable option in 1955. The next year the Navy began development of its solid-fueled ballistic missile program geared for submerged submarine launch. (An October 1956 study group concluded that a 30,000-pound missile could reach fifteen hundred nautical miles.) This weapon faced a number of serious challenges, the first from those within the Navy who opposed the new weapon as it threatened the Navy's successful World War II way of fighting wars with carrier-based aviation, surface forces, and submarines. The appointment of Adm. Arleigh Burke as chief of naval operations was significant for unlike his predecessor, he favored fleet ballistic missiles. In May 1957 the Navy planned to put a surface-launched 1,200-nm ballistic missile to sea by January 1963, followed by a submerged-launched 1,500-nm missile by January 1965. The Navy pushed schedule above all else and willingly accepting lower performance, fearing an Air Force takeover of all missile weapons. *Sputnik* accelerated the timetable.

The Navy faced two technical challenges, the ballistic missile and the submarine launch platform. The latter employing nuclear technology had already been successfully tested. To speed the process the Navy did not develop a special submarine for the missile, instead they cut an existing boat (the *Scorpion*) in two and inserted a 130-foot section to accommodate sixteen missiles. The Navy launched the renamed *George Washington* in June 1959 and declared it operational in July 1960. The first fleet ballistic submarine designed from the keel up was the *Ethan Allen*, which became operational in June 1962. The Navy built five boats in each of these two classes.

Submarine navigation was critical to the concept as the missileers had to accurately know the launch location if the missile was to hit its target. The Navy attempted a number of different devices (the inertial system from the Navaho cruise missile, celestial, radiometric from the sun or moon, sea floor maps, Loran C, and satellite). Another critical problem for the submariners was communications. Nuclear weapons demand tight control in both peace and war, which is dependent on rapid and reliable communications. There are physical problems with

The Polaris A-3 (*left*) shown alongside its successor, the larger and more capable Poseidon (*right*). The last version of the Poseidon (C-3) weighed twice as much as the Polaris, had twice the payload and twice the accuracy, and could deliver six to fourteen warheads. It was first deployed in 1971. (U.S. Naval Institute Photo Archive)

radio communications with a boat cruising beneath the ocean's surface. Antenna buoys, trailing wire antennas, various frequencies, and airborne relay stations were all employed.

The development of the missile itself was greatly aided by timing. The fear of a strategic nuclear bombardment gap opening between the United States and Soviet Union, first bomber and then missile, accentuated by *Sputnik*, broke open funding that had been monopolized by the Air Force. There were also two technical breakthroughs in the 1950s. The first, the development of smaller, lighter nuclear warheads, allowed smaller, lighter missiles. The other, solid-propellant fuel, produced a safer missile that was easier to handle, store, and maintain.

The first attempted Polaris test launch in September 1958 failed, and success was not achieved until the sixth try in April 1959. The first submerged submarine launch came in July 1960, and the first patrol began that November. The submarine was armed with sixteen of the two-stage Lockheed Polaris A-1 missiles, which could carry a 600-kiloton warhead within one mile of its aiming point at one thousand miles. The Polaris was brought into the SIOP and by the time of the Cuban Missile Crisis the Navy had nine operational Polaris boats.

The Polaris A-1 went into service quickly but was an unreliable missile that only served until September 1965. It was followed by the heavier A-2, which had a range of fifteen hundred nautical miles with its 800-kiloton (later 1-megaton) warhead. The Navy's goal was to improve missile reliability, which it did, along with improving the second-stage engine. It became operational in June 1962. Concerned by the prospect of Soviet antiballistic missile (ABM) defense, during the 1961–62 time frame the Navy considered penetration aids for the submarine-launched ballistic missile (SLBM). The A-3 quickly followed with its first flight in August 1962 and became operational in September 1964. The A-3 was heavier and longer ranged (2,500 nm) than its predecessors. The additional range had strategic consequences, giving the submarines more sea room for maneuver, and allowing the Navy in December 1964 to begin its first Pacific SLBM patrol. The A-3 featured three 200-kiloton warheads, termed multiple reentry vehicles (MRV) with a CEP of .5 nm.[11] Later the Navy substituted penetration aids for one of the warheads. The Navy retired the A-3 in November 1979. Polaris missiles armed forty-one of the first three classes of nuclear ballistic missile submarines.

The SLBM continued to grow and improve. Launch technology changed from a compressed air launch to the surface before rocket ignition to a steam system that could launch a bigger and heavier missile with increased payload and range. (The submarine had adequate volume to accommodate the larger missile.) The Poseidon C-3 had twice the payload and twice the accuracy of the Polaris missile. While it could carry six to fourteen warheads, on average it carried ten (MIRV, multiple independently targetable reentry vehicle), which could hit separate targets at a range of 2,500 to 3,200 nm with a CEP of .25 nm. It was first deployed in March 1971. The thirty-one *Lafayette*-class boats were retrofitted with Poseidon missiles.

The larger missiles required larger submarines, an effort added by President Richard Nixon's desire for leverage in dealing with the Soviets. In the early 1970s the Navy saw the possibility of fitting an ICBM aboard a nuclear submarine. The program, initially known as undersea long-range missile system (ULMS), consisted of a Trident missile and a new submarine also named Trident (*Ohio* class). It was larger than the *Lafayette* class, measuring 560 feet in length and weighing 18,700 tons, compared with 425 feet and 8,300 tons, and could carry twenty-four larger missiles. However, unlike the remarkably rapid, efficient, and successful early history of the Navy's ballistic missile nuclear submarine (SSBN) program, the Trident submarine was beset with a slew of problems, delays, rising costs, litigation, and public criticism. The Navy launched the first of the class in April 1979 and deployed it on its first patrol in September 1982. Fortunately the Navy designed the missile to fit existing submarines as the *Ohio* class struggled toward service.

The manufacturer began development of the three-stage C-4 version in November 1971 and produced a weapon with a range of four thousand miles that could deliver eight 100-kiloton warheads. Lockheed incorporated a number of innovations to allow these marked improvements in performance. The builder used new materials, lighter components, advanced propulsion, and a deployable "aerospike" that reduced aerodynamic drag. There were efforts in the 1970s to develop a common missile for both Air Force and Navy use, which collapsed by the end of 1978 when studies indicated only a relatively small cost saving associated with greater risks. Accuracy improved as Trident's stellar inertial system allowed a CEP of .12 to .25 nm. It first flew in January 1977 and was deployed in October 1979 on the existing SSBNs years before the Trident submarines went to sea.

The push for more accuracy continued despite the policy of MAD. The proliferation of weapons, the promising technology, and unease with a mutual suicide pact were factors in this continued effort. The Navy resisted but was caught up in it in the 1980s. Actually the move toward more accurate weapons was endorsed at the highest level in January 1974, when President Nixon signed a memo that called for the option of small strikes on military targets, a move that reinforced the DoD's move toward greater accuracy. President Ronald Reagan was more outspoken against MAD, advocating a policy of counterforce and ballistic missile defense.

In 1980 the Navy began a program to provide a SLBM to fully use the larger space in the Trident submarine. This led to the three-stage Trident II that would supplement the USAF's MX in a counterforce strategy. The Trident II D-5 is considerably larger than the C-4 and can reach out over four thousand miles with ten to fourteen warheads with an accuracy of .06 nm (365 feet).[12] It was initially deployed in 1990.

In contrast to the Trident submarine, the Trident missile development went well. As seen here, the missile made its first flight in January 1977. The U.S. Navy deployed the Trident missile in older, retrofitted SSBNs beginning in October 1979, and years later aboard Trident submarines. (U.S. Naval Institute Photo Archive)

The British also participated in the U.S. SSBM program. They obtained four Polaris submarines along with Polaris missiles and later upgraded to Trident IIs. The Royal Navy began SLBM patrols in June 1968.

The Navy's accomplishments with the SSBN and SLBM were remarkable. The first Polaris submarine began its first operational patrol in July 1960, an astonishing four years after its inception. This required the Navy to bring together a number of new technologies and integrate them into a weapons system. It was new, if not revolutionary, and has stood the test of time.

## The New and Improved Cruise Missile

At the same time ballistic missiles were challenging the preeminence of bombers, another contender arose. Actually, it was an old concept revived by new technologies. As already described, into the 1960s cruise missiles had a long and rather undistinguished history, including wartime German and postwar American service. Two technological breakthroughs transformed this unreliable, short-range, and inaccurate cruise missile into a formidable weapon of war. The first was an improvement in computing power that permitted a navigation system that compared digitized terrain elevations of the planned route with the actual elevations at waypoints to guide the missile: terrain contour matching (TERCOM). Whereas inertial navigational systems of the day had an accuracy of two thousand feet per hour of flight, TERCOM mated with an inertial system yielded an overall accuracy at impact many hours later of between one hundred and six hundred feet.[13] This allowed the missile to be programmed for very low altitude flight, which coupled with the device's small size made it difficult to spot and track. The second innovation was the development of a small, lightweight engine that had remarkably low fuel consumption, greatly boosting range.

At the end of 1960s the Air Force sought a replacement for its obsolete Quail decoy. In June 1970 the Air Force approved a simple, low-cost subsonic cruise armed decoy (SCAD) for internal carriage on the B-52, primarily as a decoy with a later arming option. The next month the DoD approved the unarmed decoy. The program did not go well, and some suspected that the USAF did not want too much accuracy or a lethal payload because that would threaten their bombers as well as the enemy. The USAF awarded contracts in 1972 but the project was cancelled in June 1973 only to be restored in December as the air-launched cruise missile (ALCM).

The Navy came to the idea of an underwater-launched cruise missile a little later, around 1970. One idea considered was to fit cruise missiles into the ten retiring Polaris submarines and to build a new class of nuclear-powered submarines to be armed with twenty vertical tubes to accommodate cruise missiles. In January 1972 the Navy began a program on a strategic cruise missile that soon became known as the sea-launched cruise missile (SLCM), which eventually would be capable of surface, subsurface, air, and ground launch.

The timing was fortuitous because the Strategic Arms Limitation Treaty (SALT I) agreement did not limit cruise missiles as it did other nuclear delivery systems. In

November 1972 the Navy opted for a missile launched from existing torpedo tubes, which, like the short-range air-launched missile (SRAM) rotary launcher's constraints on the ALCM, limited the SLCM in size and weight. SALT energized both cruise missile programs as decision makers saw them as bargaining chips in the arms talks. The DoD ordered the two services to cooperate, especially share the key equipment, the Navy's TERCOM and the Air Force's turbofan engine and high-energy fuel. Boeing made minor modifications to its SCAD, renamed it AGM-86A, and began flight tests in March 1976. Although there was a belief that the SLCM, the Navy's General Dynamics (GD) Tomahawk, was two years behind the ALCM, it had support in Congress and in the DoD because of Air Force reluctance to develop the ALCM, indicated by its priority behind the B-1 bomber and MX ballistic missile. Moreover the specter of a naval missile on an Air Force aircraft, "having a torpedo rammed up its bomb bay," spurred the USAF to action, an example of how civilians exploit interservice rivalry. The SLCM first flew only two weeks after the Air Force missile, had a much more successful testing program, and swept on by the AGM-86A.[14] The question arose as to whether to select one of the two cruise missiles, which would be the Tomahawk, or develop both. The Navy gained an advantage when the DoD established a joint program office in 1977 under Navy leadership. Its mandate was to stress commonality, concentrate on long-range (fifteen hundred nautical miles) versions, and develop a ground-launched cruise missile (GLCM) from the Tomahawk for USAF theater use.

(Top left) The invention of a small turbofan jet engine and a new and accurate navigation system combined to produce a very accurate, long-range, subsonic missile. The first air-launched cruise missile, the AGM-86A (*right*), was enlarged into the 1,500-nm Boeing AGM-86B (*left*), which was armed with a nuclear warhead and deployed on B-52s. (USAF)

(Bottom left) The U.S. Navy developed a sea-launched cruise missile comparable to the air-launched cruise missile, but it was more versatile as it could be launched from submerged submarines, surface ships, trucks, and aircraft. The General Dynamics Tomahawk was first fielded as a conventionally armed antiship missile in March 1983 and the next year as a nuclear-armed, land-attack missile. (USN)

The joint office ran a competition between GD and Boeing to pick the ALCM builder. Each company attempted ten flights and each had six successes, with the unanimous decision for Boeing announced in March 1980. The AGM-86B became operational in December 1982 on B-52Gs, and more than seventeen hundred were delivered to the Air Force. The missile can deliver a 200-kiloton warhead over fifteen hundred nautical miles at Mach .75. The B-52 can carry twelve ALCMs externally and eight internally. To comply with the SALT II treaty, Boeing added a strakelet (a faring connecting the wing to the fuselage) onto the B-52G and a spade antenna onto the B-52H that distinguishes the ALCM carrying B-52s from other B-52s. The Air Force also considered using converted airliners to carry 48 to 90 ALCMs, an idea that died in August 1979. The B-1 is able to carry ALCMs.

The Air Force developed other versions of its cruise missile, one conventionally armed (AGM-86C) and guided by the satellite-based global positioning system (GPS) that saw action in the 1991 Gulf War and the other a stealth version, the AGM-129, known as the advanced cruise missile (ACM). The latter had greater range, less radar cross-section (RCS), and higher cost than the AGM-86B. Studies began in 1982, and GD won the contract the following year. The AGM-129 endured technical and managerial difficulties that delayed schedules, problems attributed to General Dynamics. The missile made its maiden flight in July 1985 but was not delivered to the USAF until June 1990. While essentially the same size as the AGM-86B, the ACM is quite different in shape, distinguished by its sharp nose, flush intake, vertical tail below the fuselage, and swept forward wings. The Air Force intended to replace the AGM-86B with the ACM, but the 2,500 missile buy was reduced at the end of the Cold War to 460. In March 2007 the Air Force announced it was retiring the missile.

The Tomahawk had wide service. The Navy deployed three versions, antishipping, conventional ground attack, and a nuclear-armed missile deployed aboard both surface ships and submarines. The antiship version became operational in March 1983, while the land-attack version entered service in 1986. As the Air Force and Navy cruise missiles are powered by the same engine and weigh about the same, they have essentially the same flying performance. The nuclear-armed Tomahawk became operational in 1984 and was taken out of service by September 1991.

The GD missile also served with the Air Force. For political (reassuring NATO allies) and military (countering Soviet nuclear missiles) reasons the United States deployed the nuclear-armed Pershing II ballistic missile and ground-launched cruise missile to Europe. The GLCM was a Tomahawk, known to the Air Force as Gryphon, mounted in sets of four on a truck. It first flew in May 1980 and became operational in December 1983. In 1987 the Americans and Soviets agreed to the Intermediate-range Nuclear Forces (INF) Treaty, which banned nuclear weapons with ranges between 260 and 1,840 nm from Europe. In April 1988 the United States began withdrawing both the Pershing II and GLCM as the Russians withdrew their SS-20.

The U.S. Air Force also deployed a nuclear-armed ground-launched cruise missile version of the U.S. Navy's Tomahawk. The missile was deployed briefly to Europe until arms reduction agreements in the late 1980s led to its withdrawal and destruction. (USAF)

## Minuteman and Peacekeeper

The Air Force also made use of the new solid-fuel propulsion technology. While the Navy began its effort in 1956, formal development of the Minuteman program did not begin until February 1958 with a small $50 million research project. Nevertheless, progress was swift as the missile made its first flight in February 1961, flawlessly flying forty-six hundred miles and making impact in the target zone. In October 1962, at the time of the Cuban Missile Crisis, ten Minuteman missiles were on alert.

The Air Force developed and deployed three versions of the Minuteman, all built by Boeing, all solid fueled, and all three-stage missiles. They made good use of the improving technology, growing but 10 percent in length and 20 percent in weight yet gaining markedly better performance. (Minuteman was appreciably more effective than either Atlas or Titan.)[15] Minuteman I encountered some first-stage technical problems with its nozzles reducing its range from sixty-three hundred to forty-three hundred nautical miles. As a consequence the Air Force based these missiles, designated IA, farther north[16] and at the higher elevations and latitudes in Montana and North Dakota to somewhat compensate for the decreased range. (Later Minutemen were based as far south as Missouri.) The IA version became operational in December 1962 and the improved IB version in September of the next year. Boeing built 150 Minuteman IA missiles and 650 IBs. The Minuteman II began development in October 1963, made its first flight in September 1964, and went on alert in May 1966. It was slightly larger than Minuteman I, had a longer range, and carried a 1.2-megaton warhead to a CEP of one nautical mile. It also carried penetration aids as did the later versions of the Atlas and Titan. By decade's end the Air Force had five hundred Minuteman Is and five hundred Minuteman IIs on alert. Minuteman III was not far behind, beginning development in December 1964. It had greater range, payload, and accuracy and was the first ICBM with multiple warheads able to carry two to three 170- to 375-kiloton warheads out to eight thousand nautical miles with an accuracy of eight hundred feet. The Air Force declared Minuteman III operational in December 1970 and put it on alert the next month. By July 1975 the United States had 450 Minuteman II, 550 Minuteman III, and 54 Titan IIs in service.

The three-stage Boeing Minuteman went on alert in October 1962 during the Cuban Missile Crisis. The U.S. Air Force deployed three versions of the missile that, along with submarine-launched ballistic missiles and strategic bombers, formed the core of the United States' nuclear triad and nuclear deterrent. Shown here is a Minuteman I launch in September 1959. (USAF)

The Air Force ran into problems siting its ICBMs. As more accurate and more powerful enemy missiles became available, one possible defense was mobility that would make an enemy's effort to disarm the United States much more difficult. The USAF had studied mobility for both Atlas and Titan in 1956 but concluded it was not feasible. The concept reemerged in September 1958 when the Air Force considered placing Minuteman on rails. SAC was a strong proponent of this concept, in February 1959 advocating mounting three hundred of four hundred Minutemen on trains and sent a requirement to the Air Staff for deploying an operational railroad-based Minuteman unit no later than January 1963. Tests between June and August 1960 proved satisfactory but studies uncovered operating and logistical problems. President Kennedy deferred the mobile Minuteman concept in March 1961 and in December 1961 the DoD cancelled the project, citing high cost. Instead the Air Force emplaced all the Minutemen in hardened silos, designed to withstand 200 psi overpressure, which were less expensive. However, the railroad idea was not entirely dead.

The MX (missile experimental, later named Peacekeeper) development began in April 1972. It overcame problems involving domestic politics, arms control, and cost to enter service. In October 1981 Reagan set out a program that called for the deployment of at least one hundred MX missiles. The solid-fuel missile was very large, heavy, and capable of delivering ten independently targeted 300-kiloton warheads over fifty-two hundred nautical miles.[17] The first test flight in June 1983 went well, and ten Peacekeepers became operational in December 1986.

The MX's most serious problem proved to be its basing. The increasing accuracy of Soviet missiles posed the threat that a first strike ("bolt from the blue") might take out U.S. nuclear forces. While mobility complicated the attackers plans, it brought with it problems of expense, accuracy, security, and reaction time. In addition there was the issue of "not in my back yard" domestic politics. The planners considered forty or so options, including protection by an anti-ballistic-missile system, superhardening, deep underground basing, airborne launch, buried trench, multiple protective shelters (MPSs), closely spaced basing (dense pack), railroad basing, off-road truck basing, sea basing, and basing them in the existing Minuteman III silos. President Jimmy Carter picked the MPS option (September 1979),[18] which President Reagan cancelled in October 1981 and replaced in November 1982 with dense pack. This option was more attractive politically as it required less land despite questions about it effectiveness[19] and although Congress favored buried trench or shelter basing. Therefore the president appointed a government group (the Scowcroft Commission) to study the matter, which in April 1983 recommended the immediate deployment of one hundred MXs into resized Minuteman III silos. In addition it called for acceleration of work on an ABM and other strategic offensive systems (ALCM, Trident, and bombers) and beginning work on a small, lightweight missile.

Beginning in January 1986 the first fifty MXs went into the Minuteman III holes to replaced those missiles. This was possible because a unique feature of MX was its cold launch. Unlike other silo-based missiles, which ignited their first stage in the silo, MX was ejected in its capsule from the silo by highly pressurized steam and ignited when 150 to 300 feet above ground. However, Congress balked at deploying the second fifty in the same manner and capped

(Left) The four-stage Peacekeeper (MX) encountered considerable opposition and controversy regarding basing before being sited in former Minuteman silos. The U.S. Air Force declared the missile operational in December 1986 and fielded fifty that were deactivated in the early 2000s. This Peacekeeper launch in 1987 from a modified Minuteman silo reveals the installation's heavy hexagonal door. (U.S. Naval Institute Photo Archive)

the MX deployment at fifty until the administration developed a more survivable concept. In April 1989 President Reagan decided to limit the MX program to fifty missiles and transfer the existing missiles onto rails, a plan cancelled by President George H.W. Bush in 1991.[20] Arms-control agreements, along with maintainability issues, led to the first Peacekeeper being deactivated in October 2002 and the last in September 2005.

Meanwhile the Air Force considered the missile proposed by the Scowcroft Commission, a small intercontinental ballistic missile (SICBM) that was mobile. The three-stage Midgetman was designed to deliver a single 500-kiloton warhead sixty-eight hundred nautical miles.[21] Another proposal to meet this requirement was to upgrade the Pershing missile to a four-stage, 25,000-pound missile. Both missiles were connected with mobile and fixed siting. The United States conducted test launches of the SICBM in 1989 and 1991; however, in response to the dissolution of the Soviet Union it was cancelled in January 1992.

There was a considerable difference of opinion on how many ICBMs the United States should deploy. In 1956 President Eisenhower believed that 150 missiles would deter the Soviets, and while a 1957 report recommended 600 ICBMs, the Air Force wanted far more. In 1958 the air service proposed a force of sixteen hundred Minutemen, while Curtis LeMay wanted ten thousand. In December 1964 Secretary of Defense McNamara set the ICBM force at one thousand Minutemen and fifty-four Titan IIs, a number that stood for many years. While the U.S. ICBM numbers leveled off at 1,054, the Soviets continued to build, and by 1970 they surpassed the American numbers, fielding 1,300 ICBMs. Warhead numbers rose greatly due to the development of multiple warheads.

## Soviet Missiles

The Soviets had great incentive to pursue missile technology because of the great American strategic advantage based on technical and combat experience with strategic bombing and the U.S. lead with nuclear weapons. Just as jet propulsion offered the Russians a way to leap frog the existing bomber imbalance, missiles presented a similar opportunity. This the Soviets pursued, for while they had suffered greater losses than any other country in World War II, they possessed vast human and natural resources, and a determination to be a superpower. Russian strategic programs started with the help from stolen U.S. nuclear secrets, copied B-29s, and German technicians and V-2s. This outside boost should not be overemphasized, as Westerners tend to do, for the Soviets had a strong scientific establishment, a long and productive history with rockets, and determination. Consequently they took advantage of the new technology and were the first to launch an ICBM, space satellite, man in space, and SLBM. While this lead did not immediately translate into military superiority over the Americans, it did establish the springboard for the Soviets to reach strategic equivalence by the mid-1970s.

The Russians were quick to jump into the missile field. The Soviets overran the V-2 testing facility at Peenemünde and manufacturing plant at Nordhausen and

(Right) The Soviet SS-8 (Sasin) was a liquid-propelled missile that was first flight tested in April 1961. The two-stage ICBM became operational in 1963 and went out of service in 1976. (U.S. Naval Institute Photo Archive)

scooped up hardware along with German missile technicians, albeit not of the caliber of the von Braun group. With German parts and labor the Soviets built thirty V-2s by September 1946 and in late 1947 fired twenty of them. The large family of tactical ballistic missiles that emerged in the Soviet program are outside the scope of this study with one exception. The R-11 was a 5-ton missile that used a different fuel than the V-2, was not as efficient but easier to handle, and would later become known as the Scud. It made its first flight in April 1953 and went into production in 1955. It and the V-2 are the only ballistic missiles thus far to see combat in a strategic role.

The Soviets made a number of efforts with cruise missiles, although far fewer than the Americans. In June 1944 Stalin ordered a Soviet version of the V-1 that did not perform as well as the original. In April 1953 the Soviets began two programs that in timing and concept appear to copy the American Navaho, a supersonic cruise missile vertically launched by liquid-fuel booster and then propelled by ram jets. The larger of the two, the Buran, was cancelled in November 1957 before test flights began. The La-350 Burya achieved success in four of seventeen test flights during 1957–60 and had a range of fifty-six hundred miles. Nevertheless due to the greater success and promise of the ICBM, the Soviets cancelled the project in 1960, three years after the USAF terminated the Navaho. While the Soviets made some efforts in this area, their focus was on ballistic missiles.

The Soviets pushed the technology, as the Soviet leader, Nikita Khrushchev, liked missiles and found the Russian bomber program disappointing. The advantages of ballistic missiles were appealing as their upkeep was lower than bombers and, once launched, they were invulnerable. In December 1959 the Russians formed Strategic Missile Forces, an all-missile Soviet SAC. The Soviets developed and tested many more missiles than the Americans, some of which they did not put into service. However, despite the well publicized accomplishments of getting *Sputnik* up in October 1957, getting a dog into orbit the next month, and orbiting the moon in January 1959, the United States had a lead with military ballistic missiles.

Russia ordered its first successful ICBM, the R-7 (U.S. designation SS-6, NATO designation Sapwood), in February 1953 and began development the next year. It was designed to carry a 3-ton warhead thirty-eight hundred to forty-three hundred nautical miles. The SS-6 was developed at the same time as the Atlas and shared two common features with the American missile. First, it had the same type of guidance, radio control during its powered stage and then inertial during the remainder of the flight. And it also was a one-and-a-half-stage missile, using four (instead of two as in the Atlas) booster rockets and one sustainer motor; all five engines fired at liftoff, and after burnout the boosters fell away. The Atlas was of a lighter construction, designed to carry a lighter and more advanced warhead than the SS-6, which was much more ruggedly built and consequently more than twice the weight of the American missile. First launched in May 1957, its first successful flight in August of that year covered four thousand miles, months before the first

successful Atlas flight. The initial SS-6 had an average range of forty-five hundred nautical miles with a CEP of over two nautical miles, less range but accuracy equivalent to that of the Atlas. The major threat posed by the missile was its huge thrust, far exceeding American engines. Although it was a flawed military weapon requiring twenty hours to launch, it nevertheless became operational in December 1959, shortly after the Atlas went on alert. The Soviets improved the missile with a lighter warhead, more powerful engines, and a new guidance system that extended the SS-6's range to sixty-five hundred nautical miles. Flight tests from December 1959 to July 1960 led to the missile deployment in the early 1960s. It was taken off alert by mid-1968 and went on to serve well as the booster for *Sputnik* and more than a thousand other space shots.

The next Soviet ICBM, the R-16 (SS-7 Saddler), was approved in December 1956 and began development in 1958. The first attempted launch in October 1960 ended with an explosion that destroyed the missile and killed one hundred, including the head of Strategic Rocket Forces. The first successful test of the two-stage liquid-fueled missile occurred in April 1961, and the ICBM went on alert in November 1961. Like the later Atlas missiles, the SS-7 were first emplaced in coffins and, beginning in 1964, in silos, the first Soviet missile so emplaced. (Unlike the first silo-based Atlas, the SS-7 could be "hot launched" out of the silo.) The SS-7 was a little more than half the weight of the SS-6 yet had almost three-fifths greater range. It probably had a minimum capability to deliver a 3-megaton warhead six thousand nautical miles with a CEP of one nautical mile. However, it took almost three hours to fire the missile. At its maximum deployment in 1968 the Soviets had 69 missiles in silos and another 128 in coffins. These were gradually decommissioned starting in 1971 with the last leaving service in 1977.

The R-9 (SS-8 Sasin) was another liquid-fueled ICBM. Proposed in April 1958, it began flight tests in April 1961 that extended into February 1964 and reached an initial operating capability in November 1963 at soft sites and in April 1964 at hard ones. It probably went on alert in July 1965. The two-stage missile was able to deliver a 3,500-pound warhead more than six thousand nautical miles with a one nautical mile CEP. Later the ICBM could be hot launched from silos. The missiles were inactivated from soft positions in 1971, while those in silos served on until 1976.

The two-stage R-36 (SS-9 Scarp), also a liquid-fueled ICBM, began flight tests in September 1963. The silo-based missile (hardened to withstand 500 psi overpressures) could be fired in three to five minutes. The Soviets tested the SS-9 in four versions, two with unitary warheads (with warheads of 5 to 10 megatons), one fractional orbit (see below), and a fourth with three warheads. The single-warhead missile became operational in 1966 and the multiple-warhead version in October 1970. With the lighter-weight warhead the missile had a range of fifty-five hundred nautical miles. American intelligence saw it as a significant improvement over previous Soviet ICBMs. It was decommissioned in 1978.

While almost all of the Soviet and American strategic weapons had a counterpart in the other camp, one system did not. In 1961 the Soviets began work on a fractional

Another Soviet liquid-propelled two-stage ICBM was the SS-9 (Scarp). It began flight testing in September 1963 and was decommissioned in 1978. (U.S. Naval Institute Photo Archive)

orbital bombardment system (FOBS), an orbiting nuclear weapon that could be brought down onto a target on demand. The advantages were unlimited range and the ability to avoid known radar coverage. The threat was made loud and clear by Premier Khrushchev in an August 1961 public statement. The Soviets tested the R-36O in orbit and apparently conducted two dozen tests between December 1965 and August 1971. The 1967 Outer Space Treaty banned nuclear warheads in space but did allow the missile to be deployed. The Soviets flew two missiles a year for a time and began deployment in November 1968 to eighteen silos. SALT II banned the weapon, which was phased out in January 1983.

Development of the two-stage liquid-fuel UR-100 (SS-11 Sego) began in the early 1960s, and it was deployed in a variety of versions. The first entered service in 1966 with a CEP of 1 nm. Later versions extended range to 6,500 nm, improved accuracy to .6 nm, and employed penetration aids, as well as appearing in one version with multiple warheads (three) with some reduction in range. The Soviets began to deploy these in the early 1970s into simple silos hardened to withstand overpressures of 400 to 700 psi. More SS-11s were deployed than any other Soviet ICBM. The missile was eliminated by SALT I and was gone by 1996.

The UR-100MR (SS-17 Spanker) is another two-stage, liquid-fueled ICBM, most notable as the first Soviet missile to employ cold launch and to carry MIRVs. It was sized to use the silos of the SS-11 it replaced. The Soviets tested the first version, which carried MIRVs (four .3- to .75-megaton yield), in the early 1970s and put the missile on alert in 1975. A second version carried one warhead, and another model carried four warheads. The missile had a range of about 5,500 nm and a CEP of .5 nm. The SALT I led to its dismantlement.

The rival to the SS-17 to replace the SS-11 was the UR-100N (SS-19 Stiletto), also a two-stage, liquid-fueled ICBM. It was deployed in three versions, two with multiple warheads (six 550 kilotons each) and one with a single warhead (2.5 to 5 megatons). The missile has a range of 5,400 nm with a CEP of .5 nm. It was tested and deployed in the same time frame as the SS-17. As a consequence of the SALT II Treaty Russia must either dismantle the multiple-warhead missile or convert it to a single-warhead configuration.

The R-36M (SS-18 Satan) was the replacement for the R-36. It was a large, two-stage, liquid-fueled missile fitted with either a massive single warhead (18 to 25 megatons) or multiple warheads (eight to ten with a yield of .5 to 1.5 megatons each). It was deployed in superhardened silos (estimated to be able to withstand overpressures of 4,000 psi or possibly 6,000 psi) and used a cold launch. Testing began in the early 1970s, and it was deployed in 1975. This missile was greatly feared by the United States because of its range (eighty-six hundred nautical miles), multiple warheads (eight to ten), and accuracy (820-foot CEP) and became the focus of American arms limitations. START II banned land-based multiple-warhead systems and the SS-18, although the Russians were allowed to keep the silos, albeit downsized to prohibit accommodation of that missile.

The SS-13 (Savage) was a three-stage ICBM that served until 1996. It is significant as the first Soviet solid-propellant ICBM. (U.S. Naval Institute Photo Archive)

The Soviets also fielded a number of solid-fueled ICBMs. The first was the RT-2 (SS-13 Savage). The three-stage missile had a 5,500-nm range and could deliver its single warhead with a CEP of .7 to 1 nm. Flight tested in the mid-1960s, the SS-13 was deployed into hardened silos late in the decade. The Russians phased out the ICBM in 1996.

The RT-2PM (SS-25 Sickle) is a three-stage, solid-fueled, road-mobile ICBM. Road mobility, while more expensive than fixed sites, presents a more difficult target. The Topol was authorized in 1977, not flight tested until 1985, and went on alert in 1988, and by the end of 1996 the Russians had 360 of these missiles in service. Although tested with multiple warheads, the SS-25 was deployed with a single warhead.

The RT-2 UTTH (SS-27, no NATO designation) is a single-warhead, mobile ICBM. The solid-fueled Topol-M reportedly has a 5,700-nm range with a CEP of .2 nm and was first launched in December 1994. The missile's deployment has been restricted by both START II and the shortage of funds in the post-Soviet state. Nevertheless the SS-27 went on alert in December 1997. It was fielded in both mobile and silo versions, in a single-warhead configuration.

The RT-23 (SS-24 Scalpel) is a rail mobile and silo-based, solid-propellant, three-stage ICBM. It carries ten warheads (550 kilotons each) and is cold launched. The rail version will be eliminated under the START II agreement and the silo version will replace the SS-19. Development of the SS-24 began in January 1969 but did not achieve alert status (rail version) until October 1987. (Each train carried three missiles with their associated equipment and personnel.) The silo version was first test flown in July 1986 and began deployment in November 1989. It has a CEP of .1 nm and a range of 5,400 nm.

The first Soviet SLBM was a navalized version of the well-known Scud, the R-11FM (SS-1b Scud), deployed on the Zulu-class diesel-powered submarines. Whereas the first U.S. SSBNs mounted sixteen ballistic missiles amidships, the Soviets fitted two R-11FMs inside the sail. In September 1955 the Soviets launched the first R-11FM from a surfaced submarine. The missile's range and accuracy were limited (90 nm and .8 nm CEP), but the project was less important as a weapon than as a step toward deploying SLBMs. In any case the Soviets beat the U.S. Navy, getting the missile to sea in a long-range test patrol in August 1956, with deployment on Zulu- and Golf-class submarines beginning in February 1959. The Soviets encountered reliability problems with their submarines, ballistic missiles, and cruise missiles.

The Golf-class submarine, another diesel-powered boat, was designed from the keel up for SLBMs. Compared to the Zulu class it was one and half times larger (displacement), slightly slower, could dive one and half times deeper, had greater endurance, and was quieter. As in the Zulu class the missiles were carried in the sail and fired from the surface. The first three

In March 1994 Secretary of Defense William Perry (in dark coat and hat) visited the ICBM base at Pervomaysk, Ukraine, where he saw this SS-24 (Scalpel) silo. The missile was also deployed on rail, carried ten MIRV warheads, and was later dismantled in accordance with arms control agreements. (U.S. Naval Institute Photo Archive)

of the class (Golf I and II) were armed with three R-11FMs; later boats were armed with three R-13s. The latter (SS-N-4 Sark), another liquid-fueled missile, carried a 500-kiloton nuclear warhead out to a range of 320 nm. Submarines with reduced-range missiles became operational in 1958, with the full-range model in 1960. By the start of 1975 the Soviets had phased out both the Golf I class and the R-13.

A refitted Golf boat (Golf II) became operational in May 1963 and went to sea armed with three R-21 (SS-N-5 Serb) missiles. The R-21 had a range of 700 nm and a CEP of 1.5 nm and was most notable as the first Soviet missile that could be launched submerged. The R-13 and R-21 also armed the next Soviet submarine class (Hotel), a modification of the first Soviet nuclear-powered boat (November class). The missiles were mounted directly behind the sail and initially surface launched, however, the Soviets upgraded the boats for submerged launch by December 1963. The first of this class became operational in October 1960 within months of the much better performing American SSBN. The Hotel-class boats were later rearmed with R-21 missiles, which could be fired submerged. The first of these modified submarines (Hotel II class) became operational in April 1964. In all the Soviets built eight of these boats.

The Soviets authorized the Yankee-class submarine in 1962. It was larger than its predecessors, quieter, faster, deeper diving, and more capable, and it was the first Russian submarine comparable to its U.S. counterparts. The nuclear-powered boat was designed to carry sixteen R-27 (SS-N-6 Serb) missiles. This is another single-stage liquid-propellant missile with a range of thirteen hundred to sixteen hundred nautical miles. However, while it was fielded with a single warhead, it is notable for its multiple warheads, two or three in the upgraded R-27U, which went into service in January 1971. The latter version also was deployed with a single warhead that had

The SS-N-5 was the first Soviet missile to be launched from a submerged submarine. It was deployed on the diesel Golf II boats and later on the nuclear-powered Hotel-class submarines. (U.S. Naval Institute Photo Archive)

The Soviets deployed the SS-N-6 on Yankee boats that could carry sixteen missiles. (U.S. Naval Institute Photo Archive)

increased range and accuracy. The first boat of the Yankee class was launched in 1964 and became operational in late 1967. The Soviets built thirty-four of this class. In late 1982 some of the Yankee-class boats were enlarged and armed with SS-N-24 cruise missiles. Between 1982 and 1991 other Yankee boats were armed with twenty to forty launchers of SS-N-21 (Grenade) cruise missiles and became known as Yankee Notch subs. In compliance with arms-limitations treaties, the Yankee-class submarines were removed from operations by 1994.

The first Delta I–class submarine went into service in December 1972. This class was larger than the Yankee class and armed with a dozen R-29 missiles. The SS-N-8 Sawfly is a two-stage liquid-fueled missile armed with a single warhead. It had a range of 4,200 nm, carried decoys, and had an estimated .5 nm CEP. The system was declared operational in March 1974. The Soviets built eighteen of these boats, followed by four Delta II boats, the first of which entered service in September 1975. These had 15 percent greater displacement than the Delta I and carried four additional R-29s. The Delta III class carries sixteen R-29R (SS-N-18 Stingray) missiles and went into operation in 1976. These missiles are armed with three to seven MIRVs that can reach 3,500 to 4,300 nm, depending on the number of warheads. The Soviets built fourteen of this class. In 1985 the first Delta IV–class boat went into service. It is larger than its predecessor and carries sixteen R-29M (SS-N-23 Skif) three-stage liquid-fueled missiles armed with four MIRVs. It is larger than the R-29 and capable of exceeding forty-three hundred nautical miles. The Soviets built seven Delta IVs.

In the next decade the Soviets put the even more capable Typhoon-class submarines to sea, the largest submarine in the world. It has twice the displacement of the Delta IV and 40 percent more than the American Trident. While about the same length as each of the smaller boats, the Typhoon is much wider and built with multiple-pressure hulls. The Soviets launched the first Typhoon in September 1980 and commissioned it a little over a year later. Although notable for its size, the Typhoon had greater maneuverability, was quieter, faster, and deeper diving, and had a greater endurance than the Delta IV. It was armed with twenty R-39 (SS-N-20 Sturgeon) ballistic missiles. Each of these solid-propellant three-stage missiles can carry

**The Soviets built the Typhoon, the largest submarine in the world, which can carry twenty ballistic missiles. Although six were built, only one is in service. (U.S. Naval Institute Photo Archive)**

ten warheads out to fifty-four hundred nautical miles. It was deployed in 1984 and five years later rearmed with an improved version with increased accuracy. Although the Russians built six of these subs, at this point only one is in service, testing a new (SS-NX-30) ballistic missile.

The Russians laid the keel of Borei-class SSBN in November 1996. It is about half the size of the Typhoon yet was to be armed with twenty R-39Ms (SS-N-28 Bark) missiles. Financial problems and a redesign necessitated by repeated failures of the missile created extraordinary problems delaying the boat's launch until April 2007. Two other boats of the class are, or were, under construction. It is now designed to carry twelve or sixteen (depending on the public source) SS-NX-30 Bulava missiles with a range of fifty-four hundred nautical miles. This missile is a lighter and more sophisticated version of the land-based, three-stage, solid-propellant SS-27 (Topol-M), although with a decrease in range due to submarine launch. The Russians designed it to counter ballistic missile defenses, claiming it can perform evasive maneuvers, disperse decoys during midcourse flight, and carry shielding to protect its warhead. The missile carries six to ten MIRVs, although a tradeoff between warhead numbers and decoys and shielding is necessary. The missile was successfully test launched twice in 2005 from a Typhoon boat, with tests continuing in 2006 and 2007. Western commentators write that the Topol-M and its sibling, the Bulava, will form the core of the Russian strategic missile force in the future.

In June 2000 the Russians claimed to be operating five Typhoon-, seven Delta IV-, and thirteen Delta III-class submarines. However, it is believed that not all of these are seaworthy. The Russian navy, as with the rest of the Russian military, is facing very difficult problems of finances and manning. The press reports that the Russians have set a goal of twelve SSBNs for the period through 2010.

## Other Nations

China is thus far the last nation to have deployed ICBMs. The DF-3 program began in November 1961 before technical and economic factors forced its cancellation in 1963. (In October 1964 the Chinese exploded their first nuclear weapon.) The Chinese first tested the DF-5 in September 1971, but it did not undergo full-scale flight testing until 1978 or 1979; in May 1980 it flew six thousand miles. The two-stage liquid-propelled missile could carry a single 3-megaton warhead to a range exceeding five thousand nautical miles and entered service in 1981. The Chinese began development of the DF-5A in 1983. It had both longer range (seven thousand nautical miles) and better accuracy and entered service in the mid-1980s. The missile delivers a single 2-megaton warhead, although the Chinese are working on a multiple-warhead configuration. Initially the DF-5 was stored in mountain tunnels and moved outside for fueling and launch, all of which took about two hours. In contrast the DF-5As are sited in silos. The Chinese deployed the first in 1981 and has fielded a small number of ICBMs, with most published estimates at twenty to twenty-four. The Chinese are developing three-stage solid-propelled ICBMs as well as solid-fuel SLBMs. The Chinese have a handful of submarines and put a two-stage solid-propellant missile (JL-1) with a nine-hundred-nautical-mile range into operational service in 1988 after demonstrating a submerged launch. There is some evidence that the Chinese are developing a multiple-warhead submarine-launched ICBM.

The Chinese test fired this CSS-NX-3 from a Golf-class submarine. (U.S. Naval Institute Photo Archive)

No countries other than Britain, China, Russia, and the United States have deployed ICBMs or SLBMs with intercontinental range, although a number of others possess nuclear weapons. The French have tested nuclear weapons and deployed a nuclear triad consisting of a small number of submarines, silo-based IRBMs, and thirty-four or so Mirage IV bombers. Other countries that have tested nuclear weapons include India, North Korea, and Pakistan. Israel also has nuclear weapons and Iran is suspected of pressing forward in this area. North Korea has demonstrated missile design, development, and production ability by fielding the No-dong, a medium-range ballistic missile (MRBM), which it has sold to Iran (Shahab-3) and Pakistan (Ghauri II). It is working on a much longer range missile named Taep'o dong-2, which failed in a July 2006 test. In October 2006 the North Koreans conducted a nuclear test that also failed. The North Koreans have aided both Iran and Pakistan with nuclear and missile technology.

## Ballistic Missile Defense

The story of strategic bombardment is a tale of offensive weapons based on improving technology. Nevertheless there have been two major thrusts to deflect the offensive. A nontechnological line has been international arms-control agreements. A number of these abortive efforts have already been mentioned: the turn-of-the-century Hague agreements and the 1930s disarmament efforts. Efforts over the past four decades have been more successful. In the wake of the Cuban Missile Crisis in 1963 the United States and Soviet Union concluded an agreement that banned above-ground nuclear testing. Ten years later the two nations signed SALT, which limited the numbers of strategic weapons to the existing levels and the Anti-Ballistic Missile Treaty, which limited those systems to two sites in each country, later amended to one site each. In June 1979 the two nations concluded SALT II, which reduced the number of strategic launchers to twenty-four hundred. While never ratified because of the Soviet invasion of Afghanistan, both countries observed its restrictions. In July 1991 the Strategic Arms Reduction Treaty (START I) reduced the number of warheads to six thousand on sixteen hundred launchers. Both sides then took unilateral moves, which included the United States taking Minuteman II off of alert, curtailing bomber alert status, stopping development of MX and Midgetman, canceling the SRAM II ASM, and, in December 1993, beginning destruction of Minuteman II silos. This reduced U.S. ICBMs to 50 MX and 450 Minuteman III missiles. In January 1993 START II went further, limiting each side to 3,000 to 3,500 warheads by 2003 and banning multiple warheads.[22] Later talks led to an agreement (SALT III) in 1997 to further reduce the warhead limit to 2,000 to 2,500 by December 2007. Shortly thereafter the Russians indicated their willingness to go even lower, to 2,000 to 2,500.[23] In May 2003 the two parties agreed in the Strategic Offensive Reduction Treaty to reduce warheads to 1,700 to 2,200 by 2012. While diplomacy has reduced the nuclear arsenals of the two leading nuclear powers, it has not halted the spread of nuclear weapons.

Another effort to blunt offensive strategic weapons was to use new technology and build an anti-missile system. Following World War II a number of AAF/USAF and Army projects to develop a ballistic missile defense (BMD) system surfaced and then were cancelled. One that survived began in 1944 as a surface-to-air missile and was fielded in March 1954 as the Nike Ajax and in June 1958, in an improved version, as the Nike Hercules. The Army achieved some success with adaptations of SAMs as an ABM. The Army's Hawk SAM intercepted a short-range Honest John ballistic missile and the Hercules intercepted both a Corporal and another Hercules. In 1956 the Army proposed a new Nike missile as a defense against ICBMs. BMD became embroiled in interservice rivalry as the Army and Air Force dueled for this mission, leading the secretary of defense in November 1956 to grant the Army the terminal (point) defense role, and the USAF area defense. In January 1958 the secretary gave the Army the entire anti-ballistic-missile defense mission.

Critics of BMD have been vocal and their arguments continue to this day. Aside from the basic problem of intercepting the incoming missile, other issues include sorting out decoys and debris from warheads, defending against a saturation attack, the systems vulnerability to direct attack, and the inability to test the entire system. There are also financial, domestic political, and diplomatic objections.

The Army pushed forward with the Nike Zeus project. It began missile tests in August 1959 and in December 1961 intercepted a ballistic missile over the White Sands range. (Interception consisted of getting the two missiles within the lethal radius of the defending missile's nuclear warhead, which was not used in the test.) On two tests in December 1962 at Kwajalein atoll the Zeus intercepted Atlas Ds and was successful in seven, and partially successful in two, of the next Atlas and Titan tests. Because there were doubts about the feasibility of the system, two new technologies were advanced. In January 1963 McNamara authorized development of a new system (Nike X) based on two missiles (one new) and a new radar. The existing long-range missile, Nike Zeus, later renamed Spartan, was designed to intercept incoming missiles at the range of three hundred nautical miles and an altitude of one hundred nautical miles and kill them with x-rays from a 5-megaton warhead. It was tested between 1968 and 1973. A new short-range high-acceleration missile, the Sprint, which accelerated at 100 g's and reached Mach 10 in a mere five seconds, would use the atmosphere to sort out and intercept warheads from decoys and debris at a maximum range of one hundred nautical miles and around forty thousand feet altitude. It first flew in November 1965. The system also used a new type radar, phased array radar (PAR), that could handle more than one target at a time, was more resistant to nuclear blackout, and had greater range than existing radars.

The Soviets also were engaged in ABM efforts. The Griffon was an enlarged SAM resembling the SA-2 of Vietnam War fame that began flight tests in 1957 and reportedly intercepted a ballistic missile in March 1961. The Soviets deployed it outside of Leningrad in 1960, but by the end of 1964 they abandoned these installations. At the same time the Russians began site construction for the SA-5 Gammon, which some believe had ABM capability and which became operational in 1968. In 1966 the Soviets built sites for a much larger ABM, the Galosh, around Moscow. Although Western intelligence did not consider it an effective weapon, the Soviets put it into service in 1970. The Russians were determined to build an ABM system, according to two American secretaries of defense, spending two and a half to four times as much money in this pursuit as the United States. In the June 1967 Glassboro meeting between the heads of the Soviet Union and United States, President Lyndon Johnson attempted to cut a deal to limit defensive weapons in order to avoid an arms race, to which the Soviet premier heatedly responded, "Defense is moral, offense is immoral!"[24] As a consequence the president decided to approve ABM development.

The new Republican administration also favored ABM, albeit in a scaled-down and renamed version (Sentinel became Safeguard). The system encountered fierce domestic opposition, winning Senate approval only with a tie-breaking vote of the vice president. Playing against stereotype and with the thawing of the Cold War, the Republicans traded off the Safeguard for a treaty that limited the numbers of offensive weapons and ABMs. A 1972 Soviet-American treaty limited ABM to two sites in each country, reduced in 1974 to one each. The Air Force declared its one ABM installation operational in September 1975, but in January 1976 the JCS, judging it to be too costly and of questionable value, ordered it dismantled. The Soviets chose to maintain their one site deployed in the Moscow area.

BMD, as it later came to be called, was inoperative in the United States but not dead. It was considered for protection of the MX ICBM and then got a new lease on life when Ronald Reagan came to office. In March 1983 he proposed a space-based program called the Strategic Defense Initiative (SDI), better known to the critics and public as "Star Wars." He put forth the radical idea of making offensive nuclear weapons obsolete by relying not on arms-control agreements but on superior technology. The program faced monumental technical and financial obstacles that were overshadowed by the Soviet Union's collapse in 1991. Some credit SDI with triggering the breakup by making clear to the Soviets their permanent trailing position in technology and economics and the impossibility of closing that gap.

The push for BMD responded to the changed international situation. Although the United States faced a diminished threat from the Russians, fears of an accidental or unauthorized missile attack or an attack by another state persisted. This was fueled by the proliferation of both nuclear and missile technology to countries unrestrained by either the discipline of the Cold War or the international community. The menace of ballistic missiles to civilians was highlighted in the 1991 Gulf War Scud campaign. The success and limitations of the Patriot missile in that conflict gave new life to BMD. In 1991 President George H. W. Bush redirected SDI on three paths: defense of a theater forces, defense of the homeland, and a space-based defense. Congress lent further impetus to BMD when in November it overwhelmingly passed the Missile Defense Act of 1991. First, it wanted the deployment of a treaty compliant BMD to protect against limited threats when the technology was available or by fiscal year 1996. Second, it wanted the ABM treaty amended to permit deployment of more ground-based interceptors and space-based sensors. Under the next president, BMD was cut, and the first direction given priority.

The United States has devoted much effort to an airborne laser (ABL) system, but made more progress with the less demanding theater missile defense systems. Events in the late 1980s spurred BMD. The specter of nations such as Iran, Iraq, Libya, North Korea, Pakistan, and Syria deploying missiles and weapons of mass destruction was unnerving. Democratic president Bill Clinton attempted to derail, or at least stall, BMD, but Republican control of Congress made this difficult. BMD became a partisan issue with the Republicans putting the earliest possible deployment of a cost effective system into their 1994 political program, "Contract with America." President Clinton was sustained by a razor slim vote in the House (222 to 218) in February 1995 and by a National Intelligence Estimate in November that stated that aside from the declared nuclear powers, there was no ICBM threat against the United States over the next fifteen years. And just as the Clinton administration attempted to soothe these fears in 1998, two events changed the dynamic. First, a blue-ribbon congressional study group (the Rumsfeld commission) faulted American intelligence on its appreciation of the missile threat and posited a much greater threat by "rogue nations" to the United States. (It claimed that such a nation could field an ICBM within five years of making such a decision.) Less than two months later a second event lent credibility to these findings. On the last day in August the impoverished, isolated, and presumably technologically backward North Koreans test fired a three-stage missile that some believed had a range of thirty-seven hundred miles, which would put Alaska and Hawaii, and when lightly loaded, the western United States, at risk.

The ramifications from this test were widespread. Internationally this missile shot galvanized the Japanese to join BMD research with the United States, a position the Taiwan government wanted to follow. The domestic politics forced the Clinton administration to increase BMD spending and seek changes in the ABM treaty to permit a restricted national missile defense (NMD). In September 1999 U.S. intelligence, following the pessimistic guidelines used in the Rumsfeld report (emphasizing what could happen, not what was likely), stated that within the next fifteen years the United States faced an ICBM threat from the Russians, Chinese, Iranians, North Koreans, and possibly the Iraqis. In 1999 Congress passed legislation that called for deployment of BMD when it became technically feasible, a measure reluctantly signed by President Clinton. The George W. Bush administration went further, abrogating the ABM Treaty in June 2002 and authorizing deployment of ABMs to Alaska and California; they became operational in 2006. Midway in the first decade of the new century the United States proposed deployment of ABMs to Europe, specifically to the two new NATO members, the Czech Republic and Poland. This action was ostensibly prompted by efforts of Iran to develop nuclear weapons and a delivery system. In addition the U.S. Navy began BMD patrols with *Aegis*-class destroyers in September 2004 and activated a ballistic missile defense system in June 2006. The American BMD effort has created international criticism, principally in a resurgent Russia,[25] and unhappiness among some America's allies. At the same time other American allies feel threatened by the proliferation of missiles and weapons of mass destruction (WMD).

As the first decade of the twenty-first century draws to a close the world order is in turmoil, still adjusting to the post–Cold War realities. The demise of the Russians as a superpower leaves the United States as the world's sole superpower to face a changed world balance of power, a new international situation, and new threats. While there is only one superpower, China is growing in power and assertiveness. Further, the growing energy crisis is empowering a number of old and new powers, specifically allowing Russia to reassert itself. Meanwhile the more immediate concern is the proliferation of WMD and delivery systems, the failure of the international community to control the situation, the emergence of worldwide terrorism, and the rise of small, aggressive nation states. What role strategic weapons and BMD will play in this environment remains to be seen. However, these concerns will focus attention on the issue of strategic bombardment.

Chapter 11

# Strategic Bombardment into the Twenty-first Century

## A LIMITED FUTURE?

The implosion of the Soviet Union reordered the world balance of power and recast the threat to the United States. Nuclear deterrence and retaliation seemed less of a priority while conventional and unconventional conflicts came to the fore. Signs of these changes were the 1991 Gulf War and the phasing out of Strategic Air Command in May 1992. The next month SAC and TAC were combined into a new organization, Air Combat Command. The historic organizational distinction between strategic and tactical aircraft had officially dissolved, as the role of the strategic bomber declined. How did these events unfold?

### The B-1 Lancer

The USAF's unsuccessful experience with the B-58, FB-111, and B-70 did not stop SAC's quest for farther, faster, and higher flying bombers, what by 1964 became known as the advanced manned strategic aircraft (AMSA). The influence of politics is apparent throughout the bomber's history. The Johnson administration (and especially Secretary of Defense McNamara) was skeptical of AMSA and embraced the FB-111, a position reversed by the Nixon administration, which cut the FB-111 buy and accelerated the AMSA, which in April 1969 became the B-1. In June 1970 the Air Force ordered 244 from North American Rockwell with which to replace the entire B-52 fleet.

The new bomber, officially the Lancer but known as the Bone (B-one), made its initial flight in December 1974. It was a sleek-looking aircraft with a long fuselage and swing wing and was powered by four turbofan engines mounted in two nacelles directly under the wing root. The bomber reached a top speed of Mach 2.2 at altitude and Mach .85 at low level and could carry

12.5 tons of munitions in each of its three weapons bays. Four men manned the aircraft, which carried only ECM for defense. The bomber featured two innovations that improved performance, a blending of the wing into the fuselage, which brought the advantages of both aerodynamics and radar reduction, and two small moveable fins ("whiskers") mounted beneath the cockpit, which automatically dampened out the ride at low altitudes and thus minimized fatigue for both crew and airframe. There were problems with the engines and escape system,[1] as well as delays, increasing costs, and the threat of alternative bomber options. The bomber survived these challenges, although in December 1976 the Air Force announced a B-1 production decision that reduced the buy to between 120 and 150.

Politics was a major part of the B-1 saga. Some have asserted that the government chose North American Rockwell not for technical reasons but to preserve the industrial base (Rockwell did not have a major defense contract while its two competitors did) and electoral politics (major economic benefits would go to southern California, always a key state). To generate support for the bomber, Rockwell subcontracted lucrative contracts to five thousand companies in forty-eight states despite the costs in money, quality, and efficiency. This led to wide congressional support for the bomber in contrast to the caution, if not hostility, of the executive branch.

The B-1 had to deal with a number of challenges and problems. The first was the air-launched cruise missile, which allowed bombers to launch their weapons outside an enemy's main air defenses. While SAC had deployed the Hound Dog and SRAM on its bombers, these were conceived as aiding bomber penetration as they lacked range and accuracy. In contrast the ALCM offered both greater range and accuracy, albeit at subsonic speeds, as well as greater numbers per carrier. A standoff force that would not have to penetrate increasingly potent air defenses would cost less and be more survivable than a penetrating one. Advocates of the penetrating bomber argued it could deliver a heavier bomb load more accurately and would be able to hit mobile targets as well as targets that survived a first strike. In addition they emphasized that a penetrator would have a man in the loop and that it was recallable and reusable. Further, the penetrating bombers forced the Soviets to spend many times the bombers' cost on defense, and due to the difference in the size of the U.S. and Soviet economies, this was a much greater burden on the Soviets than on the Americans. Finally they noted that under the arms treaty counting rules there was an advantage for penetrators over standoff bombers. The second challenge was financial. Some accuse the Air Force of keeping two sets of books to conceal the cost escalation. In fact most of this increase was due to inflation and overall the B-1's cost rise was reasonable when compared to other bombers. The third issue was technical. These centered on the defensive avionics that were complicated, costly, did not work adequately, and, most of all, did not stay ahead of the threat.

The Air Force was divided on whether to procure the B-1 or wait for a stealth bomber, so by the early 1980s it pushed for both, 100 B-1s and 132 stealth bombers. In

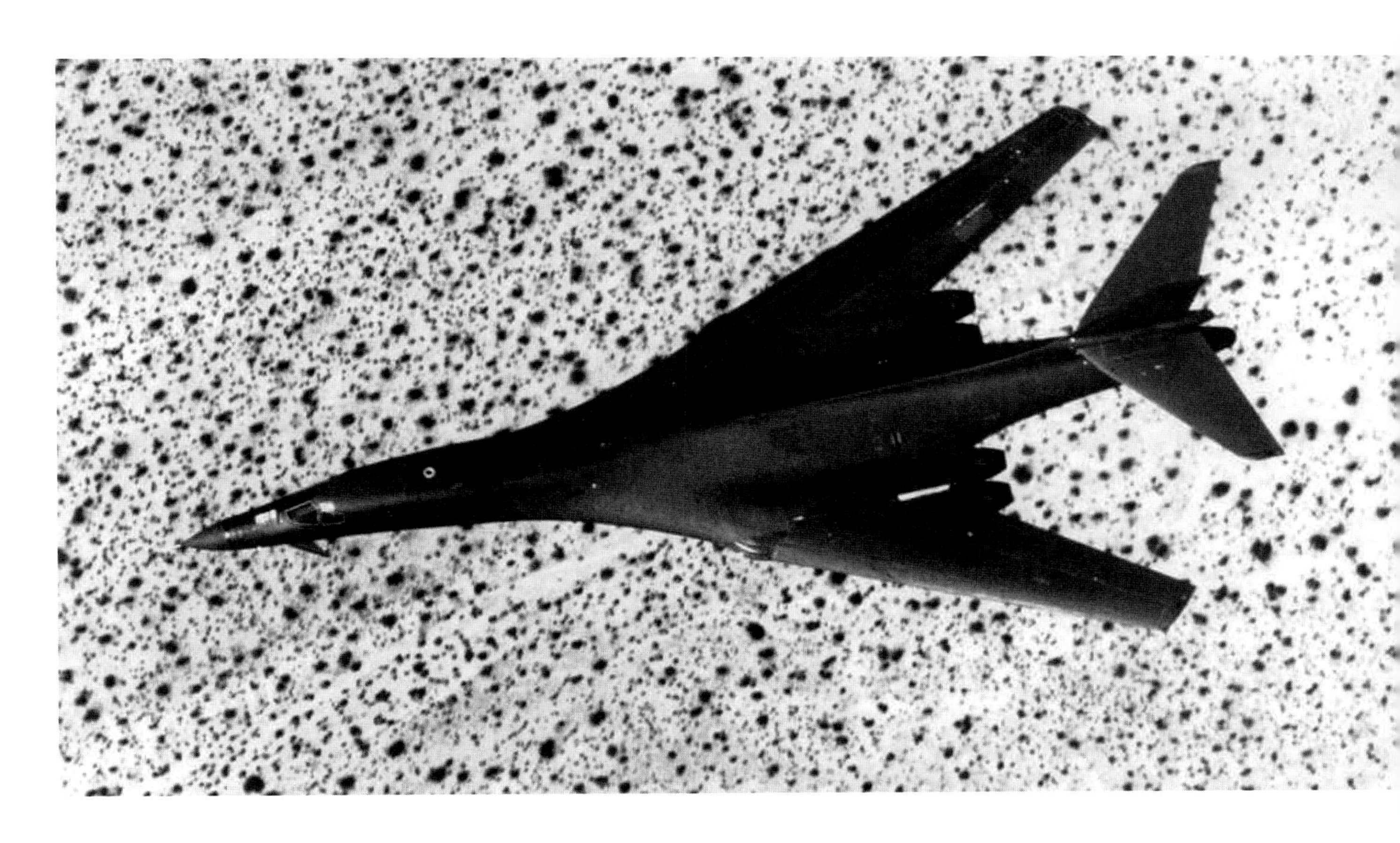

President Ronald Reagan quickly revived the B-1. The reconstituted bomber looked like, and had 85 percent common construction with, the B-1A. The U.S. Air Force accepted one hundred B-1Bs that now form the bulk of the bomber portion of the American nuclear triad. Note that the wings are swept fully back for maximum speed. (National Museum of the USAF)

June 1977 President Jimmy Carter decided to buy only the stealth bomber, although Rockwell was allowed to build and test four B-1As. Carter's decision did not end the matter, however, as lobbying by industry and the Air Force continued. In October 1981 President Ronald Reagan revisited the decision and announced a plan to significantly bolster American nuclear forces that included the USAF's two new bomber program along with one hundred MX ICBMs and Trident II SLBMs. The United States also continued to develop both the SRAM II and advanced cruise missile, which would increase the lethality of the bombers.

The B-1B looked like the cancelled "A" model with 85 percent of its airframe common. There were significant changes, however. Rockwell beefed up the airframe, which added four tons and permitted the bomber to carry additional fuel and ordnance, and that increased maximum takeoff weight from 395,000 to 477,000 pounds. The builder also simplified the engine inlets, which, coupled with a redesign of the forward engine nacelles and the use of radar-absorbing materials (RAM), reduced the aircraft's radar signature to one-tenth that of the "A" model.[2] These changes reduced top speed at high altitudes to Mach 1.25 while increasing the maximum low-level speed from Mach .85 to Mach .92, an adjustment long overdue in view of the vulnerability of flying at high altitudes in SAM country and SAC's low-level penetration tactics. Rockwell also reconfigured the bomb bays.[3] The B-1B first flew in October 1984 and became operational in October 1986. SAC received one hundred of these bombers, the last of which was delivered in April 1988.

The B-1 confronted numerous problems and was subjected to a torrent of hostile criticism. It had major difficulties with its defensive electronics that stemmed from overambitious requirements, concurrency, and poor management. Initially the bomber also suffered fuel leaks and had inadequate range when operating at low altitudes. In September 1987 a bird strike downed a low-flying bomber, indicating the aircraft was very vulnerable to natural hazards when employing its standard combat tactics. The Air Force identified five vulnerable airframe locations and Rockwell added protective shielding to these areas. Structural fatigue caused by the low-level flying was noted, and repaired, in more than a third of the fleet. By the end of the decade SAC had shaken out most of the bomber's problems as evidenced by the B-1 winning a number of awards in its first SAC bombing competition in 1988. It was converted to conventional role despite a 1988 congressional study that asserted that the bomber was too valuable to overfly its target, except for one pass over a poorly defended one. Although testing with iron bombs began in April 1988, much to the chagrin of the USAF, its most modern bomber did not see action in the 1991 Gulf War, according to the Air Force because of its inability to employ conventional weapons. The Air Force continued to have problems keeping the bomber in commission and making its ECM work properly as late as the mid-1990s. Nevertheless, one student of U.S. bomber development asserts that the B-1 was not the acquisition horror story as so often depicted but indeed a qualified success. The B-1B saw its first combat in Operation Desert Fox against the Iraqis in December 1998, delivering sixty-three tons of 500-pound bombs.

## Stealth

One of the key military aviation innovations that has emerged in the past few decades is stealth technology. Along with precision-guided munitions, stealth has altered air warfare, symbolizes the air power of the future, and dramatically demonstrates American military and technological dominance. Initially the defender had only two means of detecting aircraft, visual and acoustic, both limited in a number of ways. Thus the defender's problem appeared insoluble, as he would have great difficulty detecting and responding to a fast moving enemy attack before it did its deadly work. This reality led in the 1920s and 1930s to the belief that "the bomber will always get through." The answer to this problem was radar that gave the defender the ability to detect enemy aircraft at a distance, day or night in fair weather or foul, aim antiaircraft artillery, and guide fighter interceptors. Radar was the centerpiece of air defense systems that defeated the Luftwaffe in the Battle of Britain and later allowed the Germans to mount a stout defense against Allied bombers.

Early on engineers knew what caused large radar returns and how these returns could be increased or decreased. The ability to enhance the radar image could be used for offensive purposes as exploited in decoy missiles. The attempt to decrease the aircraft's radar return received much less attention as it faced two major obstacles. First, the engineers did not adequately understand the physical interactions between radar and aircraft and lacked analytical tools (test ranges, high-speed computers, and computer models) to study and test stealth. Second, making an aircraft less detectable to radar made it less flyable as the standard aircraft configuration, the right angles and vertical surfaces, antennas, inlets, canopies, and external stores created large radar returns. To further complicate the designers' problems, it took major reductions of radar cross-section to significantly reduce radar detection ranges.

The first operational effort to reduce radar signature was made in 1945 by the Germans, who used mats of radar-absorbing material around the snorkel tubes of their new Type XXI U boats, which cut radar detection range by 25 percent. Applying RAM to aircraft was more difficult due to the size, weight, and structural integrity of the aircraft, and the shape of the RAM, as the aerodynamics and electronics operated against each other. In the mid-1950s the USAF tested a RAM-covered T-33 (Passport Visa), which demonstrated that although the RAM degraded flying performance, such an aircraft could fly and that RCS could be reduced. The USAF also tested metal wire screens over the engine inlets on both bomber and cruise missile that cut the RCS, but as it reduced the flow of air into the engines it degraded engine and aircraft performance. The USAF attempted at least four different schemes to reduce the U-2's RCS, but none resulted in major RCS reduction, although most adversely affected aircraft performance. This reinforced the view that once an aircraft's configuration was set, little could be done to significantly reduce RCS. Lockheed's Mach 3 reconnaissance aircraft, the A-12, was the first aircraft that considered RCS from its origin. The company used RAM, composite materials, as well as "iron ball" paint, but problems of size, cost, and structural strength greatly limited this approach; the A-12 also demonstrated another RCS reduction method: shaping the structure. The aircraft's blended wing, inward canted vertical tails, and body flared outward into broad chines lessened the aircraft's radar return. Reportedly the 140,000-pound aircraft had the RCS of a tiny Piper Cub. Nevertheless there were major problems with radar returns from the engine inlets.

During this period the military and industry made numerous efforts to reduce the RCS of unmanned aerial vehicles. Lockheed's D-21 made generous use of RAM throughout its structure and had the lowest RCS of any Lockheed aircraft built to this point. Teledyne engineers mounted a wire mesh screen over the engine inlet, used RAM blankets, and later employed built-in RAM on the Ryan BQM-34 drone. Reportedly in a 1962 test it did not appear on the radar screen of the Air Force's best interceptor of the day, the F-106, and that aircraft's radar-guided Falcon air-to-air missile could not guide on the drone. The Air Force also retrofitted its Hound Dog, air-to-surface missile, with RAM that notably reduced head-on RCS.

In the summer of 1974 the USAF went one step further, initiating a program to study and demonstrate a low observable aircraft. Lockheed fought its way into the project and came up with a solution from an unlikely direction that yielded a satisfactory answer. Two Lockheed mathematicians concluded that RCS could be reduced by using a series of properly oriented flat plates, what came to be called "faceting." More convincingly, in five weeks they wrote a computer program that could evaluate the radar returns of the various designs. Resistance to the concept within Lockheed dissipated after RCS tests of models indicated that the computer program had accurately assessed the design's RCS characteristics and its amazingly low RCS.[4]

The government ran a competition for a flying stealth demonstrator (XST) between Lockheed and Northrop to be decided by an RCS test of full-scale models. The government announced Lockheed as the winner in April 1976 and began what became known as the Have Blue program. Lockheed built two manned aircraft that proved to be both flyable and essentially undetectable, although one Air Force estimated that its design reduced aerodynamic efficiency 20 to 30 percent. A new aviation technology (computer stabilized flight, "fly-by-wire") allowed this ungainly aircraft to fly. The Air Force had a new technology that promised to give it a significant advantage. Now it had to develop this jewel.

The USAF came up with specifications for two types of stealth aircraft, a single-seat attack bomber (ATA [advanced tactical aircraft] "A") and a two-seat medium bomber (ATA "B"). Some in the Air Force pressed for the larger aircraft, but as the faceted design created size and weight problems when scaled up beyond the size of the ATA "A," the Air Force shelved the "B" version.[5] In November 1978 the USAF awarded Lockheed a contract to build five test and twenty operational aircraft. The goals were an RCS similar to Have Blue, a five-thousand-pound bomb load (nominally two 2,000-pound bombs), a mission radius of four hundred nautical miles, and a maximum speed of Mach .9.

The F-117 looks like Have Blue but is about twice its size and three times its weight, or about the size of an F-15 fighter.[6] Besides the increase in size, the most noticeable difference between the two was the change from an inward to an outward-canted tail fin. Lockheed added a rotating receptacle for air-to-air refueling just behind the cockpit to give the aircraft greater range and mounted mesh grids over the engine intake and wire mesh screens over the two infrared (IR) weapons sensors to reduce RCS. As in the Have Blue aircraft, the F-117 used components designed for other aircraft to speed development and reduce cost and risk. Lockheed fitted the aircraft with two side-by-side bomb bays, each capable of carrying a 2,000-pound guided bomb. No radar was put aboard the aircraft; instead Lockheed borrowed IR sensors from the F-16 and mated them to laser designators. To diminish their RCS, they were mounted internally and protected from radar by wire mesh. The aperture proved a considerable problem until the designers could develop a transparent, low RCS window, which created the greatest impact of all the deficiencies encountered in the test program.

The aircraft made its initial flight in June 1981. Most would agree with the description that even "its own mother would have a hard time calling the F-117A pretty."[7] It truly is an absurd-looking flying machine; however, despite the perennial belief of aviators that "if it doesn't look good, it can't fly

One of the major military aviation innovations of the late twentieth century was stealth technology, first fully applied to an operational aircraft in Lockheed's F-117. Considerably different in appearance than conventional aircraft, it had two valuable military capabilities: low radar detectability and the ability to deliver precision-guided munitions. (National Museum of the USAF)

good," the aircraft handled well, thanks to fly-by-wire technology. While it lacked both flaps and speed brakes, some say it was easier to land than the F-16, was a very stable platform, and easy to fly.

As with all aircraft there were a number of problems. The engine inlets were prone to icing, a problem solved by adding a wiper with a de-icing spray device. The tailpipe caused considerable difficulties, and then in September 1985 a structural problem emerged when a fin failed in flight after a test pilot side slipped the aircraft near its maximum speed. The company mounted a new fin manufactured out of graphite composite, a fix retrofitted into the entire F-117 fleet.

Tactically some saw the stealth aircraft limited to special operations, that is, small, covert operations. The majority view, however, was that the aircraft would be valuable in a full-scale war. The Air Force remained true to its October 1979 Senior Trend Acquisition Plan, which saw the aircraft's mission to be "to confuse, disrupt and destroy the enemy's war making capability. . . . The principal targets will be command, control, communications centers, air defense facilities, airfields, logistics choke points, and other targets of high military value."[8] The Air Force procured fifty-nine F-117As. Stealth technology had granted penetrating bombers a new lease on life.

Getting the aircraft into operations was not easy. Because of the changes in the program, various difficulties, high security, and the initial lack of a ground simulator, the unit was nineteen months late becoming operational. The project was very maintenance intensive. Initially the unit averaged about 150 man-hours of maintenance for every flying hour (MMH/FH), compared with 32 for the F-15A/B, 22 for the F-15C/D, and 19 for the F-16. By 1989 MMH/FH had dropped to about 45. The unit was involved in shaking down the aircraft while developing combat tactics. Despite the public's perception of an invisible aircraft, the aircraft can be picked up by radar, but at a much-reduced range. Therefore attack planners direct the F-117s around threats. The big question was when the new weapon would be used in combat. According to one source the F-117 was primed for action following both the bombing of the Marine barracks in Beirut in October 1983 and the April 1986 bombing of a Berlin discotheque. Secretary of Defense Caspar Weinberger reportedly cancelled both strikes an hour before their planned launch.

The F-117 made its combat debut in Panama in December 1989. The aircraft was not employed because of its stealth characteristics—the Panamanians had no air defense capability—but for its night bombing accuracy. The United States aimed to minimize casualties of both the Panamanian military—a former and future ally near an important U.S. strategic interest, the Panama Canal—and, of course, civilians. Therefore the Army wanted bombs to impact near the Panamanian Guardia National barracks to disorient possible opposition. The Air Force launched six aircraft from their Nevada base, supported by five aerial refuelings, on the round-trip mission of fifty-two hundred nautical miles. The crews encountered both marginal weather and indecision and confusion as to precisely what was required, both prior to takeoff and in the air. The aiming point was changed as many as four times between takeoff and bomb delivery. Some believe, and one journalist reported, that the bombs landed somewhat farther from their target than planned. The unit commander, hardly a disinterested observer, insists the bombs landed exactly where they were supposed to land. In his words the "airplane performed precisely as it was intended to perform."[9] In any case the bombs did not kill anyone on the ground or hit a hospital located behind the barracks. Its next appearance would be much more successful.

## The 1991 Gulf War

For the four decades after World War II the United States prepared for two kinds of war, a nuclear conflict and a major ground war in Western Europe. Neither materialized, instead America engaged in two limited wars in Asia that proved less than successful. The reputation of America's military suffered greatly as a result of the loss in Vietnam, racial strife, drug problems, overpriced and underperforming weapons, and at best lackluster results in subsequent operations from Cambodia (1975) to Libya (1986). At the same time the military reinvented itself by fielding a smaller force manned by better educated and higher motivated volunteers, armed with expensive, high-technology weapons, and trained with new techniques for a swifter, more daring kind of warfare. It was this force that would engage in its next war in the Middle East against a minor power.

The Gulf War of 1991 was provoked by the ambitions of Iraqi dictator Saddam Hussein, who aspired to dominate the Arab world, dictate Middle East policy, and control the world's oil supply. Following a failed and costly war against Iran, in August 1990 he invaded and took over oil-rich Kuwait.[10] The international community, aided by the demise of the Soviet Union in 1991 and with it the end of the Cold War, called for his withdrawal. After Iraq refused, an international coalition applied diplomatic pressure as it fielded a military force. It faced a large, experienced Iraqi military armed with modern equipment that many feared included nuclear, biological, and chemical (NBC) weapons.[11]

This successful war need not detain us beyond strategic air operations. However, two major aspects deserve attention as they changed the way Americans, especially airmen, would wage war. The first, precision-guided munitions (PGMs) had a significant role in the war and have overcome the airman's major obstacle of accuracy, and thus have revolutionized air combat. Airmen made attempts with guided free-fall bombs during World War II and Korea with limited success. The breakthrough came in the 1960s with the invention of the laser, which was quickly and cheaply turned into an effective weapon. The laser-guided bomb (LGB) was combat tested in Vietnam, where it achieved phenomenal success. In 1969 LGBs scored direct hits on 61 percent of releases, and 85 percent of those bombs registered a CEP of ten feet, less than the lethal radius of the bomb. It should be noted that not only were LGBs much more accurate than conventional free-fall bombs, but the delivery altitude was considerably higher than conventional dive bombing, greatly reducing the attacker's vulnerability.

Col. John Warden III, USAF (1943– ), developed an air power theory that formed the basis of the U.S. air attack plan in the 1991 Iraq War. His concept emphasized effects, not total destruction, simultaneous not sequential attack, and targeting of enemy leadership. These ideas continue to influence the Air Force. (Air University Press)

The LGBs proved very effective during the Linebacker I campaign in which 64 percent of drops scored direct hits and accounted for 22 percent of North Vietnamese tanks that were destroyed.[12] In the Gulf War less than 10 percent of the tonnage dropped were PGMs, yet they caused over 75 percent of the damage.

In addition to some radical, if not revolutionary technologies (satellites and PGMs introduced in Vietnam and now further developed, and stealth), a second major change was the USAF's new concept of strategic operations as conceived by a serving Air Force officer, Col. John Warden. He was very critical of the graduated use of air power in Vietnam and instead advocated a short, powerful campaign to gain air superiority, "the first and most compelling task."[13] Warden called for immediate and direct attacks on the enemy's "center of gravity," the point where he is most vulnerable. Contrary to prior thinking Warden posited that partial destruction of targets might achieve the desired result, bombing for effect not for destruction. This coupled with PGMs indicated that considerably less effort would be required than in the past. In a 1988 Air Staff paper Warden illustrated his ideas of targeting priorities with five concentric rings arranged as an archery target, with leadership at the center, production next, then a nation's infrastructure, the morale of the population, and in the outermost ring, fielded military forces. As a planner at the Pentagon, Warden was a major force devising the air plan in the Gulf War.[14]

The coalition had overwhelming advantages relative to the Iraqis in terms of numbers, technology, training, and plans. Thus there could be no question as to who would win the conflict, only a question of time and cost. The air armada deployed was greater than was needed and allowed the coalition to execute the first three phases of their plan (a strategic campaign; defeating the Iraqi air defenses; and weakening the Iraqi ground forces and isolating the Kuwaiti theater) simultaneously. (The fourth phase was aiding the actual coalition ground attack.)

The F-117s helped open the attack by dropping laser-guided bombs on one-third of the strategic targets on the first night of the war and, despite poor weather, hitting three-fifths of them. Within minutes coalition airmen shattered Iraqi's extensive air defense system and obtained complete control of the air. During the course of the war the F-117s demonstrated the effectiveness of stealth as they suffered neither loss nor damage, the only combat aircraft engaged in the campaign able to make that claim. The Nighthawk also demonstrated the improvement in bombing accuracy as its bombs hit an average of forty-three feet from the target and on thirteen hundred sorties scored seventeen hundred direct hits.

Cruise missiles saw action for the first time in this war. The Navy launched 280 Tomahawks from a variety of platforms, and the Air Force employed 35 conventionally armed GPS-guided AGM-86Cs from B-52s. Seven B-52s flew a thirty-five-hour, fourteen-thousand-mile round-trip mission to and from the United States to launch conventionally armed cruise missiles on the first night, the longest combat mission of all time.

On their part the outmatched Iraqis made a strategic effort with ballistic missiles

U.S.
U.S.
WARDEN

that had the potential to unravel the alliance by drawing the Israelis into the conflict and inflict terror and casualties on allied cities. The great fear was that the missiles would deliver chemicals since the Iraqis had previously employed poison gas against both the Iranians and Iraqi Kurds. While unsuccessful, the Scud did occupy a disproportionate coalition air effort and was the one bright spot for the Iraqis, whose missileers proved innovative and effective in evading coalition countermeasures.[15] Their weapon was the obsolete Scud, an inaccurate, short-range Soviet missile, not much advanced from the

One of the few Iraqi accomplishments in the 1991 Gulf War was their employment of the Scud missile. Although a short range and inaccurate missile, it represented a significant political and military threat as it was feared that it might be armed with chemical or nuclear warheads. The Iraqis launched eighty-eight Scuds that killed fifty-three and accounted for one-fifth of U.S. combat deaths. (U.S. Naval Institute Photo Archive)

German V-2 of World War II. In all the Iraqis launched eighty-eight against Saudi Arabia and Israel, killing twenty-eight Americans (one-fifth of U.S. fatalities), thirteen Israelis, and twelve Saudis.

To meet this challenge the coalition mustered quick countermeasures. The Americans lashed together a crude system using satellites that provided a few minutes warning to both the military and civilians thus helping minimize casualties. The Army deployed its Patriot SAM and reprogrammed it to defend against the ballistic missile and announced great success, although this record was later questioned and remains one of the controversies of the war. The coalition airmen also flew some four thousand sorties against the Scuds, but their claims of eighty mobile launchers destroyed are disputed. In a similar fashion the Army's long-range missiles (Army tactical missile system) and special forces also came up dry. Nevertheless the coalition's efforts did force the Iraqis to hurry their launches, which further reduced accuracy and did cut the Iraqis launch rate in half. These efforts coupled with the inaccuracy of the Scuds, the small number involved, technical problems, and light conventional warheads reduced their effect and inflicted few casualties. Thus the Scud campaign came down to its psychological impact on the civilians and political impact on their leaders. In the end the Scud's limitations coupled with the coalition's counteractions persuaded the Israelis to stay out of the fight. This counter-effort made up 3.5 percent of strikes.

Air power claimed great success in this war while strategic operations were limited. The coalition airmen devoted only .4 percent of their total strikes to leadership targets. These attacks made Iraqi command and control more difficult, but did little more because of the coalition's limited intelligence, inadequate munitions, Iraqi countermeasures, and tough bunkers. The bombing of one of these command bunkers at Al Firdos did force a policy change. One F-117 squarely hit the facility, killed a number of Iraqi security officials and staff as intended, but also killed some two hundred to four hundred civilians who, unbeknown to the targeteers, were also sheltered in the bunker. Shrill cries from both the Iraqis and international press played up the incident as a gross, intentional American attack, typical of the bombing campaign. As a result the American commander in the theater took control over targeting against Baghdad and essentially shut down further attacks. Meanwhile the USAF quickly developed a new 4,600-pound, penetrating ("bunker buster") guided bomb (GBU-28) to destroy these buried command bunkers and used it successfully only hours before the war's end.

There were other limited efforts against strategic targets. Although only .8 percent were aimed at electricity, this effort shut down the country's power grid and reduced Iraqi's power production to 15 percent of its prewar capacity. Another 1.3 percent took out 90 percent of Iraqi's oil-refining capability, a strange target as it could have no effect in the short campaign and was counter to the aim of minimizing damage that would complicate Iraq's postwar recovery. Biological facilities were also hit, but postwar inspection found no production facilities. The airmen did better against chemical targets, destroying or heavily damaging three-quarters of these.

Once again intelligence was a major problem. For example, one source states that fifty of eight hundred strategic targets were misidentified. The attacks on nuclear facilities highlighted the severe intelligence problems. At the start of the war, intelligence identified two nuclear targets, by the war's end it found eight, and postwar UN inspection found twenty-one facilities. This striking intelligence failure undoubtedly had an impact on the later, and much more consequential, intelligence failure regarding Iraqi nuclear programs in 2003. The wartime intelligence that indicated great success against the NBC targets contrasted with the USAF's postwar inspection (*The Gulf War Air Power Survey*), which concluded that this bombing only "inconvenienced" the Iraqis. Approximately 2.3 percent of the air strikes were aimed at the NBC targets.

In all the airmen directed 11 percent of their strike sorties and 17 percent of their PGMs against strategic targets. In contrast to the overwhelming and unqualified success of the tactical bombing, these strategic attacks played essentially no role in the outcome of the war. In military terms the oil and electric attacks were efficient but had little impact, attacks on leadership harassed the Iraqis but did little more, and the efforts against the NBC and Scud targets were ineffective. In response strategic bombardment advocates argue that the war ended before the strategic effect could be felt and that the political impact of the strategic bombing was crucial on the Iraqi, Israeli, and coalition governments and their populations. The great success of air power in the war obscures the failures and shortfalls of the strategic air war.

As had been the case in the Vietnam War, the bulk of the strategic strikes were flown by fighter type aircraft, while the B-52, the symbol of strategic bombardment for decades, primarily hit ground support targets in Kuwait.[16] Aerial refueling unleashed the smaller aircraft from the constraints of range, while PGMs meant that small bomb deliveries could have the same impact as the massive bombings of previous air campaigns with much less fear of collateral damage. One discouraging note, despite the new technologies, was that weather continued to be a problem for the attackers.

There are two final points to be considered. First, in contrast to the brilliant performance of air power technology, the absence of the B-1 was conspicuous. (The most circulated explanation is that the bomber was not certified for conventional bomb delivery, and in any case was unneeded. While undoubtedly true, the aircraft was also experiencing serious technical problems at this time.) Second, the overwhelming victory and technological triumph that almost resulted in air power winning the war alone also demonstrated strategic bombardment's limitations, underscoring most clearly the importance of good intelligence.

## The Northrop B-2 Spirit

Another American bomber that did not see action in the 1991 Gulf War was the Northrop B-2 Spirit. At first glance the B-2 appears to be the result of Northrop mating stealth technology to its earlier B-49 design as both aircraft feature a tailless,

**(Right) The most expensive and possibly the last USAF strategic bomber is the B-2A Spirit. It looks like Northrop's two prior failed flying wings, but with the new stealth technology it has superior combat performance compared with existing bombers. This view reveals the bomber's distinctive double "W" trailing edge and unusual placement of the engine inlets and exhausts. (National Museum of the USAF)**

flying-wing configuration with literally the same 172-foot wing span. The company firmly rejects this notion. One major difference between the two and the earlier aircraft's flying problems and the latter's benign handling was the perfection of the fly-by-wire computer flight control system. The result is an aircraft that looks like no other. The major advantage of the design is its very low RCS, which is not based on the flat panel faceting of the F-117 but on smooth, curved lines blending the fuselage with the wing, the absence of verticals, and the use of composite materials and RAM. The bomber decreased RCS with engine intakes mounted atop the wing and a quarter of the way aft of the leading edge and decreased infrared emissions with its engine exhausts atop the wing and a quarter of the way forward of the trailing edge. The aircraft's IR signature is further reduced by mixing the engine exhaust with the air flow over the wing and using special IR-absorbent paints. Four non-afterburning engines power the bomber that is manned by two pilots. The B-2 features two bomb bays, each of which can carry up to ten tons of bombs. The bomber was initially intended for high-altitude flight, but before metal was cut, the Air Force added a low-level requirement that necessitated a redesign that cost an additional year and $1 billion. It first flew in July 1989.

The bomber, known as the advanced technology bomber (ATB), began development in a super-secret "black" program. Northrop had lost out to Lockheed in the 1970s competition for the stealth demonstrator (XST, Have Blue) but continued to work with stealth, getting a contract for a battlefield surveillance aircraft (BSAX, Tacit Blue). Unlike the F-117, which was designed mainly to minimize nose on RCS, this aircraft was to be an all-aspect stealth design. It first flew in February 1982. Meanwhile the Carter administration, which had cancelled the B-1 in June 1977, went on to authorize the development of a stealth bomber, and in September 1980 the Air Force issued a request for proposal for the ATB. In this competition Northrop beat out its old rival Lockheed with a contract awarded in October 1981 for 6 flying aircraft, 2 static test articles, and options for 127 bombers for $36.6 billion.

The B-2 is a distinctive aircraft renown for its unique configuration, its stealth characteristics, and perhaps as the last American strategic bomber. Rumors and the high costs generated considerable criticism in the press and Congress. A number of factors account for the high costs. Clearly the B-2 was on the technological cutting edge in a number of areas: stealth, new materials, new production methods, and LPI (low probability of intercept) radar.[17] Another factor that adds cost is the requirement of the avionics and stealth equipment for special maintenance buildings, and high secrecy reputedly added 10 to 15 percent to the cost and delays of eighteen to twenty-four months. Further, to reduce risk the aircraft went through extensive testing that cost both time and money. Flying tests went well, although apparently there were problems with the offensive avionics, electrical, and escape systems. The RCS tests presented the greatest difficulties, however; the decision makers decided that with the breakup of the Soviet Union the RCS requirements could be relaxed and that fixes could be retrofitted into the bomber. The results are amazing for an aircraft so large, as its RCS is a fraction of that of the B-52 and B-1B. High costs also result from the small buy. The implosion of the Soviet Union and escalating costs led the George H.W. Bush administration to lower the B-2 purchase to seventy-five and then in January 1992 to twenty-one, including the six test aircraft. The public and Congress expected a peace dividend, the incoming Democratic administration in 1993 wanted to cut costs, and the Air Force had other uses for the money. The demise of SAC and questions about the need for long-range, penetrating bombers with no superpower rival were other factors in this decision. The bomber took a long time to mature, only becoming operational in April 1997. As the total cost for the program was around $45 billion, each bomber cost $2.1 billion, an astonishing figure equal to the cost of a capital ship. This may explain why the majority of B-2s are named after states as were U.S. battleships. This was an amazing cost overrun, $45 billion for 21 aircraft compared with the $37 billion cost estimated in 1981 for 135 aircraft.

## Bombers over the Balkans and Middle East

PGMs would be even more widely used in subsequent engagements bolstered by another technology, global positioning system. GPS allowed precision accuracy despite obscured visibility (smoke, fog, cloud, or haze) that inhibited LGBs as well as gave a "launch-and-leave" capability, which further decreased vulnerability. (However, GPS-guided munitions are not as accurate as laser-guided ones and could be jammed.) In 1996 the United States deployed the GPS-aided munition (GAM), a 1-ton bomb guided by an inertial and GPS system. A more recent development is the joint direct attack munition (JDAM), a 1,000- and 2,000-pound bomb, with the same type of guidance system. It is a cheap tail kit that in tests produced a CEP of thirty feet. It went into service in the late 1990s. The joint stand-off weapon (JSOW) employs GPS and inertial guidance and has folding wings to extend its range. Whereas a free-fall bomb (the JDAM, for example) has a range of about fifteen nautical miles, JSOW can reach that distance from a low-level release and forty nautical miles from an altitude release. It is also guided by GPS and inertial systems and achieves the same accuracy. It has a modular warhead (submunitions, antiarmor, or unitary) and first saw combat in 1999. The greatly enhanced accuracy makes bombers much more lethal, changing the equation from tens and hundreds of tons of bombs required to destroy a target to numbers of targets that can be destroyed by one aircraft. Thus it allows a few, smaller fighter and attack aircraft to have great destructive capabilities.

Air power saw action in the Balkans in the 1990s after the death of the Yugoslav dictator Josef Tito in 1980 led to the disintegration of the polyglot state. In 1991 Slovenia broke away in a ten-day conflict that cost sixty-two lives; Croatia quickly followed, although it had a more difficult time. Ethnic violence in Bosnia-Herzegovina was far worse. Brutality, ethnic cleansing, and massacres prompted a UN and NATO intervention that included air patrols that began in October 1992 and some air action. Bosnian Serb ground action triggered a seventeen-day air campaign (Operation Deliberate Force) August–September 1995 that obtained acceptable results for UN/NATO forces without strategic bombardment. There was no military need for such assets, the use of heavy bombers might send the wrong diplomatic signals, and for political reasons the Italians would not allow the F-117s to use their air bases.

Action in the Balkans continued as a consequence of Serb abuse of Muslims in the former Yugoslav province of Kosovo. In March 1999 two B-2s helped open an air campaign (Operation Allied Force) against the Federal Republic of Yugoslavia, flying a thirty-one-hour mission that delivered thirty-two 2,000-pound GPS-guided JDAMs on their first combat mission. B-2 bombers dropped some 650 during the Kosovo campaign; 90 percent hit within forty feet of the aiming point. Although the Serbs had a weaker air defense than the Iraqis, they adopted a more successful, albeit losing, strategy. They did not attempt to contest the Allied air forces and remained a force in being throughout the campaign, complicating the air offensive. Casualties were low, an F-117 and F-16 downed, although both pilots were rescued, and two American helicopter pilots killed when they crashed attempting to evade a SAM. In the seventy-eight-day conflict the B-2s flew forty-nine sorties and dropped 650 JDAMs, and the B-1s flew more than one hundred sorties delivering iron bombs. B-52s were also involved, launching AGM-86C cruise missiles. However, these bomber missions were only a small proportion of the more than thirty-five thousand sorties flown by NATO forces.

Air power was very successful, but as has always been the case, not without complications. There were several instances of refugees killed when NATO aircraft attacked bridges and what they thought were Serb military convoys. More serious was a B-2 strike that accurately hit a building in Belgrade believed to be a Serbian installation that was in fact the Chinese embassy, killing three. The Chinese, and others, did not believe the embarrassed American explanation that an outdated map led to this bombing. Despite these glaring and much publicized incidents, almost all of the bombs hit their assigned targets, only twenty of twenty-three thousand noted as having gone "astray." There was criticism of the slow, graduated air campaign that was hindered by political considerations. There were also criticisms of NATO's choice of targets, specifically their attacks on oil facilities, the Serbian television headquarters, and bridges over the Danube. In the last six weeks of the campaign the bombing was effective, especially against oil and electricity and in closing the Danube. In any case, after seventy-eight days of air attack, the Serbs acceded to NATO/UN demands and pulled out of Kosovo, which declared its independence in February 2008. No ground soldiers were committed, although some special forces may have been involved; clearly air power dominated this conflict. This led some to agree with esteemed military historian John Keegan, who wrote that "the capitulation of President Milosevic proved that a war can be won by airpower alone."[18]

Air power also was important in actions in the Middle East following the terrorist attacks on the United States on September 11, 2001. That October, Operation Enduring Freedom opened against Afghanistan with strategic bomber (B-1B, B-2, and B-52), carrier aviation, and cruise missile strikes. The B-2s flew forty- to forty-four-hour missions from the United States, recovering at Diego Garcia before a direct return to the United States, while eight B-1Bs and ten B-52s were based on the island. These bombers used the GPS-guided JDAM and other munitions primarily against tactical targets as there were few strategic targets aside from initial strikes on command and control facilities. Late in the effort B-52s would loiter over the battlefield for on-call strikes. While the B-1Bs and B-52s flew only one-tenth of the total sorties, they dropped more than two-thirds of the tonnage. There was no effective Afghan air defense and no combat losses, although one B-1B went down to noncombat causes. As in all war there were a number of instances of collateral damage and friendly fire losses.

Air power played a similar role in the March 2003 campaign (Iraqi Freedom) that overthrew the Iraqi government of Saddam Hussein. An attack by cruise missiles and F-117s on the Iraqi dictator preceded the main action but failed to hit him. Again U.S. strategic bombers were important, although attacking few strategic targets. B-52s for the first time employed laser-guided bombs and overall two-thirds of the thirty thousand bombs dropped in the campaign were guided bombs. Ballistic missile defense got little notice but did better than in the 1991 Gulf War, in this case improved Patriot missiles downing all nine ballistic missiles launched by the Iraqis. Baghdad was taken in twenty days, ending the conventional phase of the war, which then devolved into a bloody and confused conflict consisting of guerrilla warfare, terrorist attacks, and civil war . . . warfare unsuited for strategic air operations.

LA

# CONCLUSION

As the United States enters the twenty-first century, its strategic bomber force has declined to fewer than one hundred B-1Bs *(lower)*, fewer than one hundred B-52s *(center)*, and a mere twenty B-2s *(upper)*. Strategic bombardment today is clearly different than that of the last century, with increasing questions about its utility and with smaller aircraft and ballistic missiles joining and sometimes supplanting the traditional large strategic bomber. (U.S. Naval Institute Photo Archive)

Air power has come a long way in the century since the Wright Brothers' first flight. That event changed both civilization and warfare as the frail wood and canvas machines evolved into vehicles enabling mass and speedy movement as well as delivery of death and destruction. Following World War I theorists forecast that the airplane would be decisive in future wars by bombing enemy cities and industries and at the same time would be more humane and cheaper than methods of conventional warfare by avoiding a repeat of the drawn out stalemate and slaughter of the western front trenches. There are two critical questions. Has strategic bombardment gotten a fair test? Has strategic bombardment delivered on the promises of its advocates?

World War II was certainly a suitable test of strategic bombardment. This was a total war between the major powers of the world allied in coalitions, that involved vast numbers of both men and the most modern military equipment. In contrast to World War I, in which aircraft numbers were limited, as was aircraft capability, in World War II America and Britain thrust enormous fleets of capable aircraft against Germany, as did the United States against Japan. Strategic bombers dropped massive tonnages of bombs, destroyed much, and killed many, but were not decisive. In the end air power proved more significant in other roles, especially in the Atlantic antisubmarine campaign and the Pacific sea war. For despite enormous investments of resources, tremendous production and great innovation, the valor and ingenuity of the airmen, and substantial military and civilian casualties, the promises of the strategic bombing theorists and advocates of a quick and cheap victory were not realized. The strategic air war over Europe was but a high-tech version of the attrition warfare of World War I, this time fought in the skies. The dramatic and expensive bombing campaigns over both Germany and Japan were not war winners.

World War II cast serious doubt, if not disproved, two key tenets of strategic bombing theory. First, the concept that bombing would break civilian morale and force a nation to capitulate was shattered despite the horrendous hardships and heavy losses inflicted on civilians. Although the fliers made great efforts at considerable cost, with a few notable exceptions (Hamburg, Dresden, Tokyo, Hiroshima, and Nagasaki), they could not inflict quick, extensive punishment, instead they meted out various degrees of death, pain, damage, and suffering that was diluted by time and geography, which lessened its impact. And while authors continue to argue whether the bombing increased or decreased the morale of the populations, this is irrelevant, for under authoritarian regimes people complied and worked to survive and thus the economies and societies in Germany and Japan continued to function.

The second tenet, which held that bombing the key points of a nation's economy would achieve victory, also floundered in the war, though certainly not for lack of trying. Poor choice of targets, duds, inadequate accuracy, poor weather, ineffective bomb damage assessment that overestimated damage and delayed or excluded restrikes, and tough defenses that made bombing expensive and frequently impractical were all factors. In addition enemy economies proved flexible and resilient. The result was that these attacks hindered the war effort but were not decisive. However, the closing shots of the war, the nuclear attacks in August 1945, seemed to portend a different future.

During the four decades of the Cold War the world was haunted by the specter of a nuclear exchange that did not occur. At first long-range strategic bombers were the only nuclear delivery system, then ballistic missiles became operational in the early 1960s. Initially missiles supplemented the bombers, then replaced them as the nuclear delivery system of choice, a position they hold today. Concurrently a deterrence system between the two superpowers emerged that kept matters under control despite some dangerous confrontations and an enormous buildup of nuclear weapons. There was neither a nuclear exchange nor a lesser war between the two nuclear superpowers as the balance of terror precluded such a conflict because the United States and Soviet Union understood and operated by the same rules. Although nuclear weapons have not been used since Nagasaki in 1945, the acquisition of nuclear weapons by states that may not follow the "rules of the road," and the instability of the international world order, does not give great confidence that this will continue.

The experience since World War II has not vindicated strategic bombardment. The wars that erupted after 1945 that involved the superpowers have been limited, hampering strategic bombing and helping render these operations indecisive. Despite marked advances in equipment and considerable investment and effort, strategic bombing has been unable to force a decision. Although technical obstacles have been largely overcome, strategic bombardment remains shackled by political restraints and intelligence inadequacies. In fact these handicaps appear greater today as modern communications make every bomb blast instantly known throughout the world and as governments and populations have become much more sensitive to the use of military force and especially to civilian casualties.

Since World War II a number of factors have altered strategic bombing. First, technology has changed. The balance between offensive and defense has shifted with the introduction of potent surface-to-air missiles that force strategic bombers to fly low, degrading their performance. While stealth technology seemed to finesse this problem, it is expensive. Another change has been the growing intrusion of smaller (tactical) aircraft into the strategic bombardment role as fighter/attack aircraft increased their capabilities to rival large, long-range bombers. Air-to-air refueling allowed smaller aircraft to overcome most of the restrictions of range. In a similar fashion precision-guided munitions reduced the effort once required to destroy a target from hundreds of tons of "dumb" bombs to the present time, when several targets can be destroyed by a single aircraft delivering a few "smart" bombs. As a result smaller fighter/attack aircraft now can fulfill most of the missions of a strategic bomber cheaper and with less vulnerability. The traditional view of strategic bombardment, of formations of large, slow aircraft carrying massive bomb loads on penetrating missions is obsolete.

Ballistic and cruise missiles, unmanned aerial vehicles, stealthy attack aircraft armed with PGMs, along with modern air defenses make the manned penetrating bomber impractical in conflicts between nations possessing major air power. For nuclear operations, ballistic missiles, both land and sea based, have emerged as a far quicker, less vulnerable, and cheaper nuclear delivery and deterrent systems than manned bombers. Conventionally armed ballistic missiles are another matter. Thus far nonnuclear surface-to-surface missiles have seen little combat and when used have lacked adequate accuracy, numbers, and destructive power to significantly influence the course of these conflicts. Practical concerns undercut such employment. Less expensive weapons can do the job and incoming ballistic missiles launched by a nuclear power could be mistaken for a nuclear attack and might trigger an excessive response. Conventional warfare against a power with weak air defenses is another matter. In this case long-range and large bomb loads coupled with expectations of low losses may well recommend the use of manned bombers.

Another factor is the change in public attitudes concerning bombardment of cities and civilians. The large casualties and massive destruction of European and Japanese cities during World War II and the threat of nuclear devastation sobered and terrorized people around the world. Attitudes changed to the extent that in Vietnam there was extraordinary concern about civilian casualties (affecting targeting because of concern over collateral damage). Today the Internet, ubiquitous camera phones, and twenty-four-hour news coverage leave no bomb blast uncovered. World War II strategic bombing is no longer acceptable, even in a worthy cause.

Since World War II the world has entered a period of limited wars. Thus far the countries that have fought these wars either had limited capabilities and targets or, in the conflicts that pitted major powers against minor powers, a lack of targets or political pressures that inhibited strategic bombardment. The use of strategic bombing by major powers against minor ones has yielded meager, mixed, and sometimes even counterproductive results as seen in Korea and Vietnam. And as the present war in Iraq has affirmed, strategic bombing has no place in civil or guerrilla wars, or against terrorists. Advocates argue that the air campaign against Serbia in 1999 and, with a little more time, the bombardment of Iraq in 1991 demonstrated that strategic bombardment alone could win wars. (Detractors, however, counter that there were other factors and, especially, that both countries were completely without outside support and thus doomed to defeat.)

A further change is the transformation of the threat to the United States following the collapse of the Soviet Union. America's present overwhelming conventional military power makes a conventional war against the United States unwinnable and thus unlikely. As recent events have indicated, the more plausible threat to the United States is a conflict at a lower level, guerrilla war or terrorism. Strategic bombardment can only play a minor or supporting role in such wars. Other nations may, and probably will, engage in conventional wars, but aside from nuclear deterrents, none have a major strategic bombardment capability. In brief, strategic bombardment has been relegated to a deterrent role and peripheral (conventional) war-fighting role.

Clearly we have entered a new era. The role of the strategic bomber, especially the manned penetrating bomber, has diminished. The thousands of bombers employed in World War II and the nearly two thousand bombers deployed by Strategic Air Command in the late 1950s are now vague memories. Today the United States fields the world's most powerful military force, the largest air force, the most strategic bombers, yet it has less than two hundred bombers in its inventory. Air power will continue to be important, perhaps even decisive, in future military operations, but not strategic bombing as waged by bombers in the mid–twentieth century. As long as the wars of the major powers remain constrained, strategic bombardment will be restrained and its successes limited. For lesser powers, the acquisition of nuclear weapons and modern strategic delivery systems may lead in a somewhat different direction.

Over the years strategic bombardment proponents have promised that their concept could win wars quickly and cheaply. Both British and Americans tested the concept in World War II, where it proved deadly and destructive but not decisive. Since then technology has greatly improved its capabilities and strategic bombardment has been employed numerous times at a considerable cost in blood and treasure. However, the shackles of inadequate intelligence combined with political restraints have prevented the concept from being fully employed. Complete application of strategic bombardment is only suitable in total war, which nuclear weapons have made unthinkable, at least for nuclear-armed nations. Thus the practitioners have fallen short of the theorists' vision. The record during its first century reveals that strategic bombardment is a case study of promises unfulfilled.

# SOURCE NOTES

## Chapter 1. The Early Years through World War I

While there is a considerable literature covering most aspects of air power, there are only two overall studies of strategic bombing. Lee Kennett's *A History of Strategic Bombing: From the First Hot-air Balloons to Hiroshima and Nagasaki* (New York: Scribner's, 1982) is especially good on the first decades but ends in 1945. R. Cargill Hall, ed., *Case Studies in Strategic Bombardment* (Washington, D.C.: Air Force History and Museums Program, 1998), like most collections of essays, varies in quality and balance but is notable for its length, scholarship, and authors. There are a number of studies on the broader subject of air power of varied value. Robin Higham's *Air Power: A Concise History* (New York: St. Martin's, 1972), which has been updated and renamed *100 Years of Air Power and Aviation* (College Station: Texas A&M, 2003), and Walter Boyne's *The Influence of Air Power upon History* (New York: Pelican, 2003) are written by two prolific and leading experts in the field. James Stokesbury's *A Short History of Air Power* (New York: Morrow, 1986) is an excellent brief history. Two scholarly studies critical of strategic bombing are Ronald Schaffer's *Wings of Judgment: American Bombing in World War II* (New York: Oxford, 1985) and Michael Sherry's *The Rise of American Air Power: The Creation of Armageddon* (New Haven, Conn.: Yale, 1987). Two recent studies are Stephen Budiansky's *Air Power: The Men, Machines, and Ideas that Revolutionized War, from Kitty Hawk to Gulf War II* (New York: Viking, 2004), and Charles Gross's *American Military Aviation: The Indispensable Arm* (College Station: Texas A&M, 2002). John Buckley's *Air Power in the Age of Total War* (Bloomington: Indiana University Press, 1999) is excellent, but his coverage ends in 1945.

Aircraft are better covered. I have used the files of the National Museum of the United States Air Force, the Boeing Company, the National Archives, the San Diego Air and Space Museum, and the U.S. Air Force Historical Research Agency, which contain both primary and secondary materials on both U.S. and foreign aircraft. Aircraft are excessively covered by the secondary sources. For British bombers throughout this study I have used Peter Lewis, *The British Bomber since 1914* (London: Putnam, 1967), along with Owen Thetford, *Aircraft of the Royal Air Force since 1918* (London: Putnam, 1968). Overall studies on U.S. bombers of especial value are Lloyd Jones, *U.S. Bombers: B1-B70* (Los Angeles: Aero, 1962); Gordon Swanborough and Peter Bowers, *United States Military Aircraft since 1908* (London: Putnam, 1971); and Ray Wagner, *American Combat Planes* (Garden City, N.Y.: Doubleday, 1982). Though rough reading, the essential work for the later period is Marcelle Knaack's *Encyclopedia of U.S. Air Force Aircraft and Missile Systems*, vol. 2, *Post–World War II Bombers* (Washington, D.C.: Office of Air Force History, 1988). German bombers are magnificently covered in William Green's *Warplanes of the Third Reich* (Garden City, N.Y.: Doubleday, 1979). For World War II bombers William Green's two volumes, *Famous Bombers of the Second World War* (Garden City, N.Y.: Hanover House, 1959 and 1960), have well stood the test of time.

The best book on the first battle of Britain remains Raymond Fredette, *The Sky on Fire* (New York: Holt, Rinehart and Winston, 1966), although Thomas Fegan's more recent book, *The "Baby Killers"* (London: Cooper, 2002), was also helpful. For details of the zeppelin campaign I have relied primarily on Douglas Robinson's wonderful *The Zeppelin in Combat* (Sun Valley, Calif.: Caler, 1971), and a useful narrative account by Ernest Dudley, *Monsters of the Purple Twilight* (London, Harrap, 1960). Clearly the best on the German R-planes is G. W. Haddow and Peter Grosz, *The German Giants* (London: Putnam, 1962). On British strategic bombing see George Williams's *Biplanes and Bombsights* (Maxwell AFB, Ala.: Air University, 1999), an update of his 1987 Ph.D. dissertation.

1. These craft had a rigid framework, an outer skin, and gas bags within providing lift. In addition to the Zeppelin company, three other companies built airships for the German military. For simplicity's sake I have used the less accurate but less confusing term "zeppelin" (lower case) to describe all German rigid airships, and capitalized "Zeppelin" when referring to the man and the company.
2. Dead reckoning uses direction flown (course) along with speed and time elapsed to determine position. The difficult variable is the wind, which will affect both speed and course.
3. Only four of the eighty-one German naval airships achieved higher speeds in their trials.
4. Fegan, *"Baby Killers,"* 12.
5. In 1917 the Germans experimented with a double-walled gas bag, the outer containing an inert gas, but the weight penalty led to its abandonment.
6. Only four large Zeppelins were built during the war.
7. Robinson, *Zeppelin in Combat*, 202, 197–202.
8. When glycerin did not solve the freezing problem, sand was substituted for ballast. In April 1917 the gun decision was modified to carrying two on overwater scouting missions.
9. Thus far the earlier model Gothas had been used to support the German army. The Germans used a prefix to indicate the function of the aircraft, "C" for observation, "D" for fighters, "G" for bombers, and "R" for large bombers.
10. Haddow and Grosz, *German Giants*, 26.
11. The Germans killed 260 Parisians with forty-five tons of artillery shells from the infamous "Big Berthas," which also caused $10 million of damage.
12. Williams, *Biplanes and Bombsights*, 5, 7–8.
13. Ibid., 115.
14. A plan from the field written in May 1918 called for attacks on industrial targets, with pride of place to iron and coal, then chemicals, followed by explosives, and finally miscellaneous production.
15. Williams, *Biplanes and Bombsights,* 176.
16. Fredette, *Sky on Fire,* 223.
17. Williams, *Biplanes and Bombsights,* 199.

18. The United States ordered 2,000 of these aircraft to be powered by two 350–400 hp Liberty engines; 107 were accepted. Some later participated in Billy Mitchell's famous battleship bombing in 1921. With more engine power, the U.S. version could carry a slightly heavier bomb load, faster.
19. Williams, *Biplanes and Bombsights*, xii.
20. Lord Northciffe quoted in Buckley, *Air Power in the Age of Total War*, 34.

## Chapter 2. The Interwar Years

On U.S. aircraft see J.V. Mizahi, *Air Corps* (Northridge, Calif.: Sentry, 1970), and the official history, Wesley Craven and James Cate, eds., *The Army Air Forces in World War II*, vol.1, *Plans and Early Operations, January 1939 to August 1942* (Chicago: University of Chicago, 1948) (hereafter cited as Craven and Cate, *AAFWWII*.) Soviet aircraft are covered in Alexander Boyd, *The Soviet Air Force since 1918* (New York: Stein and Day, 1977); Von Hardesty, *Red Phoenix* (Washington, D.C.: Smithsonian Institution Press, 1982); Robin Higham, John Greenwood, and Von Hardesty, eds., *Russian Aviation and Air Power in the Twentieth Century* (London: Cass, 1998); Robin Higham and Jacob Kipp, eds., *Soviet Aviation and Air Power* (London: Brassey's, 1977); and Robert Kilmarx, *A History of Soviet Air Power* (New York: Praeger, 1962). French bombers are dealt with in William Green, *War Planes of the Second World War*, vol.7, *Bombers and Reconnaissance Aircraft* (London: Macdonald, 1967), and Robin Higham and Stephen Harris, eds., *Why Air Forces Fail: The Anatomy of Defeat* (Lexington: University of Kentucky, 2006).

On the Chaco War, see Dan Hagedorn and Antonio Sapienza, *Aircraft of the Chaco War, 1928–1935* (Atglen, Pa.: Schiffer, 1997). The best covered of the interwar conflicts is the Spanish Civil War, well treated in Angelo Ghergo, *Wings over Spain* (Milan, Italy: GAE, 1997); E. R. Hooton, *Phoenix Triumphant* (London: Brockhampton, 1994); and Hugh Thomas, *The Spanish Civil War* (New York: Harper and Row, 1961). On the Russo-Finnish War, William Trotter, *A Frozen Hell* (Chapel Hill, N.C.: Algonquin, 1991), is useful.

The best on doctrine is Robert Futrell's classic and massive *Ideas, Concepts, Doctrine: Basic Thinking in the United States Air Force, 1907–1960* (Maxwell AFB, Ala.: Air University, 1989), and the more recent study by Tami Biddle, *Rhetoric and Reality in Air Warfare: The Evolution of British and American Ideas about Strategic Bombing, 1914–1945* (Princeton, N.J.: Princeton University Press, 2004). There are three key essays on the three most prominent air power theorists in Phillip Meilinger, ed., *The Paths of Heaven* (Maxwell AFB, Ala.: Air University, 1997). Giulio Douhet's writings are set out in his study *The Command of the Air* (New York: Coward-McCann, 1942). British doctrine is also well treated in the official Royal Air Force history, Charles Webster and Noble Frankland, *The Strategic Air Offensive Against Germany, 1939–1945*, vol. 1 (London: HMSO, 1961) (hereafter cited as Webster and Frankland, *SAOAG*). Much was written by and on Mitchell. The most important biography remains Alfred Hurley's *Billy Mitchell* (Bloomington: Indiana University Press, 1975). Mitchell wrote numerous magazine articles and three books: *Winged Defense* (New York: Putnam, 1924); *Skyways* (Philadelphia: Lippincott, 1930); and *Our Air Force* (New York: Dutton, 1931). Some of the key works on the Air Corps Tactical School and U.S. doctrine are Martha Byrd, *Kenneth N. Walker* (Maxwell AFB, Ala.: Air University, 1997); Craven and Cate, *AAFWWII*, vol. 1; Robert Finney, "History of the Air Corps Tactical School," USAF Historical Study no. 100, 1955, U.S. Air Force Historical Research Agency, Maxwell AFB, Ala. (hereafter cited as HRA); Thomas Greer, "The Development of Air Doctrine in the Army Air Arm," USAF Historical Study no. 89, 1955, HRA; and William Sherman, *Air Warfare* (New York: Roland, 1926). One of the only studies of air intelligence and targeting is the diverse collection in John Kreis, ed., *Piercing the Fog: Intelligence and Army Air Forces Operations in World War II* (Washington, D.C.: Air Force History and Museums Program, 1996).

1. These included airframe (metal construction, flaps, monoplane configuration, and enclosed cockpits), high-octane fuels, controllable pitch propellers, retractable landing gear, and increased engine power.

2. Bill Yenne, *The World's Worst Aircraft* (Greenwich, Conn.: Dorset, 1990), 32–33.
3. The B-10 was all metal, with the exception of the control surfaces, trailing edge of the wing (initially), and split flaps, which were fabric covered.
4. Craven and Cate, *AAFWWII,* 1:64.
5. Ibid., 1:65.
6. The XB-15 was powered by four engines generating 4,000 hp and had an empty weight of 38,000 pounds and a gross weight of 71,000 pounds. The B-29 had 8,800 hp available and weighed 70,000 pounds empty and 124,000 gross. The B-29 could reach 358 mph and 32,000 feet, while the XB-15's top speed was 195 mph and its service ceiling was 19,000 feet.
7. Georges Bouche, "'Grandpappy'—the XB-15," *Aerospace Historian*, Sept. 1979, 175.
8. Ibid., 173.
9. Freeman Westel, "B-15: Grandaddy of the B-17," *Air Classics*, July 1966, 15.
10. The *Maxim Gorky* had a crew of twenty-one and could carry forty-three to seventy-six others. It crashed in May 1935, killing thirty-five, when a fighter collided with it in an attempt to loop around it for a motion picture sequence.
11. John Greenwood, "The Great Patriotic War, 1941–1945," in Higham and Kipp, *Soviet Aviation and Air Power*, 70.
12. Richard Hough and Dennis Richards, *Battle of Britain: The Greatest Air Battle of World War II* (New York: Norton, 1989), 48.
13. James Corum, *The Luftwaffe: Creating the Operational Air War, 1918–1940* (Lawrence: University Press of Kansas, 1997), 89.
14. J. F. C. Fuller, *Armament and History* (New York: Scribner's, 1943), 146.
15. Douhet, *Command of the Air*, 15, 190.
16. Ibid., 110.
17. Ibid., 10.
18. Ibid., 34.
19. Ibid., 57.
20. Ibid., 61.
21. Douhet wrote that bombing would never have the accuracy of artillery; however, Douhet did not consider this significant as "bombing objectives should always be large." Ibid., 20, 19.
22. Meilinger, *Paths of Heaven*, 46.
23. Peter Gray, "Review Essay by Group Captain Peter W. Gray RAF," *Royal Air Force Air Power Review* 4, no. 4 (Winter 2001): 2.
24. Webster and Frankland, *SAOAG,* 1:47.
25. Theodore Ropp, *War in the Modern World* (New York: Collier, 1959), 308.
26. The 1899 Hague Conference accepted the U.S. proposal to ban bombing for five years, but at the Second Hague Conference in 1907 only the United States and Britain were willing to extend the prohibition.
27. The ideas of Douhet and his collaborator, aircraft manufacturer Gianni Caproni, concerning strategic bombardment were presented to the American airmen.
28. Sherman, *Air Warfare*, 5, 197. Sherman was one of the original instructors at the Tac School, where he taught between 1920 and 1923. In 1921 he wrote the first text for the school, which we can assume provided the core for his book.

29. Ibid., 214.
30. Ibid.
31. Byrd, *Kenneth N. Walker*, 36.
32. Orvil Anderson, "The Development of U.S. Strategic Air Doctrine, ETO, World War II," Air War College lecture, Sept. 1951, 8, K239.716251-9, HRA.
33. The Bolivians did not convert their three Ju 52/3ms into bombers that could have threatened the slender Paraguayan supply lines. Toward the end of the war they ordered some Curtiss Condor aircraft for that purpose that got as far as Lima, Peru, before being seized.
34. Douhet advocates countered that the attacks lacked both numbers and poison gas.
35. *Air Corps News Letter*, Aug. 1, 1937, 7.
36. Henry Arnold, "The Air Corps," Oct. 1937, 13, speech, U.S. Army War College, Carlisle, Pa.

## Chapter 3. German Strategic Bombardment

Books on the Luftwaffe in English are profuse. A number of chapters in the official German history of the war, Research Institute for Military History, *Germany and the Second World War*, vols. 1, 2, 7 [thus far] (Oxford: Clarendon, 1990, 1991), are important. I found the dated Royal Air Force, *The Rise and Fall of the German Air Force (1933–1945)* (Old Greenwich, Conn.: WE, 1969), still useful. Also see James Corum, *The Luftwaffe* (Lawrence: University Press of Kansas, 1997); E. R. Hooton, *Phoenix Triumphant* (New York: Arms and Armour, 1994); Richard Muller, *The German Air War in Russia* (Baltimore: Nautical and Aviation, 1992); Williamson Murray, *Strategy for Defeat: The Luftwaffe, 1933–1945* (Maxwell AFB, Ala.: Air University, 1983); and Hanfried Schliephake, *The Birth of the Luftwaffe* (Chicago: Regnery, 1971). On German bombers William Green asserts his authority in his extensive *The Warplanes of the Third Reich* (Garden City, N.Y.: Doubleday, 1970). For German doctrine I relied on Corum, *Luftwaffe*; and Richard Suchenwirth, "The Development of the German Air Force, 1919–1939," USAF Historical study no. 160, 1968, HRA. GAF combat operations are covered in Cajus Bekker, *The Luftwaffe War Diaries* (Garden City, N.Y.: Doubleday, 1968); and Lee Kennett, *A History of Strategic Bombing: From the First Hot-air Balloons to Hiroshima and Nagasaki* (New York: Scribner's, 1982). The literature on the Battle of Britain is immense. I found the old study by Derek Wood and Derek Dempster, *The Narrow Margin* (New York: McGraw Hill, 1961), most useful, along with the brief, more recent effort by Richard Overy, *The Battle of Britain* (New York: Norton, 2000). Also helpful were Richard Hough and Denis Richards, *The Battle of Britain* (New York: Norton, 1989); T. C. G. James, *The Battle of Britain* (London: Cass, 2005); Jon Lake, *The Battle of Britain* (London: Silverdale, 2000); and Richard Overy, *The Air War, 1939–1945* (New York: Stein and Day, 1980).

The bombing of Coventry is the subject of Norman Longmate's *Air Raid* (New York: McKay, 1976). On the He 177 see Green's *Warplanes of the Third Reich*; Manfred Griehl, *Heinkel He 177, 277, 274* (London: Airlife, 1998); R. S. Hirsch and Uwe Feist, *Heinkel 177* (Fallbrook, Calif.: Aero, 1967); and two works by Alfred Price, "He 177 Greif," *International Air Power Review*, vol. 11 (2004): 158–73, and *Heinkel He 177* (London: Aircraft Profile, n.d.). German strategic air attacks are well covered in Oleg Hoeffding, "German Air Attacks Against Industry and Railroads in Russia, 1941–1945," RAND Memo RM-2606-PR, 1970, Air University Library, Montgomery, Ala. (hereafter cited as AUL). Using archival and secondary materials I cover the V-1 in *Archie to SAM* (Maxwell AFB, Ala.: Air University, 2005), and *The Evolution of the Cruise Missile* (Maxwell AFB, Ala.: Air University, 1985). Four very useful secondary works on the V-1 are Adam Gruen, *Preemptive Defense* (Washington, D.C.: Air Force History and Museums Program, 1998); Dieter Holsken, *V-Missiles of the Third Reich* (Sturbridge, Mass.: Monogram Aviation, 1994); Benjamin King and Timothy Kutta, *Impact* (Rockville Centre, N.Y.: Sarpedon, 1998); and Stephen Zaloga, *V-1 Flying Bomb, 1942–52* (Oxford: Osprey, 2005). For details on both V-weapons (and guided bombs) the War Department's *Handbook on Guided Missiles of Germany and Japan* (Washington, D.C.: War Department, 1946) is invaluable. On the V-2 see Walter Dornberger, *V-2* (New York: Bantham, 1979); T. D. Dungan, *V-2* (Yardley, Pa.: Westholme, 2005); Gregory Kennedy, *Vengeance Weapon 2* (Washington, D.C.: Smithsonian Institution Press, 1983); Michael Neufeld, *The Rocket and the Reich* (New York: Free Press, 1995); and Steven Zaloga, *V-2 Ballistic Missile, 1942–52* (London: Osprey, 2003).

1. Green, *Warplanes of the Third Reich*, 130.
2. The DC-3 first flew in December 1935 and was the first of a new generation of aircraft, a product of the emerging technology of the 1930s, as opposed to the Ju 52/3m, which was the last of a long line of development. More than 4,800 Ju 52/3ms were built, compared with almost 13,800 DC-3/C-47s.
3. The GAF's training unit could only get 1 to 2 percent hits from 12,000 feet compared with dive bombers, which consistently got 35 percent of their bombs within 150 feet of the target.
4. Corum, *Luftwaffe*, 82, 52–53, 64, 74, 81.
5. Ibid., 139.
6. Muller, *German Air War*, 10, 14.
7. Bekker, *Luftwaffe War Diaries*, 107.
8. By October 1 the British had a numerical advantage in single-engine fighters, and during 1940 they built 4,283 fighters compared with 1,870 manufactured by the Germans. One problem with these figures is the difference between aircraft on hand and aircraft serviceable.
9. In 1939 the British completed their chain of radar stations, which had a detection range of eighty to one hundred miles against high-flying aircraft, as compared with eight or so miles with sound detectors.
10. Drew Middleton, *The Sky Suspended: The Battle of Britain* (London: Secker & Warburg, 1960), 147.
11. Between September 7 and November 13, 1940, the Luftwaffe lost only 81 aircraft on twelve thousand night sorties, mostly to flak. British night defenses improved, with fighters downing 22 of 39 GAF aircraft destroyed in March 1941, 48 of 87 in April, and 96 of 128 in the first two weeks of May.
12. During the three-month battle (July 27 to October 26), losses of Hurricanes and Spitfires exceeded production in only four weeks with a fifth week of equilibrium. Perhaps of greater significance, at this point British fighter production was more than double that of the Germans.
13. The Germans did not employ drop tanks on Me 109s, although they had been used in Spain. The pilots considered them dangerous, and their construction out of plywood proved faulty.
14. Alfred Price, *The Battle of Britain: The Hardest Day, 18 August 1940* (New York: Scribner's, 1979), 183.
15. Also helping the attackers was that the British botched the jamming of German navigational aids.
16. Green, *Warplanes of the Third Reich,* 336.
17. Hirsch and Feist, *Heinkel 177*, n.p.
18. The Hs 293 was a winged bomb powered by an underslung rocket engine. The 1,600-pound weapon had a range of nine miles when released from an altitude of eighteen thousand feet. The unpowered FX 1400 could reach almost three miles if launched from twenty-four thousand feet. The weapon's most famous exploit was the September 1943 sinking of the Italian battleship *Roma* after being launched from Do 217s.

19. The Mistel mounted a manned fighter atop an unmanned bomber (Ju 88), which flew toward the target; the latter was detached and directed toward impact while the former returned to base. The piggyback concept was tested in 1942; however, the first operational use was not until June 1944 against shipping in the Normandy invasion. Plans in January 1945 for attacks against Soviet industrial targets were thwarted when the Russians overran GAF airfields; the only other combat operations were in March against bridges on both the western and eastern front.
20. Price, "He 177 Greif," 173.
21. Circular error probable, half of the hits are within the radius of the aiming point.
22. The missile's fuselage measured 22 feet long and 17.5 feet wide, and it weighed 4,900 pounds (2,000-pound warhead). From dead astern the V-1 presented a surface area of sixteen square feet, compared with forty-five square feet on the FW 190.
23. Gruen, *Preemptive Defense*, 14.
24. The CEP of the ground-launched V-1s in 1944 was eight miles; air-launched V-1s, twenty-four miles; and the long-range V-1s in 1945, twelve miles.
25. David Johnson, *V-1, V-2: Hitler's Vengeance on London* (New York: Stein and Day, 1981), 115.
26. One source states that the failure rate on the ground and shortly after launch was 28 percent for the V-1 and 17 percent for the V-2.
27. British radar plots of eight-seven V-2 tracks indicated an average range of 184 miles, while a German source gives a range of 199 miles.
28. One source writes that twenty-seven thousand out of sixty-four thousand workers died at the facility, another states that one-third of sixty thousand died.
29. The British considered firing a barrage of antiaircraft guns into an air space upon radar warning of a missile launch. However, they calculated it would take 320,000 shells to down one missile, and that 2 percent of these shells would be duds raining down on British soil.
30. Dwight Eisenhower, *Crusade in Europe* (Garden City, N.Y.: Doubleday, 1961), 276.

## Chapter 4. British Strategic Bombing

Data on British bombers are drawn primarily from Peter Lewis, *The British Bomber since 1914* (London: Putnam, 1967); Owen Thetford, *Aircraft of the Royal Air Force since 1918* (London: Putnam, 1968); and William Green's two-volume *Famous Bombers of the Second World War* (Garden City, N.Y.: Doubleday, 1959; UK: Hanover House, 1960). On the Mosquito also see Edward Bishop, *Mosquito: Wooden Wonder* (New York: Ballantine, 1971), and Edward Shacklady, *De Havilland Mosquito* (Bristol, UK: Cerberus, 2003). The official account of Bomber Command is the four-volume history by Charles Webster and Noble Frankland, *The Strategic Air Offensive Against Germany, 1939–1945* (London: HSMO, 1961) (hereafter cited as Webster and Frankland, *SAOAG*). This is supplemented by *Report of the British Bombing Survey Unit* (London: Cass, 1998); Sebastian Cox, ed., *Dispatch of War Operations, 23 February 1942 to 8 May 1945, Air Chief Marshal Sir Arthur T. Harris* (London: Cass, 1995); Arthur Harris, *Bomber Offensive* (London: Collins, 1947); Max Hastings, *Bomber Command* (New York: Dial, 1998); and Robin Neillands, *The Bomber War* (New York: Overlook, 2001). Statistics are pulled from United States Strategic Bombing Survey, Statistical Appendix to Over-all Report (European War), Feb. 1947, Air University Library, Montgomery, Ala. (hereafter cited as AUL); and Webster and Frankland, *SAOAG*, vol. 4. The impact of the bombing is drawn from the various reports of the United States Strategic Bombing Survey, particularly Over-all Report (European War), 1945, AUL, and updated by the various writings of Richard Overy, *The Air War, 1939–1945* (New York: Stein and Day, 1980); *War and Economy in the Third Reich* (London: Oxford, 1994); and *Why the Allies Won* (New York: Norton, 1995). Also see Alfred Mierzejewski, *The Collapse of the German War Economy, 1944–1945* (Chapel Hill: University of North Carolina Press, 1988). The major raids are covered in detail in Martin

Middlebrook's trilogy, *The Nuremburg Raid* (New York: Morrow, 1973), *The Battle of Hamburg* (New York: Scribner's, 1980), and *The Berlin Raids* (New York:Viking, 1988). Of the many books on the Dresden attack, the most recent and best is Frederick Taylor's *Dresden* (New York: Harper Collins, 2004). In addition to the above works (and those noted in chapter 3) that touch on German air defenses, see Peter Hinchliffe, *The Other Battle* (Edison, N.Y.: Castle, 2001.

1. The RAF knew of this problem: In an exercise in 1937 two-thirds of the RAF bombers could not find Birmingham.
2. Interestingly, a poll of Londoners during the Blitz found them evenly split on conducting revenge bombing.
3. The British estimated that the GAF bombers carried an average of 30 percent incendiaries and on occasions 60 percent, compared with an average of 15 percent and 30 percent respectively carried by Bomber Command.
4. Webster and Frankland, *SAOAG,* 1:47.
5. Cox, *Dispatch of War Operations*, ix.
6. Horst Boog, "Harris—A German View," in Cox, *Dispatch of War Operations*, xxxvii.
7. The number of heavy bombers rose to 897 in February 1944 and 1,283 in February 1945. Lancasters comprised 46 percent of the heavy bomber force in February 1943 and 73 percent in March 1945.
8. The name came from the observation that the equipment sounded like an oboe.
9. The origin of the designation is murky and variously said to be related either to "stinky," $H^2S$ being the chemical formula of hydrogen sulfite, which is an odorous combination, or "home sweet home."
10. Cox, *Dispatch of War Operations*, 11, 10.
11. A brief note on AAF and RAF organization during World War II. Both air forces used squadrons consisting of one to two dozen aircraft as their lowest organizational unit, but while the British counted their forces in number of squadrons and organized these squadrons into groups consisting of a number of squadrons, the Americans organized four bomb squadrons (three in the case of the B-29s) into bomb groups and counted and thought in terms of groups. The AAF organized numbers of bomb groups into bomb wings and bomb wings into bomber commands.
12. Webster and Frankland, *SAOAG,* 1:433. The authors cite as evidence that Bomber Command's most famous unit, in which three of the greatest bomber pilots flew, was 617 Squadron of 5 Group. The Pathfinders were formed with one squadron from each group of Bomber Command, each flying a different type aircraft.
13. One can only speculate what impact the deletion of the copilot had on accidents and survival rates in combat.
14. While the .303s had a higher rate of fire (1,150 shots per minute versus 800), the .50s had a higher muzzle velocity (2,990 feet per second versus 2,660) and a heavier projectile (thus greater impact) as well as a longer range.
15. Green, *Famous Bombers of the Second World War,* 1:124.
16. To be clear the loss rates for Bomber Command are of aircraft that failed to return or those listed as missing compared with the number of bombers dispatched. It does not include other losses, such as accidents over friendly territory or those bombers that returned damaged and were written off as uneconomical to repair. Using the dispatched numbers obscures the fact that the bombers that actually bombed were fewer due to losses, to aborts, and to some aircraft bombing other than the designated target. The loss rate on a sortie basis was 2.36 percent for the Lancaster, 2.52 percent for the Halifax, and 3.71 percent for the Stirling.

17. The survival rate of crewmembers shot down was poorer in the Lancaster than in other Bomber Command aircraft, 11 percent in Lancasters compared with 29 percent in later model Halifaxes.
18. De Havilland claimed that wood construction required only 37 percent of the man hours compared to all-metal construction.
19. Webster and Frankland, *SAOAG,* 1:394.
20. Ibid., 2:168.
21. Ibid., 2:108.
22. Marking errors varied between six hundred yards and a mile or more to which must be added the accuracy of the main force bombers.
23. Cox, *Dispatch of War Operations*, 19–20.
24. Webster and Frankland, *SAOAG,* 2:9, 263–64.
25. Ibid., 2:55, 59. There were high RAF officers who questioned these assertions.
26. Ibid., 2:209, 207–8.
27. In addition seventy-seven bombers were damaged, twelve of which were written off.
28. Middlebrook, *Berlin Raids*, 234; Webster and Frankland, *SAOAG,* 2:201–2.
29. Prior to September 1942 GAF pilots received more flying training hours than British pilots, but after that date German flying training hours declined while Allied flying hours increased. During the period October 1942 to June 1943, the average GAF pilot was receiving about 200 flying hours, the RAF pilot about 340, and the AAF pilot about 275, a gap that widened as the war progressed.
30. Webster and Frankland, *SAOAG,* 3:57.
31. Ibid., 3:112, 117.

## Chapter 5. U.S. Strategic Bombing in Europe

The literature on the U.S. strategic air war against Germany is immense. This is demonstrated by two bibliographies, one by this author, *Eighth Air Force Bibliography: An Extended Essay and Listing of Published and Unpublished Materials, Updated and Revised* (Strasburg, Pa.: 8th Air Force Memorial Museum Foundation, 1996), and another following a tight, valuable article by Stephen McFarland and Wesley Newton, "The American Strategic Air Offensive Against Germany in World War II," in *Case Studies in Strategic Bombardment*, ed. Cargill Hall (Washington, D.C.: Air Force History and Museum Program, 1998). Materials on the B-17 and B-24 are indicated in chapter 1, along with William Green, *Famous Bombers of the Second World War*, vol. 1 (Garden City, N.Y.: Hanover House, 1959). My work on the Eighth Air Force, based on primary sources, can be found in "The Tactical Development of the Eighth Air Force in World War II," Ph.D. diss., Duke University, 1969, and on the Fifteenth Air Force in the 301st Bomb Group history, *"Who Fears?": The 301st in War and Peace* (Dallas: Turner, 1991). The official AAF history is the seven-volume, detailed, solid, and dated work by Frank Craven and James Cate, eds., *The Army Air Forces in World War II* (Chicago: University of Chicago, 1948–58) (hereafter cited as Craven and Cate, *AAFWWII*). Roger Freeman's trilogy on the Eighth is basic to any study of the subject: *The Mighty Eighth* (London: MacDonald, 1970), *Mighty Eighth War Diary* (London: Jane's, 1981), and *Mighty Eighth War Manual* (London: Jane's, 1984). On the British and German air forces, see the previous chapters. The best on GAF fighter defenses is the scholarly and excellent Donald Caldwell and Richard Muller, *The Luftwafffe over Germany: Defense of the Reich* (London: Greenhill Books, 2007). Two key books are Stephen McFarland and Wesley Newton, *To Command the Sky: The Battle for Air Superiority over Germany, 1942–1944* (Washington, D.C.: Smithsonian Institution Press, 1991), and the very detailed work by Richard Davis, *Bombing the European Axis Powers* (Maxwell AFB, Ala.: Air University, 2006).

Statistical information is drawn mainly from two wartime studies: 8AF, Statistical Summary of Eighth Air Force Operations, European Theater: 17 Aug. 1942–8 May 1945, 520.308A, HRA; and 15AF, The Statistical Story of the Fifteenth Air Force, 670.308D, HRA. Another very useful document is 8AF, Eighth Air Force Tactical Development: Aug. 1942–May 1945, 520.04-1, HRA. Ploesti is well covered by James Dugan and Carroll Stewart, *Ploesti* (New York: Bantam, 1962), and Leon Wolff, *Low Level Mission* (New York: Berkley, 1957), but superseded by Michael Hill, *Black Sunday* (Atglen, Pa.: Schiffer, 1993). For the Schweinfurt raids see Thomas Coffey, *Decision over Schweinfurt* (New York: McKay, 1977); Edward Jablonski, *Double Strike* (Garden City, N.Y.: Doubleday, 1974); and Martin Middlebrook, *The Schweinfurt-Regensburg Mission* (New York: Scribner's, 1983). On conclusions see my article "The Strategic Bombing of Germany in World War II: Costs and Accomplishments," *Journal of American History*, Dec. 1986, 702–13. I have relied on the United States Strategic Bombing Survey (USSBS) reports, especially USSBS, Overall Report (European War), Sept. 1945, Air University Library, Montgomery, Ala. (hereafter cited as AUL). On the USSBS, see David MacIssac, *Strategic Bombing in World War Two: The Story of the United States Strategic Bombing Survey* (New York: Garland, 1976), and the broader, critical, and more recent Gian Gentile, *How Effective Is Strategic Bombing? Lessons Learned from World War II to Kosovo* (New York: New York University Press, 20001).

1. Craven and Cate, *AAFWWII,* 1:149.
2. The Douglas B-18, a development of the company's DC-2, proved to be a mediocre aircraft. Although it was the most numerous AAF bomber when the war began, it did not see service in the war as a bomber.
3. Ira Eaker to Hap Arnold, Mar. 11, 1942, Film 50, HRA.
4. Bomber Command leaders were not allowed to fly combat missions, a practice one notable RAF commander called "the most deplorable mistake we made during the war." D. C. T. Bennett, *Pathfinder* (London: Muller, 1958), 231.
5. Craven and Cate, *AAFWWII,* 2:316.
6. Later, escort carriers were able to provide air cover for the convoys. Air power was involved in destroying half of the German submarines sunk by the Allies during the war.
7. Craven and Cate, *AAFWWII,* 2:305.
8. The terminology is important. "Dispatched" and "launched" refer to aircraft taking off, numbers that were reduced by crashes or aborts. "Credit" sorties were registered when an aircraft entered an area where enemy attack could be expected. An "effective" sortie was when an aircraft carried out the purpose of its mission.
9. Ira Eaker to Sorenson, Jan. 11, 1943, Film 50, HRA.
10. Bombers lost and written off relative to effective sorties.
11. The 21-centimeter rocket weighed 248 pounds and had an effective range of six hundred to twelve hundred yards. Single-engine fighters could carry two and twin-engine fighters four of these weapons. Later the GAF used the R4M, which weighed only 8.5 pounds with a 2-pound warhead. The Me 262 could carry two dozen.
12. Beginning in 1937 the British used technical experts to examine and evaluate weapons and tactics and recommend changes. The AAF adopted the concept in 1942 and by the end of the war had more than four hundred personnel, ninety in the Eighth Air Force. See Charles McArthur, *Operations Analysis in the U.S. Army Eighth Air Force in World War II* (n.p.: American Mathematical Society, 1990).
13. Efforts to convert B-24s to an escort configuration (the YB-41) also failed. The B-26 also was considered for an escort role.
14. Arnold grounded Lewis Brereton, the Ninth Air Force commander.

15. The 4th, under the command of Curtis LeMay, had practiced poor weather takeoffs and had a greater capability in these conditions than the 1st.
16. Curtis LeMay, *Mission with LeMay* (Garden City, N.Y.: Doubleday, 1965), 293.
17. In addition to LeMay and Williams, nine of the sixteen group commanders flew on this mission.
18. Anderson to all groups, message to be read at briefing, Oct. 14, 1943, Mission Folder, 520.332, HRA.
19. Ira Eaker to Hap Arnold, Oct. 15, 1943, Film 51, HRA.
20. Craven and Cate, *AAFWWII,* 2:704.
21. Swedish production was considered of such importance that both the RAF chief and the commander of AAF strategic bombers in Europe strongly advocated sabotage of Swedish production.
22. The forecasters correctly predicted the target weather 58 percent of the time, forecasting too much weather 11 percent of the time.
23. The P-38 saw brief service with the Eighth in 1942 but was shipped off to North Africa before engaging GAF aircraft. This transfer indicates the importance the AAF put on escort fighters at that time.
24. For all of the Mustang's stellar attributes, it was relatively fragile, as its underslung radiator for its liquid-cooled engine was very vulnerable and a pilot could pull the wings off in a dive.
25. The Eighth lost a number of its leading aerial aces in strafing attacks, including the top AAF ace in Europe, Francis Gabreski (twenty-eight victories).
26. Craven and Cate, *AAFWWII,* 3:8.
27. General of the Army H. H. Arnold, "Second Report of the Commanding General of the Army Air Forces to the Secretary of War," Feb. 27, 1945, 360, in *The War Reports of General of the Army George C. Marshall, General of the Army H. H. Arnold, Fleet Admiral Ernest J. King*, ed. Walter Millis (Philadelphia: Lippincott, 1947).
28. The top German ace of the war had 352 credits, compared with the top American ace in the European theater, who had 28. Approximately 107 Germans scored more than 100 victories.
29. The RAF got 15 percent of its bombs on target during its January 1945 raid.
30. On average 16 percent of the bombs inside the target area were duds. Two-fifths of the identified duds were American and three-fifths were British. The Germans are still dealing with these unexploded bombs.
31. Davis, *Bombing the European Axis Powers*, 507.
32. There are few direct comparisons. One study of attacks on three oil plants reveals that the Eighth got 27 percent hits using visual techniques (20 percent of the tonnage), 12 percent with part visual and part instrument (22 percent of the tonnage), and 5 percent with full instruments (58 percent of the tonnage), which yields an overall average of 11 percent hits compared with Bomber Command, which got 16 percent hits. This effort accounted for about one-third of the Eighth's effort against oil refineries and one-tenth of Bomber Command's.
33. Flak downed as many as half the Eighth's heavy bombers and three-quarters of its fighters.
34. USSBS, Overall Report (Europe), 107.
35. At the beginning of the war the German army employed six hundred thousand horses; at the end it used double that number.
36. Bomber Command cost the British about 7 percent of their war effort, while the AAF air offensive cost about 11 percent.
37. Davis, *Bombing the European Axis Powers*, 595.

## Chapter 6. Razing Japan

The bulk of this chapter is drawn from my *Blankets of Fire* (Washington, D.C.: Smithsonian Institution Press, 1996), which is based on primary sources. The key secondary works are Wesley Craven and James Cate, eds., *The Army Air Forces in World War II*, vol. 6, *The Pacific: Matterhorn to Nagasaki, June 1944 to August 1945* (Chicago: University of Chicago, 1953) (hereafter cited as Craven and Cate, *AAFWWII*. Alvin Coox, "Strategic Bombing in the Pacific," in *Case Studies in Strategic Bombardment*, ed. Cargill Hall (Washington, D.C.: Air Force History and Museums Program, 1998); Haywood Hansell, *Strategic Air War against Japan* (Montgomery, Ala.: Air University, 1980); and Curtis LeMay, *Mission with LeMay* (Garden City, N.Y.: Doubleday, 1965). Also see United States Strategic Bombing Survey, Summary Report (Pacific War), July 1946, Air University Library, Montgomery, Ala. (hereafter cited as AUL).

On Japanese strategic bombardment efforts, see Steve Horn, *The Second Attack on Pearl Harbor* (Washington, D.C.: Smithsonian Institution Press, 2005), and Robert Mikesh's detailed *Japan's World War II Balloon Bomb Attacks on North America* (Washington, D.C.: Smithsonian Institution Press, 1973). Information on the B-32 was found in the various materials on aircraft cited above, as well as in William Y'Blood's "The Second String," *AAHS Journal* (Summer 1968) and his "Unwanted and Unloved: The Consolidated B-32," *Air Power History* 42, no. 3 (Fall 1995): 58–71. Statistics are from 20AF, Summary of Twentieth Air Force Operations, 5 Jun 1944–14 Aug 1945, 760.308-1, HRA, while results of the bombing come mainly from the various USSBS reports. On the issue of morality see Ronald Schaffer, *Wings of Judgment: American Bombing in World War II* (New York: Oxford University Press, 1985); Thomas Searle, "'It Made a Lot of Sense to Kill Skilled Workers': The Firebombing of Tokyo in March 1945," *Journal of Military History* 66, no. 1 (Jan. 2002); and Michael Sherry, *The Rise of American Air Power: The Creation of Armageddon* (New Haven, Conn.: Yale University Press, 1987). On the atomic bomb see the subsection in my *Blankets of Fire* and the chapter in John Lynn, *Battle: A History of Combat and Culture* (n.p.: Westview, 2003).

1. The other was the Northrop P-61.
2. LeMay, *Mission with LeMay*, 321.
3. The B-29s defended themselves adequately against the Japanese but failed against jet-powered fighters in Korea.
4. Special Meeting on B-29 Project, May 27, 1944, 20AF, DF 353.01, RG 18, National Archives.
5. Its overall stateside major accident rate for the war was forty per one hundred thousand flying hours compared with thirty for the B-17 and thirty-five for the B-24.
6. Bonner Fellers, "The Psychology of the Japanese Soldiers," paper, Command and General Staff School, 1934–35, 142.041-1, HRA.
7. Claire Chennault, *Way of a Fighter* (New York: Putnam, 1949), 97.
8. Larry Bland, Sharon Ritenour, and Clarence Wunderlin, eds., *The Papers of George Catlett Marshall*, vol. 2, *"We Cannot Delay": July 1, 1939–December 6, 1941* (Baltimore: Johns Hopkins University, 1986), 675–81.
9. Carter McGregor, *The Kagu-Tshuchi Bomb Group* (Wichita Falls, Tex.: Nortex, 1981), 49.
10. Operations Analysis, Study of Incendiary Attack on Dock and Storage Area, Hankow, China, 761.310-19, HRA.
11. COMGENBMCOM 20 to COMGENAF 20, "Technical Problems," Dec. 12, 1944, 20AF, DF 425.1, RG 18, National Archives.
12. All the other AAF numbered air forces were under the command of the theater commander.
13. Neil Wemple, *Memories of a World War II B-29 Bomber Pilot* (Ashland, Ore.: IPCO, 1993), 209.

14. Commander of Air Force Pacific Fleet, Analysis of Air Operations, Jan.–Mar. 1945, vol. 3, 18, 760.310A, HRA.
15. Only 22 percent of the fighters' effective sorties were on escort missions. These AAF fighters claimed 221 Japanese aircraft destroyed in the air and 219 on the ground, and they in turn lost 157 fighters, 114 to enemy causes.
16. For a more positive view of the oil campaign see Manny Horowitz, "Were There Strategic Oil Targets in Japan in 1945?" *Air Power History* 51, no.1 (Spring 2004): 26–35.
17. These changed the bomber's bomb bay, removed all armament except the tail guns, replaced the sighting blisters with flat panels, installed reversible pitch propellers, and fitted the aircraft with fuel-injection engines.
18. Five tons, 128 inches in length, and 60 inches in diameter.
19. Craven and Cate, *AAFWWII,* 6:211, 210.
20. Y'Blood, "Unwanted and Unloved," 64.
21. Craven and Cate, *AAFWWII,* 6:332.
22. Y'Blood, "Unwanted and Unloved," 70.
23. Another eighty-seven were destroyed in training accidents and twelve were destroyed on the ground by the Japanese.
24. USSBS, Summary Report (Pacific War), 26.

## Chapter 7. The Postwar Era

Important sources on strategic bombing in the immediate postwar period are Harry Borowski, *A Hollow Threat: Strategic Air Power and Containment before Korea* (Westport, Conn.: Greenwood, 1982), and Walton Moody, *Building a Strategic Air Force* (Washington, D.C.: Air Force History and Museums Program, 1996). The Air Force–Navy fight (and much more) is well presented in the congressional hearings and in Jeffrey Barlow, *Revolt of the Admirals: The Fight for Naval Aviation, 1945–1950* (Washington, D.C.: Naval Historical Center, 1994).

Strategic Air Command annual histories have been largely declassified and are available at the Air Force Historical Research Agency and National Archives. A number of SAC monographs are valuable, such as Edward Longacre, *The Formative Years, 1944–1949* (Offutt AFB, Neb.: SAC, n.d.); and SAC, *Seventy Years of Strategic Air Refueling, 1918–1988: A Chronology* (Offutt AFB, Neb.: SAC, 1990). The best on aerial refueling is the outstanding study by Richard Smith, *75 Years of Inflight Refueling: Highlights, 1923–1998* (Washington, D.C.: Air Force History and Museums Program, 1998). Other useful studies on SAC are Norman Polmar and Timothy Laur, eds., *Strategic Air Command: People, Aircraft, and Missiles*, 2nd ed. (Annapolis, Md.: Nautical and Aviation, 1990); and Alwyn Lloyd, *A Cold War Legacy: A Tribute to Strategic Air Command, 1946–1992* (Missoula, Mont.: Pictorial, 1999).

The basic source on the USAF in the Korean War remains the dated but excellent official history authored by Robert Futrell, *The United States Air Force in Korea, 1950–1953* (New York: Duell, Sloan and Pearce, 1961). The basis for this study is his more detailed and formerly classified historical studies. A much briefer but well done contribution is USAF, *Steadfast and Courageous: FEAF Bomber Command and the Air War in Korea, 1950–1953* (Washington, D.C.: Air Force History and Museums Program, 2000). Other recent books deal with most of the other aspects of the air war with understandable lesser coverage of strategic bombing, with the notable exception of Conrad Crane's valuable *American Airpower Strategy in Korea, 1950–1953* (Lawrence: University Press of Kansas, 2000). On naval aviation during the war, see Malcolm Cagle and Frank Manson, *The Sea War in Korea* (Annapolis: Naval Institute Press, 1957); and Richard Hallion, *The Naval Air War in Korea* (Baltimore: Nautical and Aviation, 1986). About all we have on the Communist side is Xiaoming Zhang's very useful *Red Wings over the Yalu: China, the Soviet Union, and the Air War in Korea* (College Station: Texas A&M, 2002).

The principal source on the overflight missions is R. Cargill Hall and Clayton Laurie, eds., *Early Cold War Overflights, 1950–1956: Symposium Proceedings*, vol. 1 (Washington, D.C.: Office of the Historian National Reconnaissance Office, 2003). A journalistic yet well done study is William Burrows, *By Any Means Necessary: America's Secret Air War in the Cold War* (New York: Farrar, Straus and Giroux, 2001).

Information on U.S. and British bombers is taken from the archives and basic books noted above. In addition, for the postwar period the rough-reading but essential work is Marcelle Knaack, *Encyclopedia of U.S. Air Force Aircraft and Missile Systems*, vol. 2, *Post–World War II Bombers* (Washington, D.C.: Office of Air Force History, 1988). There are two fine studies on the B-36, both large and profusely illustrated: Dennis Jenkins, *Magnesium Overcast: The Story of the Convair B-36* (North Branch, Minn.: Speciality, 2002), which is footnoted, and the more detailed study by Meyers Jacobsen, *Convair B-36: A Comprehensive History of America's "Big Stick"* (Altglen, Pa.: Schiffer, 1997), which is partially footnoted. For the overall nuclear aircraft program, see Robert Little, "Nuclear Propulsion for Manned Aircraft," 1963, K168.01-11 (1959–61), HRA. The Lincoln is covered in Jon Lake, "Avro Lincoln," *International Air Power Review* 1 (Summer 2001): 176–89; and the Washington in William Suit, "Anglo-American Amity: Transferring B-29s to the Royal Air Force," *Air Power History* 41, no. 4 (Winter 1994): 30–39. For the Tu-4 and Tu-95, see Yefim Gordon and Vladimir Rigmant, *Tupolev Tu-4: Soviet Superfortress* (Hinckley, UK: Midland, 2002); Yefim Gordon and Vladimir Rigmant, *Tupolev Tu-95/-142 "Bear": Russia's Intercontinental-Range Heavy Bomber* (Leicester, UK: Midland, 1997); and David Donald, ed., *Tupolev Bombers* (Norwalk, Conn.: Airtime, 2002).

Accident statistics are drawn from USAF Statistical Digest, 1948, 134.11-6, HRA; and USAF Flying Accident Bulletin, 1952, and USAF Flying Accident Bulletin, 1954, K259.3-3, HRA. Korean War statistics are from USAF Statistical Digest: Fiscal Year 1953, K134.11-6, HRA.

1. The AAF's accident rate is one indication of this condition. The major accident rate in 1946 was 40 percent higher than 1945, and the fatal accident rate was the highest since 1941.
2. This is described so well in Borowski's *Hollow Threat*.
3. Some will argue that "Bomber" Harris is more suited for this role, but I would submit that LeMay's World War II service approximates the Bomber Command leader's record and that his Cold War service tilts the balance in his favor.
4. LeMay went on to four-star rank (October 1951) and became chief of staff of the USAF in July 1961, serving in that position until his retirement in February 1965.
5. Curtis LeMay, *Mission with LeMay* (Garden City, N.Y.: Doubleday, 1965), 433.
6. The Soviets captured some of the crew, but only the survivors of the U-2 incident (May 1960) and one B-47 incident (July 1960) were returned to the United States.
7. Moody, *Building a Strategic Air Force*, 109. Partridge was not happy with the situation and applauded when the USAF developed aerial refueling.
8. An estimate in December 1948 was that an atomic offensive might kill eight million Soviets.
9. Other changes included a five foot taller tail, different engine nacelles, hydraulic rudder boost, nose wheel steering, a lighter and stronger wing, quicker retracting landing gear, and reversible props.
10. In March 1943 Consolidated merged with Vultee, however, the term "Convair" did not become official until April 1954.
11. Borowski, *Hollow Threat*, 151.

12. The USAF considered defending the bomber with an air-to-air missile, the Dragonfly, a project that began in 1947. This was succeeded by another radar-directed rocket, the Falcon, which was switched to fighter interceptors in 1950.
13. The fighter measured fifteen feet in length with a twenty-one-foot (unfolded) wing span. The aircraft had an estimated top speed of 563 knots and a twenty-minute endurance at high speeds.
14. John Farquhar, *A Need to Know: The Role of Air Force Reconnaissance in War Planning, 1945–1953* (Maxwell AFB, Ala.: Air University, 2004), 93.
15. The *Truculent Turtle* had a gross weight of 85,600 pounds compared with the normal 61,200 pounds, was stripped down, had extra fuel tanks, and JATO (jet [actually, rocket] assisted takeoff). Its distance record was not broken until 1962.
16. Jacobsen, *Convair B-36*, 84.
17. However, combat radius decreased from thirty-seven hundred nautical miles to thirty-one hundred nautical miles.
18. The Air Force had various plans for using nuclear power; the WS 125A was cancelled in 1956 and the CAMAL in 1958.
19. Its major accident rate and fatal accident rate was two-fifths that of the B-29 and half that of the B-50.
20. The B-36 had a wing span of 230 feet and length of 162 feet. In contrast the B-52 has a wing span of 185 feet and length of 157 feet. The B-52B had a maximum takeoff weight of 420,000 pounds and the B-52H 488,000 pounds, compared with the B-36A at 311,000 pounds and B-36J at 410,000 pounds.
21. These targets were off limits to avoid the possibility of bombing either China or the Soviet Union and, initially, when the UN forces were winning, to save the facilities for the post-Communist state as well as an inducement for the Chinese to stay out of the war. Dikes were another proscribed target set.
22. The B-29s in Korea did not receive post–World War II ECM equipment until almost the end of the war.
23. The Air Force made unsuccessful attempts with direct attacks on searchlights and considered using flares and a light to temporally blind enemy fighter pilots.
24. Futrell, USAF Historical Study no. 127, HRA.
25. The United States considered, but did not conduct, attacks on Japan's food supply.
26. The Soviets claim to have downed sixty-nine B-29s with MiGs, and Chinese and North Koreans claim another four.
27. Lake, "Avro Lincoln," 176.
28. It delivered an 800-kiloton warhead at a cruise speed of Mach 1.5 to 2, a range of 350 kilometers.
29. The last B-36 model "(B-36J)" weighed 410,000 pounds gross, had a top speed of 410 mph, a 390-mph cruise speed, a 39,900-foot ceiling, and a range of 6,800 miles. The early Tu-95 weighed 344,000 pounds at takeoff, had a top speed of 550 mph, a 470-mph cruise speed, a 38,800-foot ceiling, and a 7,500-mile range.

### Chapter 8. Between Korea and Vietnam

Material on U.S. aircraft is from the sources noted above, along with invaluable studies from Air Material Command (AMC). An excellent source on the bombers tested but not procured is AMC, "Historical Data on Aircraft Developed but not Produced, 1945–Present," c. Mar 1957, K201-106, HRA. There are many books on Jack Northrop's flying wings, all covering essentially the same ground. Less accessible but more scholarly is Francis Baker, "The Death of the Flying Wing," Ph.D. diss., Claremont, 1984.

The underappreciated B-47 is covered magnificently in Alwyn Lloyd's *Boeing's B-47 Stratojet* (North Branch, Minn.: Specialty, 2005). The B-52 is the subject of numerous books and articles, two of the better books of which are Walter Boyne's still excellent *Boeing B-52: A Documentary History* (London: Jane's, 1981), and the more recent Martin Bowman, *Stratofortress: The Story of the B-52* (n.p.: Pen & Sword, 2005). From the USAF's point of view see the very well done work by Lori Tagg, *Development of the B-52: The Wright Field Story* (Wright-Patterson AFB, Ohio: Aeronautical Systems Center History Office, 2004). Richard Smith's *Seventy-Five Years of Inflight Refueling: Highlights, 1923–1998* (Washington, D.C.: Air Force History and Museums Program, 1998) is very good on that subject and the KC-135. The B-58 gets its due in the very detailed study by Jay Miller, *Convair B-58 Hustler: The World's First Supersonic Bomber* (Leicester, UK: Aerofax, 1987), and the more accessible article by Bill Yenne, "B-58 Hustler: Convair's Ultimate Delta," *International Air Power Review* 2 (2001): 116–49. Especially useful on the B-70 are U.S. Congress, Senate, *The B-70 Program*, Report of the Preparedness Investigating Subcommittee of the Committee on Armed Services, 1960; Tim Curry et al., "North American XB-70 Valkyrie Intercontinental Strategic Bomber," Aerospace Engineering Department, California State Polytechnic University, June 1994; John Sotham, "The Legend of the Valkyrie," *Air and Space*, Aug./Sept. 1999.

The Navy's attack seaplane program is well told in William Trimble's *Attack from the Sea: A History of the U.S. Navy's Seaplane Striking Force* (Annapolis, Md.: U.S. Naval Institute Press, 2005). Material on the British jet bombers can be found in the general works noted previously and in the official history, Humphrey Wynn, *The RAF Strategic Deterrent Forces: Their Origins, Roles, and Deployment, 1946–1969, a Documentary History* (London: HMSO, 1994). Clive Richards, "Short AS.4 Sperrin," *Wings of Fame* 19 (2000): 112–19, covers that bomber. David Donald does the same for the titled aircraft in "Vickers Valiant: V-bomber Pioneer," *International Air Power Review* 18 (2005): 128–57, as does Paul Jackson, in "Vulcan: Delta Force," *Wings of Fame* 3 (1996): 34–95. On operations in the Falklands, see Rodney Burden et al., *Falklands: The Air War* (London: Arms and Armour, 1986), 363–67. The Victor gets its due in Elfan ap Rees, "Handley Page Victor," *Air Pictorial*, pt. 1 (May 1972): 160–78, and pt. 2 (June 1972): 220–25. On the abortive TSR 2, see Bill Gunston, "Beyond the Frontiers: BAC TSR.2," *Wings of Fame* 4 (1996): 122–37. Soviet jet bombers are well covered and illustrated in David Donald, ed., *Tupolev Bombers* (Norwalk, Conn.: AirTime, 2002), which consists of articles from *International Air Power Review, Wings of Fame*, and *World Air Power Journal*. On the Tu-16 see Yefim Gordon and Vladimir Rigmant, *Tupolev Tu-16 Badger: Versatile Soviet Long-Range Bomber* (Hinckley, UK: Aerofax, 2004). A quick summary of Soviet bombers, complete with citations, can be found on the website of the Federation of American Scientists.

1. Tagg, *Development of the B-52*, 15.
2. AMC, YB-35, YB-49 Supplemental Case History, vol. 1, 10, 202.1-19, HRA. While the longest B-29 bomb run was forty-five seconds, the B-49's shortest time was over four minutes.
3. The B-45 had the range and bomb-load capabilities of a World War II heavy bomber.
4. Dennis Jenkins, "North American B-45 Tornado," Aug. 1989, 1, National Museum of the United States Air Force, Dayton, Ohio.
5. On the ground the wing tip drooped sixteen inches down from the horizontal, and in flight it flexed forty-five inches above the horizontal. In static tests the wing tips were deflected a total of 17.5 feet. This gave a more comfortable ride, although observers were somewhat startled by the sight.
6. The crews might have been uneasy, but the B-47B registered only one fatal accident (with three deaths) without ejection seats through the end of 1955 on forty-nine major accidents, twelve of which were fatal.
7. These consisted of two rows of small, perpendicular airfoils mounted along the top span of the wing that improved the characteristics of the air flow.

8. The seats had dependability problems as late as 1952; there were four seat failures in ten attempted ejections in forty-two fatal B-47/RB-47 accidents through 1955.
9. In July 1960 the Soviets downed one off the coast of Russia, killing four of the crew and imprisoning the other two, who were later released. Another B-47 was shot up by the North Koreans in April 1965, but it landed safely without casualties.
10. Earl Peck, "B-47 Stratojet," in *Flying Combat Aircraft of the USAAF-USAF*, ed. Robin Higham and Carol Williams, (Ames: Iowa State University Press, 1978), 2:82.
11. Boyne, *Boeing B-52*, 27.
12. Over their lifetimes the B-45, B-57, and B-58 had higher major accident rates and the B-52, FB-111, and B-1 had lower rates than the B-47.
13. The P6M-2's wing span was 100 feet and its length was 134 feet. The Sea Master was also considered for photo reconnaissance and air-refueling roles.
14. History of the Air Force Flight Test Center, Jan.–June 1953, vol. 5, 20, K286.69-28, HRA. The testers noted that stall warning was indefinite and the stall "so violent that the YB-60 airplane should not be intentionally stalled at any time". Ibid., 26.
15. Warren Greene, "The Development of the B-52 Aircraft, 1945–1953," May 1956, 3, K243.042-6, HRA.
16. Tagg, *Development of the B-52*, 63.
17. The USAF was concerned about battle damage to the fuel tanks because an analysis of World War II bomber losses indicated that 80 percent were directly caused by fire, usually from the fuel systems. Initially the airmen wanted 70 percent of the B-52's fuel cells to be self-sealing, but they reconsidered because of the cost in money and weight. The Air Force compromised with tear-proof fuel cells, which would resist direct hit causing a tank to burst but not contain fuel leaks.
18. The forward gear already had the ability to pivot for ground movement, and the rear gear was now fitted with the same capability.
19. The USAF looked at installing Falcon air-to-air missiles, suggested by Hughes in August 1951, which would add two and a half tons in weight and considerable expense ($1.5 billion). In view of the Falcon's miserable combat record in Vietnam, this was not a good idea.
20. Knaack, *Post–World War II Bombers*, 230.
21. SAC launched five B-52s, three slated to complete the flight and two spares (one aborted with an iced-up refueling receptacle, the other landed as planned). The Command spotted 78 KC-97s to get the B-52s to their record, with additional tankers on alert at four bases as a backup. One human interest note: Lt. Col. James Morris, aircraft commander of the lead ship, had been the copilot on the *Lucky Lady II* on its record-breaking flight.
22. Five B-52s were lost between June 1959 and February 1964 to severe turbulence. Gusts at eight hundred feet are two hundred times as frequent as at thirty thousand feet
23. The shorter vertical tail and spoilers tended to lead to a "Dutch roll," which if not properly managed could overstress the structure. A stability augmentation system was added to lessen the stress on both the crews and structure.
24. Convair worked on a 160-mile range, rocket-powered ASM between 1950 and its cancellation in May 1957. A pod the next year had fuel in a lower part and fuel and a warhead in an upper portion.
25. Miller, *Convair B-58 Hustler*, 37.
26. Strategic Air Command, USAF Historical Study no. 76, June 1958 to July 1959, 206, K416.01-76, HRA.

27. Donald McVeigh and Cecil Uyehara, "History of the Air Research and Development Command, Jan.–June 1959," vol. 2, "The B-70 Story," 3, K243.01 (Jan.–June 59), HRA.
28. U.S. Congress, B-70 Program, 4.
29. The B-70 had a span of 105 feet and length of 186 feet. It weighed 231,000 pounds empty with a maximum takeoff weight of 521,000 pounds.
30. Bernard Nalty, "The Quest for an Advanced Manned Strategic Bomber: USAF Plans and Policies, 1961–1966," Aug. 1966, 4, K168.8613-2, HRA.
31. Andrew Brookes, *Force V: The History of Britain's Airborne Deterrent* (London: Jane's, 1982), 73.
32. Peter Lewis, *British Bomber since 1914* (London: Putnam, 1967), 371.
33. The British scrambled to get the long-disused system back into operation and even borrowed a probe from a Vulcan displayed in the National Museum of the United States Air Force in Dayton, Ohio, for this task.
34. Gunston, "Beyond the Frontiers," 122.
35. In one of the ironies of history, after the cancellation of the TSR.2 the RAF adopted the Buccaneer in 1969.
36. One unusual feature was that the crew entered the aircraft by getting into their ejection seats on the ramp underneath the aircraft and were then hoisted, elevator fashion, into the aircraft.
37. While we do not have accident statistics on the Tu-22, more than seventy were lost prior to 1975, about 23 percent of production. It certainly had a higher accident rate than the docile Tu-16 and a higher one than its rival, the B-58, of which 22 of the 116 built were lost (19 percent).
38. The Tu-160 has a length of 178 feet and wingspan (fully extended) of 183 feet, and it weighs 258,000 pounds empty and 606,000 pounds at takeoff, compared with 146 feet and 137 feet, and 192,000 pounds and 477,000 pounds for the B-1B. The Tu-160 has a maximum speed of Mach 2.05 (versus the B-1B's Mach 1.2), and at sea level, 556 knots (versus 520 knots).

## Chapter 9. The 1960s and 1970s

There are a number of excellent books on the bombing of North Vietnam, for example, Mark Clodfelter, *The Limits of Air Power: The American Bombing of North Vietnam* (New York: Free Press, 1989), and Earl Tilford, *Crosswinds: The Air Force's Setup in Vietnam* (College Station: Texas A&M, 1993). However, I found the most useful to be Wayne Thompson's more recent *To Hanoi and Back: The U.S. Air Force and North Vietnam, 1966–1973* (Washington, D.C.: Smithsonian, 2000). An important study of Rolling Thunder that is particularly good on targets and intelligence is Paul Berg's "Assessing U.S. Air Force Bombing Effectiveness During Rolling Thunder," Ph.D. diss., Auburn University, 2001. A view from the Communist side that challenges U.S. claims is Dana Drenkowski and Lester Grau, "Patterns and Predictability: The Soviet Evaluation of Operation Linebacker II," *Journal of Slavic Military Studies* 20, no. 4 (Dec. 2007): 559–607. The SAC histories are quite detailed, although portions remain classified. For a critical and excellent view of the B-52 operations, see Marshall Michel, *The Eleven Days of Christmas: America's Last Vietnam Battle* (San Francisco: Encounter, 2002).

Both the F-111 aircraft and controversy are well covered in Robert Art, *The TFX Decision* (Boston: Little, Brown, 1968); Robert Coulam, *Illusions of Choice* (Princeton, N.J.: Princeton, 1977); Jay Miller, *General Dynamics F-111 "Aardvark"* (Fallbrook, Calif.: Aero, 1981); Anthony Thornborough and Peter Davis, *F-111* (London: Arms and Armour, 1989); and "General Dynamics F-111: The 'Earth Pig,'" *World Airpower Journal* 14 (Fall 1993): 42–101.

The primary source for the treatment of winged standoff and decoy missiles is my study, *The Evolution of the Cruise Missile* (Maxwell AFB, Ala.: Air University, 1985). Soviet missiles are covered by David Donald, ed., *Tupolev Bombers* (Norwalk, Conn.: AirTime, 2002), and the FAS databook. The ASD and SAC histories are very detailed and helpful, but some remain classified. Another valuable source on SAC is a history of one of its subordinate commands, Second Air Force Historical Data, from January 1 to June 30, 1960, K243.01, HRA. Useful on the alert system is Henry Narducci, *Strategic Air Command and the Alert Program: A Brief History* (Offutt AFB, Neb.: Office of the Historian, SAC, 1988).

1. At the end of 1962 SAC had more than twenty-six hundred bombers and 142 Atlas, 62 Titan, and 20 Minuteman missiles. In addition there were IRBMs stationed in Europe and nuclear weapons on U.S. Navy ships and submarines. The Soviets had a few bombers and fifty ICBMs that could reach the United States.
2. Robert Dorr and Chris Bishop, eds., *Vietnam Air War Debrief* (London: Aerospace, 1996), 135.
3. Seymour Hersh, *The Price of Power: Kissinger in the Nixon White House* (New York: Summit, 1983), 506; Richard Nixon, *RN: The Memoirs of Richard Nixon* (New York: Grossett and Dunlap, 1976), 606–7.
4. Clodfelter, *Limits of Airpower*, 184.
5. Clearly the SAC chief, John Meyer, was no LeMay. See particularly Michel, *Eleven Days of Christmas*, 131–34.
6. While these attacks had some limited effect, they could have been much more effective if cluster bombs had been used. The A-6s and F-111s could not use these weapons because of their low flying, but no explanation is given for non-use by the B-52s.
7. The USAF flew 36 percent of its sorties against railroad yards, 25 percent against storage facilities, 14 percent against radio communications facilities, 12 percent against power facilities, 10 percent against airfields, and 2 percent against SAM sites.
8. The SAC staff projected a 2 percent loss rate, which was astonishingly accurate. The SAC commander applied his judgment and for political reasons chose to overstate the estimate to arrive at 3 percent.
9. These losses were later attributed to a failure of a weld in the tailplane. More than half of the bombers inspected stateside showed faulty welds emanating from a faulty design.
10. Another source used 1,750 feet CEP for thirty-one missions. It states that on eleven runs under ideal conditions, the error was 233 feet, very close to the 220 feet achieved in training.
11. The F-111's gun also was removed.
12. The B-1B can carry twenty-four internally.
13. Reflex ended in March 1965, and ground alerts ended for the KC-97 in November 1965 and for the B-47 in February 1966.
14. Between 1950 and 1980 there were thirty-two incidents involving nuclear weapons, the majority involving SAC bombers.
15. The introduction of new radars that could sort out low-flying aircraft from the terrain mounted on interceptors and airborne radar aircraft did not appear until later.
16. AAF experience in World War II indicated that bombing accuracy in combat was only a fifth of that achieved in stateside training.

## Chapter 10. Missiles

The basic published sources on the ICBM are the older study by Edmund Beard, *Developing the ICBM: A Study in Bureaucratic Politics* (New York: Columbia University Press, 1976), and Jacob Neufeld's essential *Ballistic Missiles in the United States Air Force, 1945–1960* (Washington, D.C.: Office of Air Force History, 1990); however, I found most useful the more accessible study by John Lonnquest and David Winkler, *To Defend and Deter: The Legacy of the United States Cold War Missile Program* (Washington, D.C.: Department of Defense, 1996). A general overall of the subject can be found in Alan Levine, *The Missile and the Space Race* (Westport, Conn.: Praeger, 1994). On the Titan see the detailed and footnoted study by Warren Greene, "The Development of the SM-68 Titan," Aug. 1962, K243.012-7, HRA. A documented history of the cruise missile can be found in my study *The Evolution of the Cruise Missile* (Maxwell AFB, Ala.: Air University, 1985). Most valuable on the Navy's ballistic missiles are Graham Spinardi, *From Polaris to Trident: The Development of U.S. Fleet Missile Technology* (Cambridge: New Cambridge University Press, 1994), and Strategic Systems Program Office, *FBM Facts/Chronology: Polaris, Poseidon, Trident* (Washington, D.C.: Navy Department, 1996). Information on the Soviet missile program is scattered and difficult, at least for non-Russian readers; however, Steven Zaloga, *Target America: The Soviet Union and the Strategic Arms Race, 1945–1964* (Novato, Calif.: Presidio, 1993) was most useful. Also helpful are the Internet guides posted by the Federation of Atomic Scientists (FAS). Material on ballistic missile defense is drawn mainly from my *Hitting a Bullet with a Bullet: A History of Ballistic Missile Defense* (Maxwell AFB, Ala.: Air University Press, 2000). For a more detailed look at the politics of BMD during the 1990s and into the 2000s, see Bradley Graham's *Hit to Kill: The New Battle over Shielding America from Missile Attack* (New York: Public Affairs, 2001).

1. The key players included Bernard Brodie, Herman Kahn, Thomas Schelling, and Albert Wohlstetter.
2. Perhaps the most remembered aspect of this program was the death of Navy lieutenant Joseph Kennedy in a premature explosion of one of these aircraft.
3. Later the supersonic requirement was dropped and CEP loosened to eight thousand feet.
4. Neufeld, *Ballistic Missiles*, 142.
5. The Hiroshima bomb weighed 10,000 pounds and had a yield of 13 kilotons. The later thermonuclear weapons weighed 1,500 pounds and had a 1-megaton yield.
6. The deployment modes had significant vulnerability differences. The original Atlas D was vulnerable to 5 psi overpressure, in a covered coffin to 25 psi, and the "F" in a silo to 100 psi.
7. While the Titan I silo could withstand an overpressure of 150 to 200 psi, the Titan II had more protection, up to 300 to 350 psi. One example: Whereas the doors on the Titan I each weighed 125 tons, the one on the Titan II weighed 720 tons and was designed to withstand an overpressure of 1,000 psi.
8. Four Atlas and 3 Titan missiles were destroyed in their silos.
9. France, Greece, and Spain refused IRBM basing.
10. The Navy played up two differences: its system required two, not three, guidance stations and its booster was stowed with the missile and did not have to be attached just prior to launch, which reduced the men and equipment required and cut the time the submarine had to remain on the surface.
11. The warheads were programmed to impact in a triangular pattern to maximize destruction and minimize the overkill of a single large explosion. The three 200-kiloton warheads inflicted damage equivalent to 1 megaton.
12. The Poseidon C-3 and Trident I C-4 have the same dimensions (thirty-four feet long and seventy-four inches in diameter), although different weights, 65,000 pounds and 73,000 pounds,

respectively. The Trident II D-5 is bigger: forty-four feet long, eighty-three inches in diameter, and much heavier at 130,000 pounds.

13. The TERCOM maps of hostile territory were made possible by satellite imagery.
14. One major advantage of the GD missile was that, unlike the Boeing cruise missile, it was recoverable and reusable.
15. A 1964 study found the best Atlas effectiveness (measured by percentage on alert along with reliability of launch, inflight, and warhead reliability) to be in the 30 percent range, the best Titan in the low 40 percent range, and the early Minuteman in the 53–68 percent range.
16. The shortest distance between the United States and Russia is over the North Pole.
17. The MX was seventy-one feet in length and weighed about 195,000 pounds.
18. The scheme involved building twenty-three shelters for each missile and moving the one missile and many decoys between these shelters on mobile transporter-erector-launchers. The Air Force wanted forty-six hundred shelters located over five thousand square miles, believing the system could survive an assault by ten thousand warheads.
19. The one hundred MX missiles would be based in superhard capsules spaced eighteen hundred feet apart, relying on the fratricide of incoming missiles to preclude all from being destroyed. Without realistic testing, the concept was unproven and questioned, although the Air Force believed that 50 to 70 percent of the missiles would survive an attack.
20. The Peacekeeper Rail Garrison concept consisted of twenty-five trains, each carrying two missiles, that would deploy from their bases during periods of international tension.
21. Protection came from the large numbers and basing, initially 3,350 on trucks or in silos hardened to withstand 7,000 to 8,000 psi overpressures.
22. The 2002 Moscow Treaty nullified START II and eliminated the de-MIRVing of ICBMs; hence a number of Minuteman IIIs retain their multiple warheads.
23. One setback in this movement to lessen nuclear armaments came as a result of the U.S. decision in December 2001 to withdraw from the ABM treaty. As a consequence, the de-MIRVing of ICBMs stopped. The more significant issue is the number of warheads, not launchers.
24. Donald Baucom, *The Origins of SDI, 1944–1983* (Lawrence: University Press of Kansas, 1992), 34.
25. Observers note the discrepancy between the critics of BMD, who claim it will not work, and the protests of the Russians, who claim it will threaten them. Some believe the latter are merely attempting to reassert their position in world politics.

## Chapter 11. Strategic Bombardment into the Twenty-first Century

The sources on the B-1, F-117, and B-2 are the standard ones noted above. Bomber acquisition is covered in detail in Michael Brown, *Flying Blind: The Politics of the U.S. Strategic Bomber Program* (Ithaca, N.Y.: Cornell, 1992). On the B-1 see the detailed and excellent article by Bill Gunston, "Rockwell B-1B Lancer," *World Airpower Journal* 24 (Spring 1996): 52–113, updated in Ted Carlson, "B-1B Lancer in the EAF," *International Air Power Review* 23 (2007): 38–45; for the political aspects see Nick Kotz's *Wild Blue Yonder: Money, Politics, and the B-1 Bomber* (Princeton, N.J.: Princeton University Press, 1988). A detailed government view in Congressional Budget Office, "The B-1B Bomber and Options for Enhancements," Aug. 1988, Washington, D.C. For stealth and the F-117, see my *Chasing the Silver Bullet: U.S. Air Force Weapon Development From Vietnam to Desert Storm* (Washington, DC: Smithsonian, 2003). On this subject I found the most valuable "published" source to be David Aronstein and Albert Piccirillo, "Have Blue and the F-117A," final draft, ANSER, Mar. 1997. For the B-2 I relied heavily on Bill Sweetman, "B-2 'Stealth Bomber,'" *World Air Power Journal* 31 (Winter 1997): 46–95.

My treatment of the 1991 Gulf War also draws on *Chasing the Silver Bullet* and the excellent, concise treatment by Richard Davis, *Decisive Force: Strategic Bombing in the Gulf War* (Washington, D.C.: Air Force History and Museums Program, 1996), which covers considerably more than the title promises and is also found as a chapter in Cargill Hall, ed., *Case Studies in Strategic Bombardment* (Washington, D.C.: Air Force History and Museums Program, 1998). All of these works draw heavily on the five-volume *Gulf War Air Power Survey* (Washington, D.C.: [GPO], 1993) edited by Eliot Cohen. John Warden laid out his ideas in *The Air Campaign: Planning for Combat* (Washington, D.C.: Pergamon-Brassey's, 1989). On PGMs I've used *Chasing the Silver Bullet* and the more recent work by Paul Gillespie, *Weapons of Choice: The Development of Precision Guided Munitions* (Tuscaloosa: University of Alabama, 2006). The action in Bosnia-Herzegovina is the subject of the excellent and extensive study, Robert Owen, ed., *Deliberate Force: A Case Study in Effective Air Campaigning* (Maxwell AFB, Ala.: Air University, 2000). For the 1999 Serbian campaign, I relied primarily on Headquarters United States Air Force, Initial Report, The Air War over Serbia: Aerospace Power in Operation Allied Force, Apr. 2000; and Anthony Cordesman, *The Lessons and Non-Lessons of the Air and Missile Campaign in Kosovo*, CSIS Sept. 1999 (Westport, Conn.: Praeger/Greenwood, 2000). On the Afghan War David Donald's "Enduring Freedom: Fighting the War against Terrorism," *International Air Power Review* 4 (Spring 2002): 16–29, was most helpful.

1. This led to a change from a capsule to ejection seats.
2. Radar returns are measured in terms of a flat plate area, and RCS is measured in square meters. The RCS of the B-52 is said to be over one hundred square meters; of the B-1A, ten square meters; of the F-15, about five square meters; and of the B-1B, one to two square meters. Bear in mind that the RCS figures remain classified and that these figures vary by orientation and usually refer to only the head-on RCS.
3. The B-1's three bomb bays had been sized for the SRAM, not for the longer AGM-86B. The B-1B's moveable bulkhead in the forward two bomb bays can accommodate the AGM-86B. Fuel tanks (18,900 pounds capacity) can be fitted into each of the three bomb bays.
4. The importance of a mathematical paper written by Soviet mathematical Pyotr Ufimtsev in the formulation of the Lockheed computer program is controversial, although surely a great story.
5. The aircraft's performance was limited by its low thrust-to-weight ratio and low lift-to-drag ratio. The former was dictated by an engine without an afterburner (to minimize size and IR signatures) and the losses due to the inlet and exhaust design, a compromise that emphasized low observables over engine performance. The low lift-to-drag ratio was a consequence of the faceting.
6. Although the F-117 is clearly an attack aircraft, the Air Force designated it a fighter. There is no explanation for this mischaracterization, although the USAF had done this earlier with the F-111. The F-117 has a length of sixty-six feet, a span of forty-three feet, an empty weight of twenty-nine thousand pounds, and a maximum takeoff weight of fifty-two thousand pounds, compared with the Have Blue, which measured thirty-eight feet in length and twenty-two feet in span and weighed nine thousand pounds empty and twelve thousand pounds maximum.
7. Bill Sweetman and James Goodall, *Lockheed F-117A* (Osceola, Wisc.: Motorbooks, 1990), 31.
8. Aronstein and Piccirillo, "Have Blue and F-117A," 99.
9. "Gulf War Air Power Survey Interview with BG Anthony Tolin," Jan. 30, 1992, 11–12, NA-211, HRA.
10. Iraq controls 20 percent of the world's oil reserves, Kuwait 14 percent, and Saudi Arabia 25 percent.
11. The Iraqis had almost a million men under arms with the ability to mobilize another two million. They had 750 combat aircraft, including first line equipment, and hundreds of ballistic missiles. Baghdad's air defenses were considered twice as dense as the most heavily defended Eastern European target during the Cold War and seven times that which defended Hanoi in 1972.

12. The dramatic impact of PGMs is highlighted by the fact that F-105s bombing in the Hanoi-Haiphong area at the end of Rolling Thunder achieved a CEP of 447 feet with 5.5 percent direct hits, whereas in 1972 in the same area, guided bombs achieved a 23 foot CEP with 48 percent direct hits.
13. Warden, *Air Campaign*, 14.
14. Although Warden was sent home after he briefed his plan in the theater, three of his subordinates stayed on and were key in the eventual air plan. Warden's concept became the center focus, although he was denied a direct role in its execution.
15. The Iraqis cut the time required to setup, launch, and depart the firing area, made excellent use of decoys, and proved well disciplined in night operations and emissions control.
16. Only 15 percent of the bombs dropped by the B-52s were aimed at strategic targets, whereas 40 percent of the F-117's bomb loads fell on strategic targets. The B-52 bombing had a great psychological impact on Iraqi troop morale.
17. A legacy of the Tacit Blue program, as the name implies, LPI radar allows the offensive use of radar while lending little assistance to the defender.
18. USAF, Air War over Serbia, 34.

# INDEX


A

aerial refueling, 155, 160-62, 290, 299: American, 196; B-17/B-24, 160; B-29, 161; British, 160, 202, 203; cross-over, 161; flying boom, 161; probe and drogue, 161; Soviet, 177, 178, 207, 208

Air Corps Tactical School, 24, 33, 38, 41-42, 99, 132, 137, 139

air defenses, British, World War I, 5, 12: antiaircraft guns, 11; barrage balloons, 9, 11; fighters, 5, 8; ground observers, 5; sound detectors, 5

air defenses, German daylight, 107-8: air-to-air bombing, 108; antiaircraft artillery, 75, 91, 92, 108, 120, 124; dual fuses, 124; GCI, 76, 92, 124; head on attacks, 108; proximity fuses, 124; radar, 75, 76, 91; ramming, 121; rockets, 108; surface-to-air missiles, 65, 124; smoke screens, 108; sound detectors, 124

air defenses, German night: Anna Marie, 76; *Lichtenstein*, 93; *Schrage Musik*, 93; SN-2, 93; Tame Boar, 92; Wild Boar, 92

air war over Europe, targets: Antheor viaduct, 86; Dortmund-Ems canal, 85; Eder, Mohne, Sorpe dams, 85; Linoges, 86; Peenemunde, 89; Ploesti, 104, 106, 107, 110-11, 118. *See also* cities bombed

air war over Germany, campaigns: bearings, 106, 111, 113; Big Week, 118; communications, 98, 126; oil, 71, 94-95, 98, 120, 121, 126; Ruhr, 86, 87, 91, 96, 98

air war over Iraq (1991), 285-90: B-1, 280; B-52, 286-88, 290; bombing accuracy, 286; bombing targets, 289-90; cruise missiles, 286; F-117, 286, 289; PGMs, 286, 290; intelligence, 290; Patriot, 289; Scud, 286-89

air war over Japan, 138; Kyoto, Japan, 144; vulnerability, 131. *See* cities bombed, Japanese

air war over Korea, targets: irrigation dams, 174-75; Sinuiju, 173; Sui-Ho, 174

air war over Vietnam: goals, 215-16, 220, 221; Linebacker I, 219-21; Linebacker II, 221-25; PGMs, 280, 285-86, 293; Rolling Thunder, 215, 217-19, 224, 323n. 12

aircraft accident rates, 314n. 1, 317n. 12: B-1, 229; B-2, 229; B-17, 312n. 5; B-24, 312n. 5; B-29, 131, 184, 312n. 5, 315n. 19; B-45, 184; B-36, 315n. 19; B-47, 188; B-50, 315n. 19; B-52, 229; B-58, 196, 208; FB-111, 229; Tu-22, 208

aircraft, American Navy: A3J, 242; A-6, 319n. 6; AD, 174; AJ, 249; AJ-1, 169, 188, 172; F-4, 225; F9F, 174; F-14, 226; P2V, 169, 188, 193, 249; P6M, 188, 249, 317n. 13; PBY, 132

aircraft designation systems: American, 24, 322n. 6; German, 302n. 9
aircraft engines, American: Allison, 30; J 35, 185, 186; J 33, 249; J 47, 186; J 57, 189, 191; Liberty, 22, 24; Pratt and Whitney, R 4360, 165; Wright R 1820, 24; Wright R 3350, 30, 130, 137
aircraft engines, British: Merlin, 79; Vulture, 79
aircraft engines, Soviet, AM-3, 206, 212
aircraft engines, turbofan, 226: B-1, 277; B-52H, 193, F-111, 227
aircraft engines, turboprop: American, 179, 180, 190; Soviet, 177
aircraft, named: *Bockscar*, 148; *Enola Gay*, 146; *Joltin' Josie*, 139; *Lucky Lady II*, 161, 192; *Lucky Lady III*, 192; *Maxim Gorky*, 31, 304n. 10; *Question Mark*, 160; *Tokyo Rose*, 139; *Truculent Turtle*, 169, 193, 315n. 15
aircraft, parasite, 166-67, 192, 194-95
aircraft production, German, 118, 125-26
aircraft, Silverplate, 146, 313n. 17
airships: accidents, 4; advantages, 2; combat, 4, 5-6, 7; difficulties, 4; "Height Climbers," 7; losses, 5; modifications, 6
Anderson, Frederick, 112, 113
antiaircraft artillery, 34: Spanish Civil War, 43; World War I, German: 311n. 33. *See also* air defense, German, daylight
antisubmarine operations, 105-6
arms control, 37, 208, 210, 211, 212, 256, 269, 261, 272, 394n. 26: anti-ballistic missile treaty, 272, 274, 275, 321n. 23; INF, 257; Outer Space Treaty, 265; SALT I, 255, 265; SALT II, 257, 265, 266, 272, 321n. 22; SALT III, 272; START I, 272; Strategic Offensive Reduction Treaty, 272
Arnold, Henry, 24, 27, 44, 104, 114, 115, 116, 117, 118, 131, 132, 133, 136, 138, 141, 150, 153, 242
around-the-world flights: B-50, 161, 192; B-52, 192, 193, 317n. 21
atomic bomb, 97, 146-48, 152, 237, 272: American, 155, 156, 159, 162, 165, 181, 182, 183, 199, 201; objective, 146; Soviet, 176; targets, 146
Atomic Energy Commission, 157, 165
AWPD/1, 99, 133, 139
AWPD/42, 133, 139

B
Baldwin, Stanley, 32
bases, American: Pacific, World War II, 134-35, 137, 139; post World War II, 159, 160; unprepared airstrips, 160
Battle of Britain, 52-54, 90: aircraft claims, 53; aircraft losses, 53; broken codes, 54; civilian casualties, 53; GAF aircraft deficiencies, 53; German intelligence, 53; miscalculations, 53; radar, 52
Battle of the Bulge, 120
Bolling, Raynal, 38
bomb duds: World War I, 14, 19; World War II, 105, 120, 311n. 30
bombardment of Britain, World War I, 8-11: civilian casualties, 19; damage, value, 19; impact, 11, 12
bombardment of Britain, World War II: civilian casualties, 67; statistics, 67. *See also* cities bombarded, British
bombardment of France, World War I, 11-12; civilian casualties, 12
bombardment of Germany, World War I: civilian casualties, 19; damage value, 19; statistics, 127
bombardment of Germany, World War II. *See* cities, bombed, German
bombardment of Japan: impact, 152; morale, 153 statistics, 152, 153. *See also* cities, bombed, Japanese
Bomber Command: casualties, 98; difficulties, 70; loss rate, 308n. 16
bomber defenses: missiles, 192; Monica, 77
bomber escort, 14, 37, 38, 39, 41, 44, 50, 51, 52, 53, 69, 93, 99, 104, 106, 109, 112, 113, 114, 116, 117, 119, 123, 135, 142, 144, 173, 175, 221, 222, 313n. 15
bomber exercises: Dayton, 158; New York, 157; Oklahoma City, 158; Omaha, 158; Sky Shield, 236
bomber tactics, post World War II, low level, British, 202
bomber tactics, SAC: airborne alert, 215, 234; airborne command post, 234; communications missiles, 234; dispersal, 233; ground alert, 215, 233, 233; low level, 187, 192, 235-36, 279; takeoff interval, 234; unprepared bases, 233

bomber tactics, World War II, American: formations, 108; group bombing, 108-9; low altitude, 110; routing, 109
bomber tactics, World War II, British, 75: compression, 91, 92; master bomber, 89, 90; Shaker, 75
bomber training, SAC, 236
bomber aircraft, American: B-1, 193, 211, 212, 228, 232, 257, 277-80, 290, 292, 294, 322n. 2, 322n. 3; B-2 (Condor), 24; B-2 (Spirit), 290-93, 294; B-6, 24; B-7, 25; B-8, 25; B-10, 25-27, 44, 69, 304n. 3; B-15, 29, 30, 101, 304n. 6; B-17, 27, 29, 30, 41, 58, 100-2, 107, 130, 133, 160, 166, 174, 177, 226; B-18, 26, 310n. 2; B-19, 29; B-24, 30, 101, 102-4, 110, 111, 136, 148, 150, 166, 174, 177; B-26, 310n. 13; B-29, 29, 30, 80, 129-31, 136, 145, 148-49, 183, 239, 261, 304n. 6, 312n. 3, 315n. 22, 315n. 26; B-32, 148-50; B-35, 162-64, 182, 204; B-36, 165-72, 173, 179, 182, 183, 189, 192, 230, 315n. 20, 315n. 29; B-41, 310n. 13; B-42, 164-65, 180-81; B-43, 180-81, 182; B-45, 173, 181, 183-84, 316n. 3; B-46, 181, 185; B-47, 166, 170, 173, 179, 182, 183, 184-88, 191, 194, 196, 206, 226, 229, 230, 233, 234, 235, 236, 316n. 5, 316n. 6; B-48, 182, 185; B-49, 182-83, 191, 204, 290; B-50, 161, 165, 166, 173, 192, 314n. 9; B-52, 179, 189-94, 196, 197, 208, 217, 218, 220, 221, 222, 223, 226, 228, 230, 231, 232, 233, 234, 236, 255, 257, 277, 286-88, 290, 294, 295, 315n. 20, 317n. 17, 317n. 22, 317n. 23, 319n. 6, 322n. 2; B-58, 194-96, 208, 227, 229, 236, 277, 317n. 12, 318n. 37; B-60, 189, 317n. 14; B-70, 195, 196, 197-99, 204, 228, 277, 318n. 29; Barling, 22-24; F-111, 319n. 6, 319n. 9, 319n. 10; F-117, 282, 286, 292, 293, 294, 295, 322n. 5, 322n. 6, 323n. 16; FB-111, 197, 228-29, 232, 277, 322n. 6; Keystone, 22, 25; MB-1, 2; MB-2, 2-3; stealth, 278; XB-40, 110; XO-27, 25; XO-35, 25
bomber aircraft, British: 1½ Strutter, 14; Buccaneer, 205, 318n. 35; Canberra, 184, 201, 205; DH-4, 16, 17, 18, 26; DH-9, 17; F.E. 2b, 16, 18; Halifax, 77, 78-79, 82, 87, 308n. 16, 309n. 17; Hampden, 69, 81; Hendon, 81; Heyford, 69, 81; Lancaster, 58, 72, 79, 80-82, 85, 86, 87, 89, 101, 175-76, 308n. 7, 308n. 16, 309n. 17; Lincoln, 175-76; Manchester, 80, 81; Mosquito, 72, 75, 82-83, 85, 86, 88, 91, 116, 309n. 18; O/100, 16, 17; O/400, 303n. 18; Sperrin, 199; Stirling, 77-79, 82, 87, 308n. 16; TSR.2, 204-5, 318n. 35; V/1500, 16, 18; Valiant, 199-201; Victor, 199, 201, 203; Vulcan, 199, 201, 202-4, 231, 318n. 33; Washington, 176; Wellington, 69, 87; Whitley, 69
bomber aircraft, French: Bloch 162, 31; C.A.O. 700, 31
bomber aircraft, German: Do 11, 47, 48; Do 13, 48; Do 17, 43, 48, 92; Do 19, 49, 55-56; Do 23, 47, 48; Do 217, 306n. 18; Do F, 47-48; Do P, 47, 49; Gotha bombers, 7-9; He 111, 43, 48, 52, 62; He 177, 49, 54-58; Ju 52, 43, 47, 48, 51, 305n. 33, 306n. 2; Ju 86, 47, 48, 49; Ju 87, 43, 53; Ju 88, 48, 49, 92, 307n. 19; Ju 89, 49, 55-56; Risen-type, 9-11; Rohrbach Roland, 47, 49
bomber aircraft, Soviet: Il-46, 206; M-4, 177, 208; M-50/M-52, 208; Pe-8, 31; SB, 43; TB-3, 31; TB-7, 31; Tu-4, 177, 206; Tu-16, 177, 206-8, 232, 233, 315n. 37; Tu-22, 208-10, 318n. 36, 318n. 37; Tu-22M, 207, 210-11, 212, 213; Tu-95, 177-79, 212 -13, 232, 233, 315n. 29; Tu-160, 212-13, 233, 318n. 38
bombing accuracy, 120-21, 155, 158, 222, 236, 293-94, 306n. 3, 319n. 16: American, Europe, 104, 105, 109, 114, 116, 121; American, Pacific, 137-38, 140, 141, 145; American, Vietnam, 224; British, 71, 74, 75, 87, 96, 311n. 29, 311n32; German, 49, 56; Spanish Civil war, 43; theoretical, 42
bombing doctrine: American, 38-42; German 50-51
bombing retaliation: Allies consider, 65; British, 70, 71; theory, 51; World War I, 13, 15, 16; World War II, 59, 308n. 2
bombing surveys: Iraq (1991), 290; World War I, 14, 16; World War II, 125-27, 163, 174
bombing World War I, limits, 20
bombs: Bunker Buster, 289; Grand Slam, 77, 81; incendiary, 131, 132, 133; PGM, 218, 220, 222, 280, 282, 285-86, 290, 293, 294, 295, 299, 323n. 12; Tallboy, 77; Wallis bomb, 85
Burke, Arleigh, 170, 251
Bush, George W., 275
Bush, George H.W., 261, 274

C
Carter, Jimmy, 260, 279, 292
Casablanca conference, 106
Chennault, Claire, 132
Churchill, Winston, 37, 66, 71, 96, 97, 106
cities bombarded, British: London, 52-53, 57, 61, 62, 66, 70, 89, 127; Coventry, 54
cities bombed, Chinese, Hankow, 137
cities bombed, Czech, Pilsen, 86
cities bombed, Dutch, Rotterdam, 52, 89
cities bombed, French: Lille, 104; Rouen, 104
cities bombed, German: Berlin, 53, 70, 89-90, 91, 93, 96, 97, 119; Bremen, 71, 84, 107, 113; Chemnitz, 96, 97; Cologne, 84; Dresden, 89, 96-97, 118, 153, 296; Dusseldorf, 84; Essen, 84; Gelsenkirchen, 70; Hamburg, 84, 87-89, 91, 97, 118, 153, 296; Leipzig, 90, 96, 97, 118; Lubeck, 84; Mannheim, 70; Marienburg, 113; Munich, 86; Nuremberg, 90-91, 93, 118; Rostock, 84; Schweinfurt, 91, 112, 118; Stettin, 86; Stuttgart, 113; Vegesack, 113
cities bombed, Japanese: Hiroshima, 146-48, 153, 296; Kobe, 142; Nagasaki, 137, 148, 153, 296; Nagoya, 143, 144; Osaka, 143; Ota, 141; Tokyo, 143, 144, 153, 296
cities bombed, Polish, Warsaw, 51, 89
cities bombed, Spanish, 43, 89
cities bombed, Thai, Bangkok, 136
cities bombed, Vietnamese: Haiphong, 217, 222; Hanoi, 217, 222, 223
civilian casualties, 154: British, 12, 19, 53, 67; French, 105, 127; German, 19, 127; North Vietnamese, 224
Clarion, 121
Clark, Mark, 174
Clinton, William, 275
coast defense, 27
commonality: B-1, 279; B-36/B-60, 189; ballistic missile, 247; cruise missile, 256; F-111, 225, 226; KC-135, 194
Cuban missile crisis, 196, 214, 248, 258

D
Doenitz, Karl, 105
Doolittle, James, 117, 139
Douhet, Giulio, 31, 33-34, 37, 38, 44, 50, 51, 304n. 10, 305n. 34
Dowding, Hugh, 54
drones: D-21, 281; BQM-34, 281

E
Eaker, Ira, 104, 105, 106, 107, 114, 116, 121, 122
Eisenhower, Dwight, 66, 196, 198, 225, 237, 245, 261
electronic countermeasures, American: Carpet, 109; Chaff, 109, 233, 230; unit, 109
electronic countermeasures, British: Airborne Cigar, 76; Airborne Grocer, 76; Boozer, 76; Chaff, 76; Cigar, 76; Corona, 76; Grocer, 76; IFF, 76, 93; Shiver, 76; Tinsel, 76; Window, 76, 87, 89, 91
Ent, Uzal, 110-11
escape systems, 278, 292: capsule, 195, 200, 227, 322n. 1; ejection seats, 183, 186, 200, 209, 227, 316n. 6, 317n. 8, 322n. 1

F
fighter aircraft, American: F-4, 217; F-15, 282, 284, 322n. 2; F-16, 282, 284, 294; F-80, 174; F-84, 162, 167, 169, 173, 174-75; F-85, 166, 315n. 13; F-86, 173-74; F-102, 195, 281; F-105, 217, 323n. 12; F-108, 198; F-111, 204, 205, 211, 222, 225-29; P-12, 25; P-38, 105, 109, 116, 117, 311n. 23; P-47, 107, 113, 114, 116, 117; P-51, 116, 117, 311n. 23; P-80, 124; XFM-1, 110
fighter aircraft, British: Beaufighter, 77; Hurricane, 52, 53, 306n. 12; Meteor, 124; Spitfire, 52, 53, 83, 104, 107, 109, 116, 306n. 12
fighter aircraft, German: FW 190, 307n. 22; Bf/Me 109, 31, 43, 52, 53, 110, 306n. 13; He 51, 43; He 100, 56; He 112, 31; He 162, 124; He 219, 93; Ju 88, 77, 91, 93: Me 110, 53, 91; Me 163, 123, 124; Me 210, 93; Me 262, 123, 124
fighter aircraft, Soviet: I-16, 43; MiG-15, 173, 174, 175, 184, 232, 315n. 26; MiGs, 222
flying safety, 158
Foch, Ferdinand, 19, 39
Frankland, Noble, 128

G
Gaither Report, 245
gas, poison, 34, 37, 38, 44, 46, 65, 288, 305n. 34
German air force defeat: day, 94, 125-26; night, 93, 94
Goering, Hermann, 117
Green, William, 80
Groves, Leslie, 176

H
Haig, Alexander, 220
Haig, Douglas, 71
Hansell, Haywood, 139, 140, 143, 153
Harris, Arthur, 72-73, 75, 77, 79, 84, 89, 90, 95, 96, 98, 314n. 3
Hitler, Adolph, 47, 52, 53, 60, 92, 123
Hump, The, 136
hydrogen bomb, 237, 244: American, 170-71; British, 201; Soviet, 178

I
Independent Force, 17, 18: statistics, 19
intelligence, 296: Allied, 95; balloons, 158; British, 70; overflights, 158-59, 184
internment: Russia, 177; Sweden, 113, 119; Switzerland, 113, 114
interservice rivalry, 244, 247, 256
intruder tactics: British, 77, 83; German, 92
Iwo Jima, 142, 146

J
Japanese attacks: Aleutians, 151; balloon, 150-52; Hawaii, 151; Oregon, 151
JCS, 135, 139, 159, 174, 176, 217, 220, 221, 222, 274
jets, World War II, 155, 180: combat statistics, 124; German, 119, 123-24, 126
Johnson, Lyndon, 195, 215, 226, 274, 277

K
Keegan, John, 294
Kennedy, John, 198-99, 214, 231, 234, 238, 259
Kenney, George, 44, 150, 157, 160, 166
Khrushchev, Nikita, 210, 211, 262, 265
Killian Committee Report, 247
Kissinger, Henry, 221
Knauss, Robert, 50

L
LeMay, Curtis, 29, 72, 112, 129, 135, 138, 139, 141, 142, 143, 153, 157, 158, 159, 166, 191, 195, 196, 197, 217, 228, 230, 233, 244, 261, 311n. 15, 314n. 3, 314n. 4, 319n. 5
Lindbergh, Charles, 157
Lipetsk, Russia, 47
Ludendorff, Erich, 7

M
MacArthur, Douglas, 133
Marshall, George, 133
McMahon Act, 176
McMullen, Clements, 157
McNamara, Robert, 197, 216, 225, 226, 228, 229, 238, 261, 277
McNaughton, John, 216, 219
Medal of Honor, 111
Meyer, John, 319n. 5, 319n. 8
mining, 57, 134, 139, 142, 144, 152, 159
missile, ballistic, basing/launch systems, American, 260: above ground, 245; coffin, 245, 320n. 6; dense pack, 260; MPS, 260, 321n. 18; rail mobile, 259, 321n. 20; silo: 245, 246, 259, 320n. 6, 320n. 7
missile, ballistic, basing/launch systems, Soviet: coffin, 263; rail mobile, 266; road mobile, 265; silo, 263, 264, 265, 266
missile, ballistic, defense, American: 260, 270, 272-76, 307n. 29: airborne laser, 275; critics, 273; Iraq War (1991) 274; Iraq War (2003), 295; Nike X, 273; Nike Zeus, 273; Rumsfeld Commission, 275; Safeguard, 274; SDI, 274; Sentinel, 274; Spartan, 273; Sprint, 273; theater missile defense, 275
missile, ballistic, defense, Soviet: Galosh, 273; Griffon, 273; SA-5, 273
missile, ballistic, fractional orbit: 264-65
missile, ballistic, warning, 235
missiles, air-to-surface, American, 193, 198, 230: Hound Dog, 192, 193, 231, 242, 278, 281; Rascal, 166, 230; Skybolt, 199, 202, 231-32; SRAM, 193, 231-32, 272, 278, 279

missiles, air-to-surface, British, Blue Steel, 202, 203, 232
missiles, air-to-surface, German: FX 1400, 56-57, 58, 306n. 18; Hs 293, 293, 306n. 18
missiles, air-to-surface, Soviet: AS-1, 232; AS-2, 232; AS-3, 232; AS-4, 232; AS-5, 232, 233; AS-6, 233; AS-15, 233; AS-16, 233; Kh-15, 212; Kh-20, 178; Kh-22, 178; Kh-45, 210, 212; Kh-55, 179
missiles, anti-radiation: Crossbow, 230; Shrike, 202, 203; Soviet, 233-34
missiles, ballistic, American, 299: Atlas, 242, 244, 246, 248, 258, 262-63, 273, 320n. 8, 321n. 15; Midgetman, 261, 272; Minuteman, 245, 258, 261, 272, 321n. 15; Peacekeeper (MX), 254, 260, 261, 272, 274, 279, 321n. 17; Polaris, 252-53, 255; Poseidon, 253, 320n. 12; SRAM, 322n. 3; Titan, 246, 248, 258, 261, 273, 320n. 8, 321n. 15; Trident, 253-54, 255, 260, 279, 320n. 12
missiles, ballistic, Chinese, 271
missiles, ballistic, German, V-2, 26, 59, 63-67, 155, 199, 242, 262, 289, 307n. 26, 307n. 27
missiles, ballistic, North Korean, 271
missiles, ballistic, Soviet: R-7, 262; R-9, 264; R-11FM, 267, 268; R-13, 268; R-16, 262; R-21, 268; R-27, 268; R-27U, 268; R-29, 269; R-29M, 269; R-29R, 269; R-36, 264; R-36M, 265; R-39, 269; R-39M, 270; RT-2, 266; RT-2PM, 266; RT-2UTTH, 266; RT-23, 266; Scud, 262, 267, 286-88; SS-27, 270; SS-N-8, 269; SS-NX-30, 270; UR-100, 265; UR-100MR, 265; UR-100N, 265
missiles, cruise, American, 238, 286, 295, 299: accuracy, 239; ACM, 257, 279; AGM-86, 212, 256-57, 260, 294, 321n. 14, 322n. 3; ALCM, 193, 278; Aphrodite, 238-39; General Motors A-1, 238; GLCM, 256, 257; Gorgon, 249; JB-2, 239-40; Loon, 240, 249; Matador, 242, 248, 249; Navaho, 240-41, 244, 248, 251, 262; Pollux, 249; Regulus, 249, 251, 320n. 10; Rigel, 251; Snark, 240; 241, 244; Tomahawk, 256, 257, 321n. 14; Triton, 251
missiles, cruise, German, V-1, 58-63, 155, 239, 249, 307n. 22, 307n. 24, 307n. 26
missiles, cruise, Soviet: Buran, 262; Burya, 262; Kh-22, 209, 211; Kh-55, 212; SS-N-21, 269; SS-N-24, 269
missiles, decoy, 193, 229: Buck Duck, 166, 229; Bull Goose, 229; Quail, 193, 230, 231, 236, 255; SCAD, 255
missiles, intercontinental, 198, 205, 213, 215
missiles, intermediate range American, 320n. 9: Jupiter, 247, 248, 251; Pershing, 257, 261; Thor, 247, 248
missiles, multiple warheads: American, 253, 259, 260; Chinese, 271; Soviet: 264, 265, 266, 268, 269, 270
missiles, penetration aids: American, 229-30, 253, 259; Soviet, 265, 269, 270
missiles, sea launched ballistic, 198, 205, 213
missiles, surface-to-air, 155, 159, 205, 299: American, 272, 273; German, 65 124; Soviet, 187, 191, 235, 273; Vietnamese, 217, 218, 221, 222, 223-24
Mitchell, William, 22, 38, 131
Moorer, Thomas, 221
morale, 54, 61, 63, 64, 71, 72, 96, 99, 106, 121, 134, 153, 218, 219, 289, 296: impact, 20; Spanish Civil War, 44; theoretical target, 32, 34, 36, 37, 38, 39, 50; World War I, 13, 15, 17
morality of bombing, 153-54

N

NACA, 150, 195, 198
NASA, 199, 225
navigational/bombing aids, American, 115: Gee, 115; H2S, 115, 122; Oboe, 115, 122
navigational/bombing aids, British: Gee, 74; G-H, 74; H2S, 74, 81; Oboe, 74, 87, 90, 91
navigational/bombing aids, German, *Knickebein*, 53
Nixon, Richard, 220, 221, 224, 253, 254, 277
Normandy invasion, 91, 93, 94, 107, 117, 119, 120, 123, 307n. 19
Northrop, Jack, 162, 183
nuclear accidents, 234, 319n. 14
nuclear aircraft: American, 170, 198, 315n. 18; Soviet, 178
nuclear strategy, 237-38

nuclear war plans, 159, 202: SIOP, 238, 252
nuclear warheads, miniaturized, 155

O
Okinawa, 143, 146, 148
Operations Torch, 105
Ostie, Ralph, 169

P
parachutes: approach, 187, 191; braking, 187, 191
Partridge, Earle, 160, 314n. 7
pathfinders: American, 115; British, 72, 74, 75, 83, 87, 92, 308n. 12; German, 53
Peirse, Richard, 70
pilot training, 115, 119, 309n. 29
Portal, Charles, 96
Potsdam Declaration, 146
Power, Thomas, 234

Q
Quebec conference, 134

R
radar, ground, 41, 155, 306n. 9
range extension: drop tanks, 117; Ficon, 167, 169; wing tip attachment, 167
Reagan, Ronald, 254, 260, 261, 274, 279
reconnaissance aircraft: A-12, 281; RB-45, 162; RF-80, 162; RF-84, 167, 169; SR-70, 198; U-2, 158, 159
reconnaissance, British, 70
reconnaissance, satellite, 159
Revolt of the Admirals, 169
Ridgway, Matthew, 174
Roosevelt, Franklin, 132

S
Sandys, Duncan, 63
Scowcroft Commission, 260, 261
Seaplane Strike Force, 188
Sherman, William, 39-40, 304n. 28
shuttle missions to Russia, 122-23
Smuts, Jan, 15
Spaatz, Carl, 53, 98, 132
Speer, Albert, 89
*Sputnik*, 214, 231, 244-45, 246, 248, 251, 252, 262, 263
Stealth, 280-85, 299: Have Blue, 282, 292; RAM, 212, 279, 281, 291; RCS, 279, 281, 282, 291, 292, 322n. 2; Tacit Blue, 292, 323n. 17; targets, 284
Stilwell, Joseph, 134
submarine classes, American: *Lafayette*, 253; *Ohio* (Trident), 253, 269
submarine classes, Soviet: Borei, 270; Delta, 269, 270; Golf, 267, 268; Hotel, 268; November, 268; Typhoon, 269-270; Yankee, 268, 269; Zulu, 267
submarines, named American: *Ethan Allen*, 251; *George Washington*, 251; *Nautilus*, 242; *Scorpion*, 251
submarines, Polaris, 206, 232, 252, 255
Sweeney, Charles, 148
Sykes, Frederick, 16

T
tanker aircraft, American: B-24, 160; B-29, 161-62; KC-97, 162, 194, 317n. 21; KC-135, 193-94, 233, 234
tanker aircraft, British, Victor, 202, 203
target sets: American, 99, 106, 134, 139, 158, 159; British, 69, 71; German, 50; theoretical, 42; World War I, 13, 302n. 14
Tibbets, Paul, 146
Timberlake, Patrick, 150
transports, American: C-109, 134, 136; C-130, 179; DC-3, 48, 306n. 2
Trenchard, Hugh, 12, 14, 16, 17, 19, 34, 36-37, 38
Truman, Harry, 158, 174
Twining, Nathan, 192, 195

U
units, American, bomb groups: 100th, 113; 509th, 146, 156, 161
units, American, bomb wings: 58th, 135, 139; 73rd, 139, 140, 141; 315th, 145
units, American, bomber commands: XXth, 135, 136, 138, 140, 141, 144; XXIst, 138, 139, 141, 142, 143, 144
units, American, named: ACC, 277; SAC, 156, 157, 161, 186, 193, 197, 202, 215, 221, 222, 223,

228, 277, 292; TAC, 225, 277; United States Air Forces in Europe, 115
units, American, numbered air forces: Eighth, 88, 104, 107, 115, 117, 118, 150; Ninth, 115; Fifteenth, 115, 118, 150
units, American, organization, 308n. 11
units, British, 308n. 11: 617 Sqdn, 85, 308n. 12

V
Vandenburg, Hoyt, 157
victory credits, 105, 117
Von Braun, Wernher, 64, 242, 262
von Tirpitz, Alfred, 3
von Zeppelin, Ferdinard, 2

W
Walker, Kenneth, 41
Wallis, Barnes, 84
war: Afghanistan, 207, 210, 211, 294-95; Bosnia-Herzegovina, 293; Chaco, 42, 305n. 33; Chechnya, 211; China, 44-45; Ethiopia, 42; Falklands, 202; Finland, 45; Iran-Iraq, 208, 210; Iraq (1991), 208, 210, 285-90, 323n. 15, 323n. 16; Iraq (2003), 295; Italo-Turkish, 1; Korea, 173-75; Kosovo, 212, 293-94, 300; Lake Khasan/Nomonhan, 45; Middle East (1967), 207; Panama, 294; Spanish Civil War, 31, 43-44; Suez, 200; Vietnam, 215-25
Warden, John, 286, 323n. 14
weather forecasting, 115, 311n. 22
Weinberger, Casper, 284
Westover, Oscar, 27
Wever, Walther, 49, 55
Weyland, Otto, 174-75
Whitehead, Ennis, 150
Williams, Robert, 112
Wolfe, Kenneth, 138, 141

Y
Yalta Conference, 97

# ABOUT THE AUTHOR

**Kenneth P. Werrell** graduated from the U.S. Air Force Academy in 1960 and went on to earn his pilot wings the following year. He was stationed in Japan and flew weather reconnaissance missions over the northwestern Pacific, first as a pilot and later as an aircraft commander of the WB-50. He resigned his commission in 1965 and went on to earn a master's degree and doctorate degree in history from Duke University.

Werrell taught at Radford University, with brief stints at the Army's Command and General Staff College and the Air Force's Air University, until his retirement. He has published numerous articles in professional journals and several books on aviation history. He lives with his wife, Jeanne, and two cats, Fritz and Kit Kat, in Christiansburg, Virginia.

**The Naval Institute Press** is the book-publishing arm of the U.S. Naval Institute, a private, nonprofit, membership society for sea service professionals and others who share an interest in naval and maritime affairs. Established in 1873 at the U.S. Naval Academy in Annapolis, Maryland, where its offices remain today, the Naval Institute has members worldwide.

Members of the Naval Institute support the education programs of the society and receive the influential monthly magazine *Proceedings* or the colorful bimonthly magazine *Naval History* and discounts on fine nautical prints and on ship and aircraft photos. They also have access to the transcripts of the Institute's Oral History Program and get discounted admission to any of the Institute-sponsored seminars offered around the country.

The Naval Institute's book-publishing program, begun in 1898 with basic guides to naval practices, has broadened its scope to include books of more general interest. Now the Naval Institute Press publishes about seventy titles each year, ranging from how-to books on boating and navigation to battle histories, biographies, ship and aircraft guides, and novels. Institute members receive significant discounts on the Press's more than eight hundred books in print.

Full-time students are eligible for special half-price membership rates. Life memberships are also available.

For a free catalog describing Naval Institute Press books currently available, and for further information about joining the U.S. Naval Institute, please write to:

Member Services
**U.S. Naval Institute**
291 Wood Road
Annapolis, MD 21402-5034
Telephone: (800) 233-8764
Fax: (410) 571-1703
Web address: www.usni.org